Springer Series in Geomechanics and Geoengineering

Editors: Wei Wu · Ronaldo I. Borja

Richard Wan, Mustafa Alsaleh, and Joe Labuz (Eds.)

Bifurcations, Instabilities and Degradations in Geomaterials

Springer

Professor Wei Wu, Institut für Geotechnik, Universität für Bodenkultur, Feistmantelstraße 4, 1180 Vienna, Austria, E-mail: wei.wu@boku.ac.at

Professor Ronaldo I. Borja, Department of Civil and Environmental Engineering, Stanford University, Stanford, CA 94305-4020, USA, E-mail: borja@stanford.edu

Editors

Prof. Richard Wan
Department of Civil Engineering
Schulich School of Engineering
University of Calgary
2500 University Dr NW
Calgary, AB, T2N 1N4
Canada
E-mail: wan@ucalgary.ca

Prof. Joe Labuz
Department of Civil Engineering
University of Minnesota
Minneapolis, MN 55455
USA
E-mail: jlabuz@umn.edu

Dr. Mustafa Alsaleh
Engineering Specialist - Research and
Development, Virtual Product Development
Technology, Product Development Center of
Excellence, TC-E G6, Caterpillar Inc.,
14009 Old Galena Rd., Mossville, IL 61552
E-mail: Alsaleh_Mustafa_I@cat.com

ISBN 978-3-642-18283-9 e-ISBN 978-3-642-18284-6

DOI 10.1007/978-3-642-18284-6

Springer Series in Geomechanics and Geoengineering ISSN 1866-8755

Library of Congress Control Number: 2011921009

Type Design and Cover Design: Scientific Publishing Services Pvt. Ltd., Chennai, India.

Printed on acid-free paper

5 4 3 2 1 0

springer.com

Preface

Geomaterials are endowed with microstructures that make them heterogeneous even at relatively large scales. Hence, failure in geomaterials has the characteristic feature of occurring in a variety of modes, with a continuous transition from diffuse to localized deformations depending on the stress, density, and type of loading. Indeed, material failure and degradation are local material instability phenomena that can be studied within the framework of bifurcation theory.

Interests in localization and related instabilities in the field of geomechanics date back in the early 1980's when the first International Workshop on Localization of Soils was organized in Karlsruhe, Germany, February 1988. This aroused so much enthusiasm and interest in the fundamental aspects of bifurcation theory for soils that the second workshop followed soon after in Gdansk, Poland, September 1989. The topic was then extended to rock mechanics at the third international workshop in Aussois, France, September, 1993. Interests grew steadily and the scope was expanded to instabilities and degradations in geomaterials at the fourth, fifth, sixth and seventh workshops that were held in Gifu, Japan, September 1997; Perth, Australia, November 1999; Minnesota, USA, June 2002; and Crete, Greece, June 2005.

Following tradition, the eighth international workshop continued on the central theme of bifurcations and degradations in geomaterials, with further extensions into new and challenging application areas such as petroleum geomechanics and terramechanics, in particular, soil-machine interaction. The eighth workshop was thus named the International Workshop on Bifurcation and Degradations in Geomaterials (IWBDG 2008): Applications to Soil-Machine Interaction and Petroleum Geomechanics. The workshop was held in Lake Louise, Alberta, Canada. The venue offered an ideal setting for discussing the science and engineering in a relaxed atmosphere in the natural beauty of the Canadian Rockies.

IWBDG 2008 was attended by 71 participants representing 12 countries; 59 presentations were given over three days. Caterpillar Inc., USA provided generous financial support to IWBDG 2008 as a major co-sponsor to the workshop. Additional support was provided by (in alphabetical order), Chevron, Houston, USA; Gifu University, Japan; JACOS Calgary; Kyoto University, Japan; MITACS, Canada; and University of Calgary, Canada.

This special volume contains a sampling of papers as extended versions of the various presentations given at IWBDG 2008. It captures the state-of-the-art in the specialized field of geomechanics and contemporary approaches to solving the central issue of failure.

Professor Ioannis Vardoulakis, who tragically passed away in September, 2009, played an eminent role as one of the founders of this series of bifurcation workshops. His substantial contributions, both technical and as a mentor to young researchers, were pivotal to the success of the IWBDG series. We thus dedicate this special volume in Ioannis' memory for his pioneering contributions, not only to the field of geomechanics, but also to the world of science and engineering. We trust this is a fitting tribute to an outstanding man and scientist.

Richard Wan
Mustafa Alsaleh
Joe Labuz

Contents

Failure in Granular Materials: Macro and Micro Views

F. Nicot[1], L. Sibille[3], and F. Darve[2]

[1] Cemagref, Grenoble, France
e-mail: Francois.nicot@cemagref.fr
[2] Institut National Polytechnique de Grenoble, Laboratoire L3S, Grenoble, France
e-mail: Felix.darve@inpg.fr
[3] IUT de Saint Nazaire, GEM, Université de Nantes, Nantes, France
e-mail: Luc.Sibille@univ-nantes.fr

Abstract. Failure is one of the most debated notions since many decades in geomechanics. On the one hand, the discrete nature of granular materials does not make it easy to define the notion from a phenomenological point of view. On the other hand, this notion is essential for civil engineers since projects have to be designed so as no failure is expected to occur. We herein consider the failure mode related to the creation of kinetic energy, without change in the control parameters. The general framework relating the existence of bursts of kinetic energy to the vanishing of the second-order work is first recalled. Then, the second-order work is investigated from a micromechanical point of view. First, a micromechanical model (micro-directional model of Nicot and Darve, 2005) is considered. The macroscopic second-order work is shown to be the sum of microscopic second-order works, defined on each contact, extended to all the existing contacts. Then, this result is generalized without referring to any constitutive model. This basic relation between both micro and macro second-order works is used to investigate the microstructural origins of the vanishing of the second-order work. Analytical relationships are first derived, highlighting the bridge between both micro and macro scales, and then numerical simulations based on a discrete element method are presented to confirm the relevance of this multiscale approach of failure.

1 Introduction

The notion of failure can be encountered in many fields, irrespective of the scale considered. This notion is essential in material sciences where failure can be investigated at the specimen (the material point) scale. It is also important in civil engineering to prevent or to predict the occurrence of failure on a large scale.

For geomaterials, known as non-associated materials, several failure modes can be encountered strictly within the plastic surface. From a mathematical point of view, this feature is essentially related to the non symmetry of the tangent constitutive tensor. Whereas the localized mode describes a failure corresponding to a

R. Wan et al. (Eds.): Bifurcations, Instabilities and Degradations in Geomaterials, SSGG, pp. 1–12.
springerlink.com

discontinuous displacement field, the diffuse mode is associated with a homogeneous kinematic field with no localization pattern.

In geomechanics, usually it is assumed that failure occurs whenever a material specimen is no longer able to sustain any deviatoric load increment. This condition is associated with a sudden change in the specimen microstructure, namely the sharp decrease in the number of grain contacts. The significant increase in the number of degrees of freedom implies the possibility of rapid relative displacements between grains, leading in some cases to the collapse of the specimen. Because of these rapid relative displacements between grains, the failure is therefore related to a sudden increase in kinetic energy. As a consequence, an ideal way to describe failure consists in describing how the kinetic energy of a given material system can increase.

It was shown that the above-mentioned failure mode can be predicted by the vanishing of the second-order work (Nicot and Darve, 2007; Nicot *et al.*, 2007a). Introduced by Hill (1958), this quantity (hereafter denoted by W_2) is defined as the inner product of the incremental first Piola-Kirchoff stress tensor ($\overline{\overline{\Pi}}$) with the incremental displacement gradient tensor:

$$W_2 = \int_{V_o} \delta\Pi_{ij} \left(\frac{\partial(\delta u_i)}{\partial X_j} \right) dV_o \tag{1}$$

The interest of this semi-Lagrangian formulation lies in that all variables are reported to the fixed initial configuration defined by the volume V_o and the coordinates X_i. For a material point corresponding to a Representative Element Volume (REV) of a granular material, Eq. (1) simplifies into the following expression:

$$W_2 = V_o\, \delta\Pi_{ij} \left(\frac{\partial(\delta u_i)}{\partial X_j} \right) \tag{2}$$

The second-order work can also be expressed under an Eulerian formulation, introducing the Cauchy stress tensor $\overline{\overline{\sigma}}$:

$$W_2 = V \left(\delta\overline{\overline{\sigma}} + div(\delta\vec{u})\,\overline{\overline{\sigma}} - \overline{\overline{\sigma}}\left(\overline{\overline{L}}\right)^t \delta t \right) : \overline{\overline{L}}\,\delta t \tag{3}$$

where L_{ij}, such as $L_{ij}\ \delta t = \partial(\delta u_i)/\partial x_j$, is the general term of the velocity gradient tensor $\overline{\overline{L}}$. It follows that the second-order work is the combination of three terms: $V\ \delta\overline{\overline{\sigma}} : \overline{\overline{L}}\,\delta t$ is a material term, $\delta V\ \overline{\overline{\sigma}} : \overline{\overline{L}}\,\delta t$ is related to the change in volume, and $V\,\overline{\overline{\sigma}}\left(\overline{\overline{L}}\right)^t : \overline{\overline{L}}\,\delta t^2$ is associated with the change in the texture (Nicot and Darve, 2007; Nicot *et al.*, 2007b).

According to the small strains approximation, $\overline{\overline{D}}\,\delta t = \delta\overline{\overline{\varepsilon}}$, where $\overline{\overline{D}} = \frac{1}{2}\left(\overline{\overline{L}} + \overline{\overline{L}}^{t}\right)$ denotes the strain rate tensor, and $\delta\overline{\overline{\varepsilon}}$ is the incremental small strain tensor. From the symmetry of this tensor, it follows that:

$$W_2 = V\,\delta\overline{\overline{\sigma}} : \delta\overline{\overline{\varepsilon}} + \delta V\,\overline{\overline{\sigma}} : \delta\overline{\overline{\varepsilon}} - V\,\overline{\overline{\sigma}} : \overline{\overline{L}}^{2}\,(\delta t)^2 \tag{4}$$

The different formulations of the second-order work introduce macroscopic tensorial variables that represent both the complex force and displacement distributions within the granular specimen. The vanishing of the second-order work stems therefore from microstructural origins for which the local variables (contacts forces and relative displacements between adjoining particles) become relevant. As a consequence, since the vanishing of the second-order work is a proper criterion for detecting the occurrence of a certain failure mode in geomaterials, it makes sense to track the microstructural origin of this macroscopic criterion. This analysis will be first carried out by considering our micro-directional model (Nicot and Darve, 2005), which is a micromechanically based constitutive relationship. Then, this approach will be extended based on a general micromechanical derivation.

2 Microstructural Origin of the Vanishing of the Second-Order Work

2.1 *The Micro-directional Model*

The micro-directional model is a multi-scale relationhip between the Cauchy stress tensor $d\overline{\overline{\sigma}}$ and the strain tensor $d\overline{\overline{\varepsilon}}$ by taking micro-mechanical characteristics into account. In this approach, the granular assembly is described as a distribution of contacts within adjoining particles. Each contact is associated with a given direction of the physical space, corresponding to the normal direction to the tangent contact plane. The texture is therefore described by the distribution of contacts along each direction of the physical space. The probability that some contacts exist in a given direction is investigated and local variables are averaged in each direction, so that directional variables are introduced. Fundamentally, this model is based on a homogenization procedure within a representative volume element (RVE) that can be resolved in the three following basic stages (for more details, see Nicot and Darve, 2005):

The stress average corresponds to the Love formula (Love, 1927; Weber, 1966; Christoffersen *et al.*, 1981; Mehrabadi *et al.*, 1982):

$$\sigma_{ij} = \frac{1}{V}\sum_{c=1}^{N_c} F_i^c\, l_j^c \tag{5}$$

where $\overrightarrow{l^c}$ is the branch vector joining the centers of particles in contact on contact c, $\overrightarrow{F^c}$ is the contact force, and the sum is extended to all the N_c contacts occurring in the RVE of volume V. The norm of the branch vector $\overrightarrow{l^c}$ is assumed to be a constant parameter (equal to the mean diameter of the grains) whose evolution over the loading programs is ignored. This ensures that the terms $\overrightarrow{F^c}$ and $\overrightarrow{l^c}$ are uncoupled. The discrete summation given in Eq. (5) can be replaced with a continuous integration over all the contact directions in the physical space. This scheme confers the directional character to the model:

$$\sigma_{ij} = 2r_g \iint_D \hat{F}_i \, n_j \, \omega \, d\Omega \tag{6}$$

where ω is the density of contacts along each space direction $\vec{n}$, r_g denotes the mean radius of the sphere-shaped grains, $\hat{\vec{F}}$ is the average of all contact forces $\overrightarrow{F^c}$ associated with contacts oriented in the direction $\vec{n}$, and $d\Omega$ is the elementary solid angle. After differentiation it follows that:

$$\delta\sigma_{ij} = 2r_g \iint_D \delta\hat{F}_i \, n_j \, \omega \, d\Omega + 2r_g \iint_D \hat{F}_i \, n_j \, \delta\omega \, d\Omega \tag{7}$$

The kinematical projection relation is given by:

$$\delta\hat{u}_i(\vec{n}) = 2r_g \, \delta\varepsilon_{ij} \, n_j \tag{8}$$

where $\hat{\vec{u}}(\vec{n})$ is the directional kinematic variable linked to $\hat{\vec{F}}(\vec{n})$ along the contact direction $\vec{n}$.

The local behavior is described by introducing a constitutive relation between both average normal force $\hat{F}_n$ and tangential force $\hat{F}_t$ and both average relative normal displacement $\hat{u}_n$ and tangential displacement $\hat{u}_t$. An elastic–plastic model is introduced, and the following local constitutive incremental relations can be inferred:

$$\delta\hat{F}_n = k_n \, \delta\hat{u}_n \tag{9a}$$

$$\delta\hat{\vec{F}}_t = \xi \frac{\hat{\vec{F}}_t + k_t \, \delta\hat{\vec{u}}_t}{\left\| \hat{\vec{F}}_t + k_t \, \delta\hat{\vec{u}}_t \right\|} - \hat{\vec{F}}_t \tag{9b}$$

where $\xi = \min\left\{ \left\| \hat{\vec{F}}_t + k_t \, \delta\hat{\vec{u}}_t \right\|, \tan\phi_g \left(\hat{\vec{F}}_n + k_n \, \delta\hat{\vec{u}}_n \right) \right\}$, k_n is the normal elastic stiffness, k_t is the tangential elastic stiffness, and ϕ_g is the local friction angle.

2.2 Microstructural Expression of the Macroscopic Second-Order Work

Starting from Eq. (7), and noting that the density of contact ω along each direction expresses as $\omega = \omega_e / V$, where ω_e is the number of contacts along the considered direction, it follows that the differentiation of the Cauchy stress tensor is the sum of three terms, i.e

$$\delta\sigma_{ij} = \frac{2r_g}{V} \iint_D \delta\hat{F}_i \; n_j \; \omega_e \; d\Omega + \frac{2r_g}{V} \iint_D \hat{F}_i \; n_j \; \delta\omega_e \; d\Omega - \frac{\delta V}{V} \sigma_{ij} \quad (10)$$

which also writes as:

$$\delta\sigma_{ij} + \frac{\delta V}{V} \sigma_{ij} - \frac{2r_g}{V} \iint_D \hat{F}_i \; n_j \; \delta\omega_e \; d\Omega = \frac{2r_g}{V} \iint_D \delta\hat{F}_i \; n_j \; \omega_e \; d\Omega \quad (11)$$

Now, taking advantage of the kinematical projection relation yields:

$$V \; \delta\sigma_{ij} \; \delta\varepsilon_{ij} + \delta V \; \sigma_{ij} \; \delta\varepsilon_{ij} - 2r_g \iint_D \hat{F}_i \; \delta\varepsilon_{ij} \; n_j \; d\omega_e \; d\Omega = \iint_D \delta\hat{F}_i \; \delta\hat{u}_i \; \omega_e \; d\Omega \quad (12)$$

Interestingly, Eq. (12) can be compared to Eq. (4). The term $V \; \overline{\overline{\sigma}} : \overline{\overline{L}}^2 \; (\delta t)^2$, which is shown to be related to the change in texture (Nicot *et al.*, 2007b), can be assimilated to the term $2r_g \iint_D \hat{F}_i \; \delta\varepsilon_{ij} \; n_j \; d\omega_e \; d\Omega$ which also accounts for textural change. In these conditions, it can be established that the macroscopic second-order work can be expressed in a very straightforward manner with respect to microscopic variables:

$$W_2 = \iint_D \delta\hat{F}_i \; \delta\hat{u}_i \; \omega_e \; d\Omega \quad (13)$$

The integral $\iint_D \delta\hat{F}_i \; \delta\hat{u}_i \; \omega_e \; d\Omega$ corresponds to the summation of scalar product $\delta F_i^c \; \delta u_i^c$ over all the contacts contained within the assembly. As demonstrated by Nicot and Darve (2007 and 2007b), the term $\delta F_i^c \; \delta u_i^c$ can be interpreted as the microscopic second-order work associated with the contact '*c*' between two given adjoining particles. As a consequence, Eq. (13) states that the macroscopic second-order work is equal to the sum of the microscopic second-order works associated with all the contacts existing within the assembly.

This basic result was inferred by considering a given constitutive relation, namely the micro-directional model. The purpose of the next section consists in generalizing this result without referring to any constitutive model.

2.3 From Micro to Macro Second-Order Work

Let us consider a granular assembly containing N grains 'p', with $1 \le p \le N$. Each grain 'p' is in contact with n_p other adjoining grains 'q', with $1 \le q \le N$. Boundary particles ($p \in \partial V$) are subjected to an external force $\vec{F}^{ext,p}$ directed by the external medium. We introduce the Galilean reference frame $\Re$, together with the local reference frame $\hat{\Re}\{\vec{n}, \vec{t}_1, \vec{t}_2\}$ attached to the considered contact whose normal to the tangent contact plane is $\vec{n}$. $\hat{\delta}\psi$ denotes the differentiation of any variable ψ with respect to this reference frame.

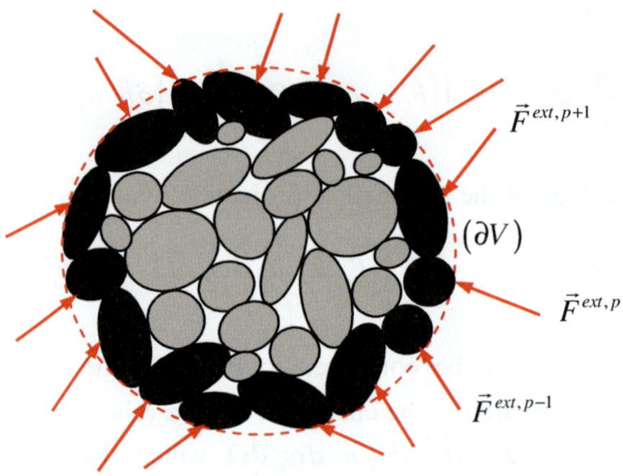

Fig. 1. Granular assembly: boundary particles and external forces

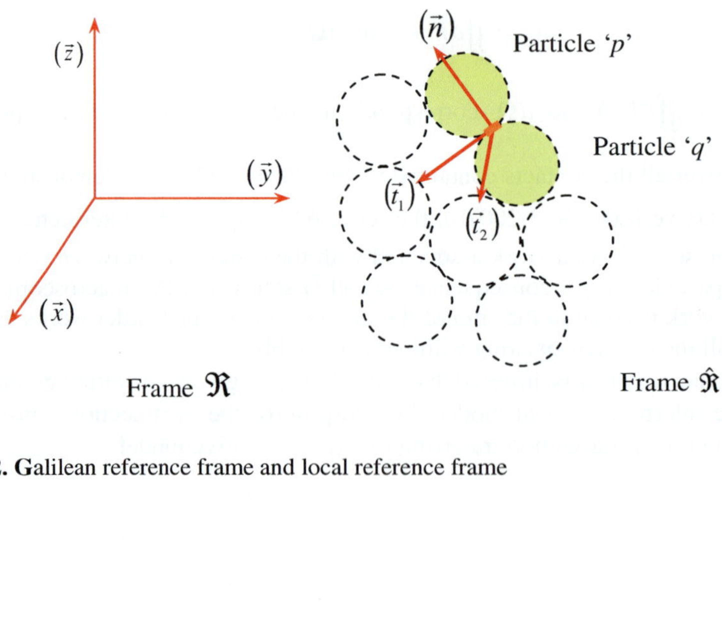

Fig. 2. Galilean reference frame and local reference frame

The microscopic second-order work attached to the contact ‘*c*’ between particles ‘*p*’ and ‘*q*’ is given by the relation (Nicot and Darve, 2007; Nicot *et al.*, 2007b):

$$W_2^{p,q} = \hat{\delta}\vec{F}^{p,q} \cdot \hat{\delta}\vec{u}_c^{p,q} \tag{14}$$

where $\hat{\delta}\vec{F}^{p,q}$ denotes the incremental contact force exerted by particle ‘*p*’ on particle ‘*q*’, and $\hat{\delta}\vec{u}_c^{p,q}$ is the incremental relative displacement of particle ‘*p*’ with respect to particle ‘*q*’.

On the granular assembly scale, the macroscopic second-order work can be related to the second-order time derivative of the kinetic energy as:

$$W_2 = \sum_{p\in\partial V} \delta\vec{F}^{ext,p} \cdot \delta\vec{u}^p - \delta^2 E_c(t) \tag{15}$$

Taking into account the expression of the kinetic energy,

$$\delta E_c(t) = \sum_{p=1}^{N}\left(\vec{F}^p \cdot \delta\vec{u}^p + \vec{M}^p \cdot \delta\vec{\omega}^p\right) \tag{16}$$

it follows, after some algebra (Nicot and Darve, 2007):

$$W_2 = \sum_{p=1}^{N}\sum_{q=1}^{p-1}\left(\hat{\delta}\vec{F}^{p,q} \cdot \hat{\delta}\vec{u}_c^{p,q}\right) - \sum_{p\in\partial V}\left(\vec{F}^{ext,p} \cdot \delta^2\vec{u}^p\right) \tag{17}$$

This basic relation indicates that the macroscopic second-order work is the sum of the microscopic second-order works extended to all the contacts of the whole assembly, $\overline{W}_2 = \sum_{p=1}^{N}\sum_{q=1}^{p-1}\left(\hat{\delta}\vec{F}^{p,q} \cdot \hat{\delta}\vec{u}_c^{p,q}\right)$, minus a boundary complementary term $\sum_{p\in\partial V}\left(\vec{F}^{ext,p} \cdot \delta^2\vec{u}^p\right)$. This last term seems to be negligible from simulations based on a discrete element method (Sibille, 2006; Sibille *et al.*, 2007). This relation, that connects the macroscopic second-order work to microstructural elements embedded in the term $\overline{W}_2$, provides insight into the microstructural origins of the vanishing of the second-order work. The next section is concerned with examining this feature.

2.4 Micromechanical Analysis of the Vanishing of the Second-Order Work

Let us consider the term $\overline{W}_2 = \sum_{p=1}^{N}\sum_{q=1}^{p-1}\left(\hat{\delta}\vec{F}^{p,q} \cdot \hat{\delta}\vec{u}_c^{p,q}\right)$. The vanishing of $\overline{W}_2$ requires that the quantities $\hat{\delta}\vec{F}^{p,q} \cdot \hat{\delta}\vec{u}_c^{p,q}$ vanish for a certain number of contacts. But

$\delta\vec{u}_c^{p,q}$ and $\delta\vec{F}^{p,q}$ are related through constitutive equations such as those given in (9). Considering any contact 'c', $\delta\vec{u}_c$ splits into a normal component δu_c^n and a tangential component δu_c^t. When the contact behaves in the plastic regime, the microscopic second-order work W_2^c is a quadratic form that can be positive or negative:

$$W_2^c = k_n \left(\delta u_c^n\right)^2 + \tan\phi_g \cos\alpha \, k_n \, \delta u_c^t \, \delta u_c^n + k_t \sin^2\alpha \left(\delta u_c^t\right)^2 \tag{18}$$

where α is the angle between both vectors $\vec{t}_1 = \vec{F}_c^t / \left\|\vec{F}_c^t\right\|$ and $\delta\vec{u}_c^t$. For axisymmetric conditions, $\alpha = 0$, and Eq. (18) yields:

$$W_2^c = k_n \left(\delta u_c^n\right)^2 + \tan\phi_g \, k_n \, \delta u_c^t \, \delta u_c^n \tag{19}$$

The vanishing of W_2^c requires that both following conditions are fulfilled (Nicot and Darve, 2006 and 2007), i.e. $\delta u_c^n \leq 0$ (unloading along normal direction) and $\delta u_c^t \geq -\delta u_c^n / \tan\phi_g$ (the amplitude of the tangential displacement is sufficient so as the contact behaves plastically). It is worth noting that the microscopic second-order work is always positive when the contact undergoes a normal compression.

As in plastic regime, $\delta F_c^t = k_n \tan\phi_g \, \delta u_c^n$, condition $\delta u_c^n \leq 0$ also means that both components δF_c^n and δF_c^t are negative. Locally, at the contact scale, the stress state slides down the Coulomb line, as seen in Fig. 3. This result can be regarded as the microstructural origin of the fact that the vanishing of the macroscopic second-order work is essentially observed within the third quadrant, corresponding to $\delta\sigma_1 < 0$ and $\delta\sigma_3 < 0$ (in the stress incremental space), as seen for instance in Fig. 4 (in some cases, negative values of the second-order work can also be observed within the first quadrant; Darve *et al.*, 2004).

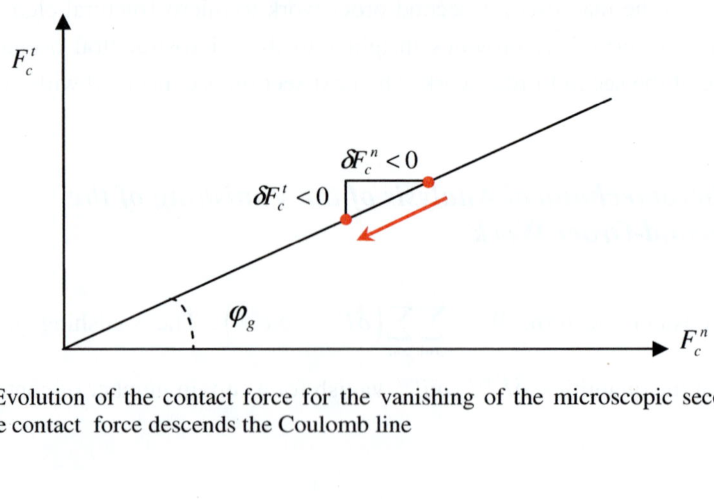

Fig. 3. Evolution of the contact force for the vanishing of the microscopic second-order work: the contact force descends the Coulomb line

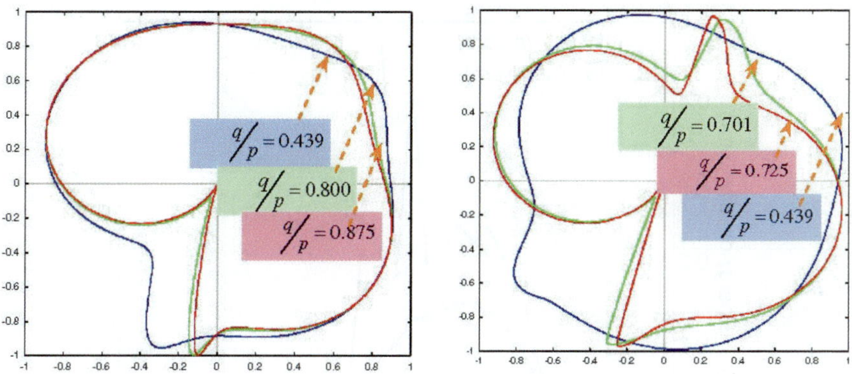

Fig. 4. Polar representation of the second-order work along incremental stress direction (octo-linear model on the left side, micro-directional model on the right side) for different deviatoric ratios (after Nicot and Darve, 2006)

3 Some Remarks on the Basic Micro-Macro Relation for the Second-Order Work

Let us come back to further discuss the relation (17). This relation was investigated from discrete element simulations. Considering a cubic granular specimen, at rest after an initial axisymmetric drained triaxial loading, a series of stress probes was imposed, along all the directions of the incremental stress space, and both quantities W_2 (macroscopic second-order work) and $\overline{W}_2$ (sum of the microscopic second-order works) were compared.

As seen in Fig. 5, Eq. 17 is perfectly verified within the elastic tensorial zone (the zone gathering loading directions leading to no plastic dissipation), when contacts behave essentially elastically, and in a part of the plastic tensorial zone where plastic dissipation is related to sliding on contacts (Fig. 6). As soon as loading directions are characterized by contact opening and/or creation (which corresponds to the central part of the plastic tensorial zone), a significant shift between W_2 and $\overline{W}_2$ exists (Fig. 6). Should the validity of Eq. (17) be questioned? It is our conviction that this basic relation is valid, irrespective of the tensorial zone considered. Nevertheless, it is worth noting that Eq. (17) applies to an equilibrium state; on the contrary, discrete element simulations require considering a finite time interval to compute both quantities W_2 and $\overline{W}_2$. For loading directions belonging to the central part of the plastic tensorial zone (characterized by contact opening and creation), grain rearrangements continuously take place, so that the medium is (at least locally) no longer in equilibrium. As a consequence, for such loading directions, discrete element simulations do not constitute an appropriate way to check a relation valid at the equilibrium but involving (force and displacement) rates.

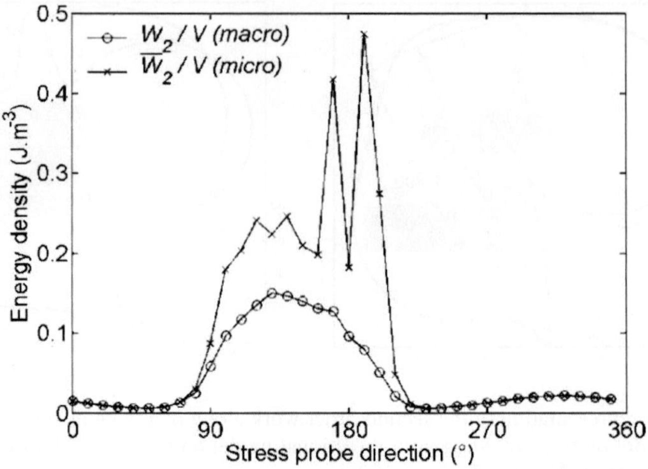

Fig. 5. Microscopic and macroscopic second-order work densities along different stress loading directions (after Nicot et al., 2007b)

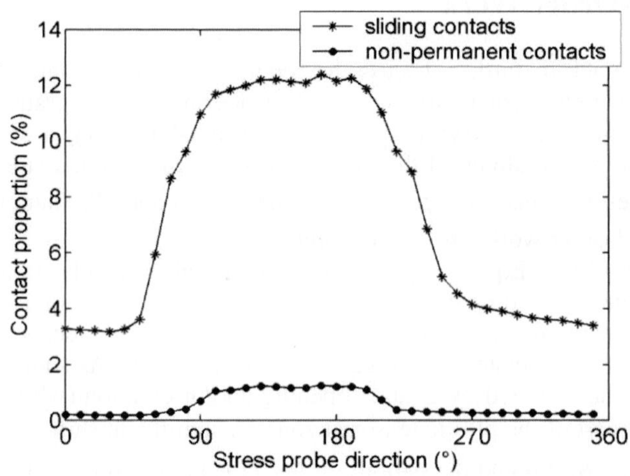

Fig. 6. Grain rearrangement by sliding and opening/creation of contacts along different stress loading directions (after Darve et al., 2007)

4 Conclusion

This paper was devoted to the micromechanical investigation of the vanishing of the second-order work. As this quantity was shown to play a fundamental role to detect the occurrence of a certain failure mode (diffuse failure mode, related to the spontaneous burst of kinetic energy), it is of a great interest to understand what are the microstructural conditions that lead to the vanishing of the second-order work.

First, by considering our microdirectional model, it was inferred that the macroscopic second-order work of a given granular assembly is equal to the sum of the microscopic second-order works extended to all the contacts within the assembly. Then, the validity of the relation was extended based on general micromechanical arguments. This relation is fundamental since it bridges both (between) microscopic and macroscopic worlds. The analysis was pursued by introducing an elastoplastic (frictional) model at the contact scale. The conditions for the vanishing of the microscopic second-order work (which is quadratic form with respect to the relative displacement) were examined, and an interpretation of the fact that the "unstable cones" containing the loading directions of the incremental stress space corresponding to negative values of W_2 are contained in the third quadrant ($\delta\sigma_1 < 0$ and $\delta\sigma_3 < 0$) was provided. The microstructural ingredient of the analysis is essentially related to the local sliding condition. An important aspect remains to be considered in relation with the sudden (and brutal) deletion of contacts on the (mesoscopic) force chain scale. This geometrical aspect should be considered in addition of the former material aspect (sliding condition) considered in this paper.

References

Christoffersen, J., Mehrabadi, M.M., Nemat-Nasser, S.: A micro-mechanical description of granular material behavior. Journal of Applied Mechanics 48, 339–344 (1981)

Darve, F., Servant, G., Laouafa, F., Khoa, H.D.V.: Failure in geomaterials, continuous and discrete analyses, Comp. Methods Appl. Mech. Engrg. 193, 3057–3085 (2004)

Darve, F., Sibille, L., Daouadji, A., Nicot, F.: Bifurcations in granular media, macro- and micro-mechanics. Compte-Rendus de l'Académie des Sciences – Mécanique 335, 496–515 (2007)

Hill, R.: A general theory of uniqueness and stability in elastic-plastic solids. J. Mech. Phys. Solids 6, 236–249 (1958)

Love, A.E.H.: A treatise of mathematical theory of elasticity. Cambridge University Press, Cambridge (1927)

Mehrabadi, M.M., Oda, M., Nemat-Nasser, S.: On statistical description of stress and fabric in granular materials. Int. J. Num. Anal. Meth. Geomech. 6, 95–108 (1982)

Nicot, F., Darve, F.: A multiscale approach to granular materials. Mechanics of Materials 37(9), 980–1006 (2005)

Nicot, F., Darve, F.: Micro-mechanical investigation of material instability in granular assemblies. Int. J. of Solids and Structures 43, 3569–3595 (2006)

Nicot, F., Darve, F.: A micro-mechanical investigation of bifurcation in granular materials. Int. J. Solids and Structures 44, 6630–6652

Nicot, F., Darve, F., Khoa, H.D.V.: Bifurcation and second-order work in geomaterials. Int. J. Num. Anal. Methods in Geomechanics 31, 1007–1032 (2007a)

Nicot, F., Sibille, L., Donzé, F., Darve, F.: From microscopic to macroscopic second-order works in granular assemblies. Mechanics of Materials 39(7), 664–684 (2007b)

Sibille, L.: Modélisations discrètes de la rupture dans les milieux granulaires. Ph-D, Grenoble, INPG (2006)

Sibille, L., Nicot, F., Donze, F., Darve, F.: Material instability in granular assemblies from fundamentally different models. Int. J. Num. Anal. Methods in Geomechanics 31, 457–481 (2007)

Weber, J.: Recherches concernant les contraintes intergranulaires dans les milieux pulvérulents. Bulletin de Liaison des Ponts et Chaussées (20), 1–20 (1966)

Instability in Loose Sand: Experimental Results and Numerical Simulations with a Microstructural Model

A. Daouadji[1], P.-Y. Hicher[2], C.S. Chang[3], M. Jrad[1], and H. Algali[1]

[1] Laboratoire de Physique et Mécanique des Matériaux, UMR CNRS 7554, Université Paul Verlaine Metz, France
[2] Research Institute in Civil and Mechanical Engineering, UMR CNRS 6183, Ecole Centrale de Nantes, France
[3] Department of Civil and Environmental Engineering, University of Massachusetts, Amherst, MA 01003, USA

Abstract. Under certain loading conditions, loose sand can develop instability at a shear stress level much lower than the critical state failure line. To analyze these types of problems, we have adopted the micromechanics model developed by Chang and Hicher for modelling granular material behaviour. The stress-strain relationship for a granular assembly is determined by integrating the behaviour of the inter-particle contacts in all orientations. The constitutive model is applied to simulate undrained triaxial, constant-q and proportional strain tests on loose Hostun sand. Experimental results are used to evaluate how well the model can capture the modes of instability at the assembly level. The notion of control variables is discussed according to these different loading conditions.

Keywords: Granular material, Instability, Micromechanics, Stress-strain relationship, Sand.

1 Introduction

The instability of granular materials is an important topic in geotechnical engineering because it may lead to catastrophic events such as the collapse of earth structures. There are two aspects in the study of instability, namely, material instability (also known as intrinsic/constitutive instability) and geometrical instability (see, for example, Goddard 1993). In this paper, we concentrate on material instability and, more specifically, on a mode of instability called diffuse failure (Darve et al. 1998, 2004; Nova 1994). Experimental results support these theoretical approaches. For example, in loose sand under undrained conditions, an unstable condition can be obtained at a low shear stress level and, subsequently, strength is reduced to almost zero, which corresponds to a material state known as static liquefaction.

The present study is based on the micromechanical approach developed by Chang and Hicher (2004), whereby the stress-strain relationship for a granular assembly can be determined by integrating the behaviour of the inter-particle

R. Wan et al. (Eds.): Bifurcations, Instabilities and Degradations in Geomaterials, SSGG, pp. 13–31.
springerlink.com

contacts in all orientations, based on a static hypothesis which relates the average stress of the granular assembly to the mean field of particle contact forces. Model simulations are compared to experimental results obtained on loose Hostun sand along different loading paths: undrained triaxial tests, constant – q tests and proportional strain paths. The notion of control variables will be discussed according to these different loading conditions.

2 Stress-Strain Model Based on Micromechanical Approach

In this section, the microstructural model developed by Chang and Hicher (2004) is briefly described. In this model, a granular material is viewed as a collection of particles. The deformation of a representative volume of the material is generated by the mobilization of particle contacts in all orientations. On each contact plane, an auxiliary local coordinate can be established by means of three orthogonal unit vectors $\{\mathbf{n},\mathbf{s},\mathbf{t}\}$. The vector $\mathbf{n}$ is outward normal to the contact plane. Vectors $\mathbf{s}$ and $\mathbf{t}$ are on the contact plane.

2.1 Inter-particle Behaviour

Elastic Stiffness: The contact stiffness of a contact plane includes normal stiffness, k_n^α, and shear stiffness, k_r^α. Thus, the elastic stiffness tensor relates contact forces f_i^α to displacements $\delta_j^{\alpha e}$ as

$$f_i^\alpha = k_{ij}^{\alpha e}\delta_j^{\alpha e} \tag{1}$$

which can be related to the contact normal and shear stiffness, i.e.

$$k_{ij}^{\alpha e} = k_n^\alpha n_i^\alpha n_j^\alpha + k_r^\alpha\left(s_i^\alpha s_j^\alpha + t_i^\alpha t_j^\alpha\right) \tag{2}$$

The value of the stiffness for two elastic spheres can be estimated from Hertz-Mindlin's (1953) formulation. For sand grains, a revised form was adopted (Chang et al., 1989), given by

$$k_n = k_{n0}\left(\frac{f_n}{G_g l^2}\right)^n \quad ; \quad k_t = k_{t0}\left(\frac{f_n}{G_g l^2}\right)^n \tag{3}$$

where G_g is the elastic modulus for the grains, f_n is the contact force in normal direction, l is the branch length between two particles, k_{no}, k_{ro} and n are material constants.

Plastic Yield Function: The yield function is assumed to be of Mohr-Coulomb type, defined in a contact-force space (e.g. f_n, f_s, f_t), i.e.

$$F(f_i,\kappa)=T-f_n\kappa\left(\Delta^p\right)=0 \tag{4}$$

where $\kappa(\Delta^P)$ is a hardening/softening parameter. The shear force T and the rate of plastic sliding Δ^p are defined as

$$T=\sqrt{f_s^2+f_t^2} \quad \text{and} \quad \Delta^p=\sqrt{\left(\delta_s^p\right)^2+\left(\delta_t^p\right)^2} \tag{5}$$

The hardening function is defined by a hyperbolic curve in the $\kappa-\Delta^p$ plane, which involves two material constants: ϕ_p and k_{p0} such that

$$\kappa=\frac{k_{p0}\tan\phi_p\,\Delta^p}{\left|f_n\right|\tan\phi_p+k_{p0}\Delta^p} \tag{6}$$

Plastic Flow Rule: The plastic sliding often occurs along the tangential direction of the contact plane with an upward or downward movement; thus shear-induced dilation/contraction takes place. The dilatancy effect can be described by

$$\frac{d\delta_n^p}{d\Delta^p}=\frac{T}{f_n}-\tan\phi_0 \tag{7}$$

where the material constant ϕ_0 can be considered in most cases equal to the inter-particle friction angle ϕ_μ. On the yield surface, under a loading condition, the shear plastic flow is determined by a normality rule applied to the yield function. However, the plastic flow in the direction normal to the contact plane is governed by the stress-dilatancy equation in Eq. (7). Thus, the flow rule is non-associated.

Elasto-plastic Relationship: With the elements discussed above, the incremental force-displacement relationship of the inter-particle contact can be obtained. Including both elastic and plastic behaviours, this relationship is given by

$$\dot{f}_i^\alpha=k_{ij}^{\alpha p}\dot{\delta}_j^\alpha \tag{8}$$

The detailed expression of the elasto-plastic stiffness tensor is not given here since it can be derived in a straightforward manner from the yield function and flow rule, among others.

2.2 *Interlocking Influence*

The resistance against sliding in a contact plane is dependent on the degree of interlocking arising from neighbouring particles. This resistance can be related to the packing void ratio e by

$$\tan\phi_p=\left(\frac{e_c}{e}\right)^m\tan\phi_\mu \tag{9}$$

where m is a material constant (Biarez and Hicher, 1994) and e_c corresponds to the critical void ratio for a given state of stress.

For a dense packing, (e_c / e) is greater than 1 and therefore the apparent inter-particle friction angle ϕ_p is greater than the internal friction angle ϕ_μ. When the packing structure dilates, the degree of interlocking and the apparent frictional angle are reduced, which results in a strain-softening phenomenon. For a loose packing, the apparent frictional angle ϕ_p is smaller than the internal friction angle ϕ_μ and increases during contraction of the material.

The critical void ratio e_c is a function of the mean stress applied to the overall assembly and can be written as follows:

$$e_c = \Gamma - \lambda \log(p') \qquad \text{or} \qquad e_c = e_{ref} - \lambda \log\left(\frac{p'}{p_{ref}}\right) \tag{10}$$

where Γ and λ are two material constants, p' is the mean stress of the packing, and (e_{ref}, p_{ref}) is a reference point on the critical state line.

2.3 *Micro-Macro Relationship*

The stress-strain relationship for an assembly can be determined by integrating the behaviour of inter-particle contacts in all orientations. In the integration process, a micro-macro relationship is required. Using the static hypothesis, we obtain the relation between the global strain and inter-particle displacement

$$\dot{u}_{j,i} = A_{ik}^{-1} \sum_{\alpha=1}^{N} \dot{\delta}_j^\alpha l_k^\alpha \tag{11}$$

where the branch vector l_k^α is defined as the vector joining the centres of two particles, and the fabric tensor is defined as

$$A_{ik} = \sum_{\alpha=1}^{N} l_i^\alpha l_k^\alpha \tag{12}$$

The mean force on the contact plane of each orientation is

$$\dot{f}_j^\alpha = \dot{\sigma}_{ij} A_{ik}^{-1} l_k^\alpha V \tag{13}$$

The stress increment can be obtained by the contact forces and branch vectors for all contacts (Christofferson et al., 1981; Rothenburg and Selvadurai, 1981), as follows

$$\dot{\sigma}_{ij} = \frac{1}{V} \sum_{\alpha=1}^{N} \dot{f}_j^\alpha l_i^\alpha \tag{14}$$

2.4 Stress-Strain Relationship

Using Eqs. (11-14), the following relationship between stress increment and strain increment can be obtained:

$$\dot{u}_{i,j} = C_{ijmp}\dot{\sigma}_{mp} \quad \text{where} \quad C_{ijmp} = A_{ik}^{-1}A_{mn}^{-1}V\sum_{\alpha=1}^{N}\left(k_{jp}^{ep}\right)^{-1} l_k^{\alpha} l_n^{\alpha} \tag{15}$$

When the number of contacts N is sufficiently large in an isotropic packing, the summation of compliance tensor in Eq. (15) and the summation of fabric tensor in Eq. (12) can be written in integral form, given by

$$C_{ijmp} = A_{ik}^{-1}A_{mn}^{-1}\frac{NV}{2\pi}\int_0^{\pi/2}\int_0^{2\pi} k_{jp}^{ep}(\gamma,\beta)^{-1} l_k(\gamma,\beta) l_n(\gamma,\beta)\sin\gamma d\gamma d\beta \tag{16}$$

$$A_{ik} = \frac{N}{2\pi}\int_0^{\pi/2}\int_0^{2\pi} l_i(\gamma,\beta) l_k(\gamma,\beta)\sin\gamma d\gamma d\beta \tag{17}$$

The integration of Eqs. (15) and (16) in a spherical coordinate system can be carried out numerically using Gauss integration points over the surface of the sphere.

3 Experimental and Numerical Evidences of Instability in Sand

3.1 Undrained Triaxial Tests

Undrained triaxial tests on Hostun sand with various initial relative densities, D_r, are presented in Fig. 1 (Hicher, 1998). These results show two distinctive trends corresponding to either a contractive or a dilative behaviour. For loose sand (small values of D_r), a maximum strength is reached in the $q-\varepsilon_1$ plane, and the peak in the stress-strain curve is followed by a rapid decrease of the deviatoric stress down to a minimum strength. This peak corresponds to the development of material instability as will be analyzed later on. The minimum strength can be almost zero for a relative density close to zero. This represents the phenomenon called static liquefaction. For medium dense sand (medium values of D_r), the tendency of softening still occurs at the beginning of the loading, but it is followed by an increase of the deviatoric stress up to the ultimate strength, corresponding to the critical state at large deformations. For dense sand (high values of D_r), the material is strongly dilatant and no strain softening can be observed. Instead, the deviatoric stress continuously increases up to the ultimate strength achieved at large deformations. These different evolutions can be related to the stress paths followed in a $p'-q$ plane; a continuous decrease of the mean effective stress is observed for strongly contractant materials; a decrease followed by an increase of the mean effective stress up to the critical state is observed for dilatant materials.

If we examine the position of the peak in the stress plane $p'-q$, we can see that instability occurs at a stress state below the critical state failure line. Depending on the density of the sand, the position of the peak stress can be more or less distant from the failure line.

3.2 Numerical Simulations of Undrained Tests

The model needs a set of input parameters, such as mean particle size, particle stiffness, inter-particle friction, initial porosity, and parameters defining the critical state of the sand. The mean size of the particle for fine Hostun sand is d = 0.4 mm. The inter-particle elastic constant k_{n0} is assumed to be equal to 61000 N/mm.

The total number of contacts per unit volume changes during the deformation. Using the experimental data by Oda (1977) for three mixtures of spheres, the total number of contacts per unit volume can be approximately related to the void ratio by the following expression:

$$\frac{N}{V} = \left(\frac{N}{V}\right)_0 \frac{(1+e_0)e_0}{(1+e)e} \tag{18}$$

where e_0 is the initial void ratio of the granular assembly. This equation is used to account for the evolution of the contact number per unit volume. The initial number of contacts per unit volume can be obtained by matching the predicted and experimentally measured elastic modulus for specimens with different void ratios (Hicher and Chang, 2007).

The value of k_{t0} / k_{n0} is commonly about 0.4, corresponding to a Poisson's ratio for Hostun Sand $\nu = 0.2$ and the exponent $n = 0.5$. From drained triaxial test results, we were able to derive the values of the two parameters corresponding to the position of the critical state in the $e-p'$ plane: $\lambda = 0.2$ and $p_{ref} = 0.01$ MPa for $e_{ref} = e_{\max} = 1$. In Eq. (9), the value of $m = 1$ was determined from the test results. The values of k_{p0} are assumed to be same as the elastic stiffness $k_{p0} = k_n$. The set of parameters for fine Hostun sand is presented in Table 1. The model performance will be demonstrated in the following sections by comparing the predicted and measured stress-strain behaviours.

Table 1. Model parameters for fine Hostun Sand

e_{ref}	p_{ref} (MPa)	λ	ϕ_μ(°)	ϕ_0(°)	m
1	0.01	0.2	30	30	1

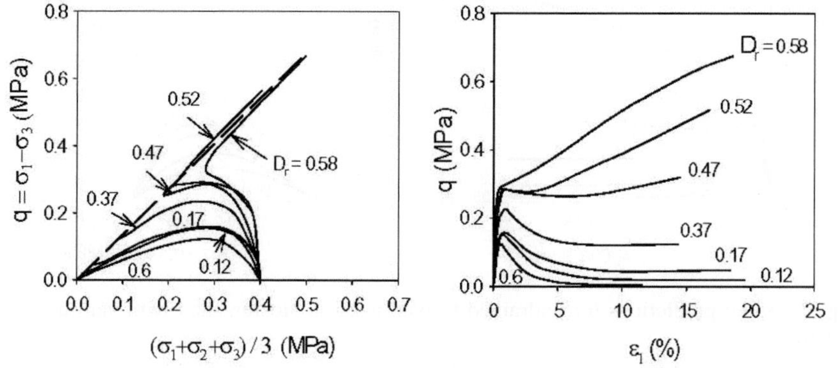

(a) (b)

Fig. 1. Experimental results for undrained triaxial tests on Hostun sand with various densities: (a) stress- strain curves, and (b) stress paths (Hicher, 1998)

Figure 2 presents numerical results for undrained triaxial tests on Hostun sand with various initial void ratios, corresponding to relative densities between 0.05 and 1. Both predicted stress-strain curves and stress paths are in agreement with the experimental curves in Fig. 1. Results indicate that the model is capable of capturing the general trend observed for contractive and dilative sands. In order to examine the inception of instability, the predicted shear stress and second-order work are plotted against the shear strain in Fig. 3 for a test on loose Hostun sand with an initial confining stress $p'_0 = 300$ kPa .

For triaxial tests, the second order work is given by:

$$d^2W = dq\,d\gamma + dp'd\varepsilon_v \tag{19}$$

On the other hand, for undrained conditions ($d\varepsilon_v = 0$), the second-order work is reduced to $d^2W = dq\,d\gamma$ where γ is the deviatoric strain. Since the deviatoric strain increases continuously, $d\gamma$ is always positive and the second-order work can become non-positive, if and only if $dq \le 0$ (i.e., decrease in *q*). Figure 3 shows that instability begins at the shear stress peak. The second-order work is positive before the peak stress. After the peak, the second-order work remains negative and approaches zero at critical state. The numerical simulations can be carried out after the peak because the control variable is the vertical strain ε_1 and not the deviatoric stress q .

It should be noted that the undrained condition imposed in the tests is taken into account in the modelling by the condition of no volume change, i.e., isochoric condition. The instability is therefore not triggered by the pore pressure built up, but rather tests on a dry loose sand specimen have also led to the same instability mode (Lanier et al., 1989).

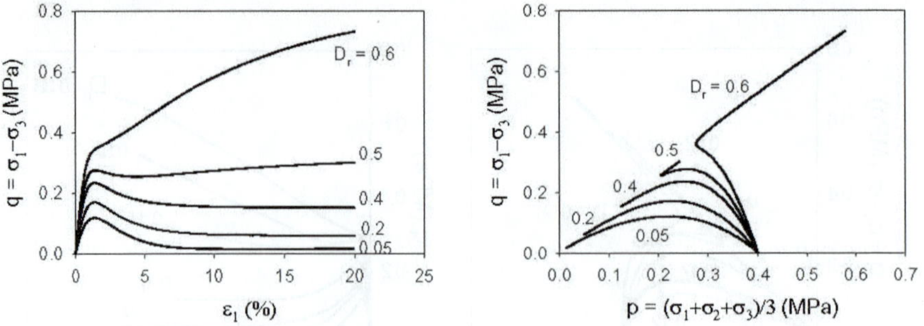

Fig. 2. Model predictions for undrained triaxial tests on Hostun sand with various relative densities

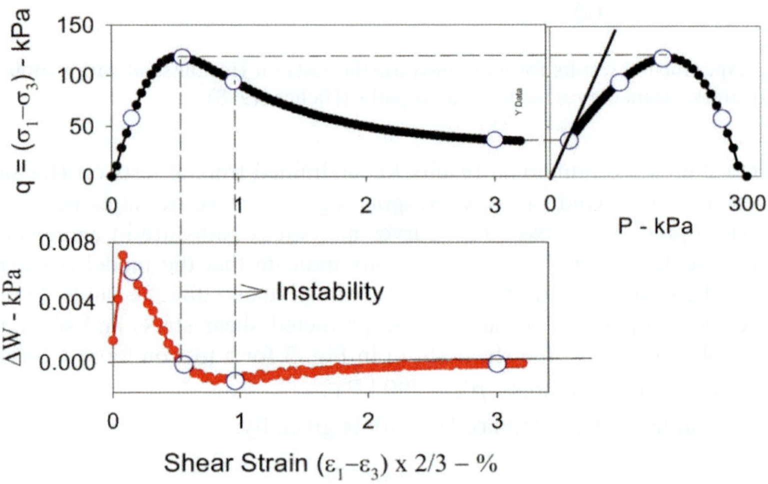

Fig. 3. Predicted second-order work and stress-strain curves for undrained triaxial test on loose Hostun sand

3.3 Constant-*q* Tests

This type of test consists of shearing the specimen to a prescribe stress ratio along a drained compression triaxial path, and then in decreasing the mean effective stress while keeping the deviatoric stress constant. This stress path can simulate the loading condition of a soil element within a slope when a progressive increase in pore pressure occurs.

Several investigations have demonstrated that instability can occur in loose sand during a constant-q stress path (Sasitharan et al. (1993), Nova and Imposimato (1997), Gajo et al. (2000), Lade (2002), Chu et al. (2003). Darve et al. (2007) have presented similar test results on Hostun sand and observed a sudden collapse of the sand specimen for stress states located well below the critical state failure line. Typical results of constant-q tests on loose Hostun sand are presented in Fig. 4. After an isotropic consolidation stage to a desired initial effective mean pressure p'_0, a drained triaxial compression test was applied to the sample up to a prescribed value of the deviatoric stress q. Then, while keeping q constant, a decrease of the mean effective stress p' was applied by increasing the pore water pressure and maintaining constant the total stresses.

At a given point during the test, the axial strain rate started to increase very rapidly and the deviatoric stress could no longer be kept constant. The test is no longer controllable (as defined by Nova, 1994) in the sense that the imposed loading program cannot be maintained. As will be shown later, this point corresponds to a loss of stability, since any small change of one control variable, as defined below, will lead to a catastrophic failure.

Similarly, we also used the parameters in Table 1 to predict the results of constant-q tests on loose Hostun sand. The predicted and measured results for the confining stress $p'_0 = 300$ kPa are presented in Fig. 5. The initial part of the $p-\varepsilon_v$ curve shows that, as the mean stress p decreases, the volume increases. This trend continues until a certain point where the volume starts to decrease. For constant-q tests $(dq = 0)$, according to Eq. 19, the second-order work is reduced to $d^2W = dp'd\varepsilon_v$. Since the mean stress is progressively decreased (i.e., $dp' < 0$), the second-order work becomes negative, if and only if $d\varepsilon_v \leq 0$ (i.e., the volume contracts). Thus the onset of instability corresponds to the peak of the $p-\varepsilon_v$ curve, which is well reproduced by the model simulation.

As for the undrained tests discussed previously, it is found that numerical simulations can be carried out after the instability condition because the control variable is the mean effective stress p' and not the volume change ε_v.

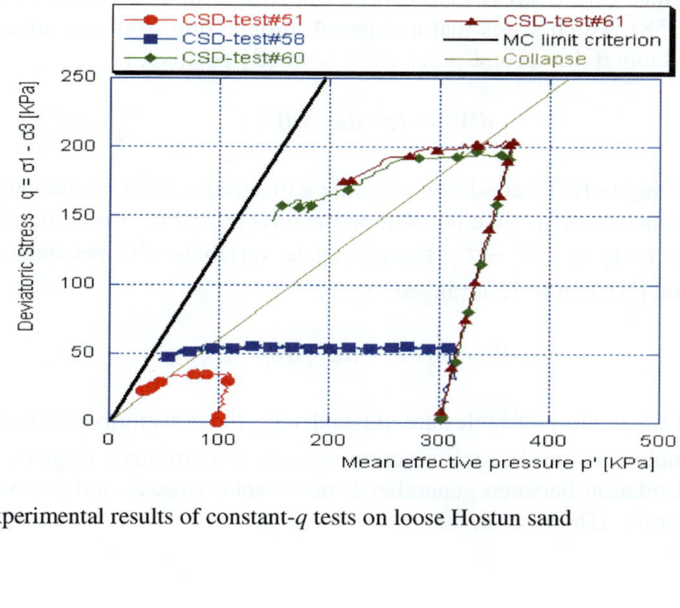

Fig. 4. Experimental results of constant-q tests on loose Hostun sand

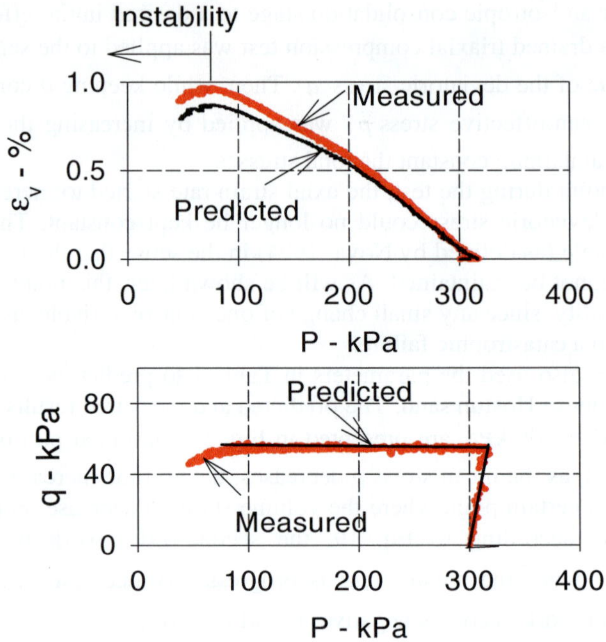

Fig. 5. Comparison of predicted and measured results for constant-*q* test on Hostun sand

3.4 Analysis of Undrained Compression and Constant-q Tests in Terms of Instability Condition

The importance of the stress path can be linked to Hill's sufficient condition of stability (1958), which states that a material, progressing from one stress state to another, is stable if the second-order work is strictly positive, i.e.

$$d^2W = d\sigma_{ij}\, d\varepsilon_{ij} > 0 \tag{20}$$

Thus, according to Hill's condition, whether a material is stable or not depends not only on the current stress state but also on the direction of the stress increment.

Because $q-\varepsilon_1$ and $\varepsilon_v-\sigma'_3$ are conjugate variables with respect to energy, Eqs. (19) and (20) can be re-arranged to give

$$d^2W = dq\, d\varepsilon_1 + d\sigma'_3\, d\varepsilon_v \tag{21}$$

Noting that the control variables are dq and $d\varepsilon_v$, the constitutive relation linking the stress increments to the strain increments can be rearranged to give a generalized mixed relation between generalized incremental stresses and generalized incremental strains (Darve et al. 2004):

$$\begin{Bmatrix} dq \\ d\varepsilon_v \end{Bmatrix} = \begin{bmatrix} E_1 & \frac{2E_1}{E_3}\nu_{13} - 1 \\ (1-2\nu_{31}) & \frac{2(1-\nu_{33}-2\nu_{13}\nu_{31})}{E_3} \end{bmatrix} \begin{Bmatrix} d\varepsilon_1 \\ d\sigma_3 \end{Bmatrix} = [P] \begin{Bmatrix} d\varepsilon_1 \\ d\sigma_3 \end{Bmatrix} \tag{22}$$

where E_i are pseudo Young moduli and ν_i are pseudo Poisson coefficients. As no volumetric variation is allowed during undrained tests, Eq. (22) is modified as follows

$$\begin{Bmatrix} dq \\ 0 \end{Bmatrix} = \begin{bmatrix} E_1 & \frac{2E_1}{E_3}\nu_{13} - 1 \\ (1-2\nu_{31}) & \frac{2(1-\nu_{33}-2\nu_{13}\nu_{31})}{E_3} \end{bmatrix} \begin{Bmatrix} d\varepsilon_1 \\ d\sigma_3 \end{Bmatrix} \tag{23}$$

Equation (21) indicates that instability will occur only if $dq = 0$ (peak shear stress); thus:

$$\begin{Bmatrix} 0 \\ 0 \end{Bmatrix} = \begin{bmatrix} E_1 & \frac{2E_1}{E_3}\nu_{13} - 1 \\ (1-2\nu_{31}) & \frac{2(1-\nu_{33}-2\nu_{13}\nu_{31})}{E_3} \end{bmatrix} \begin{Bmatrix} d\varepsilon_1 \\ d\sigma_3 \end{Bmatrix} \tag{24}$$

Therefore, instability can take place only if the determinant of the constitutive matrix **P** in Eq. (24) becomes equal to zero.

Turning to constant-q tests ($dq = 0$), no variation of the shear stress is imposed and as such, Eq. (22) can be rewritten as

$$\begin{Bmatrix} 0 \\ d\varepsilon_v \end{Bmatrix} = \begin{bmatrix} E_1 & \frac{2E_1}{E_3}\nu_{13} - 1 \\ (1-2\nu_{31}) & \frac{2(1-\nu_{33}-2\nu_{13}\nu_{31})}{E_3} \end{bmatrix} \begin{Bmatrix} d\varepsilon_1 \\ d\sigma_3 \end{Bmatrix} \tag{25}$$

Taking into account Eq. (24), instability will occur in this case only if $d\varepsilon_v = 0$, because $|d\sigma'_3| > 0$ is imposed by the loading program. In conclusion, the condition is the same as the one obtained for undrained tests (Eq. 24) and both correspond to

$$\det(\mathbf{P}) = 1 - 2\nu_{31} + 2(1-\nu_{33}-\nu_{13})\frac{E_1}{E_3} \tag{26}$$

Equation (26) is verified for non trivial solutions if $\det(\mathbf{P}) = 0$. As described by Darve et al. (2004), the condition of instability for constant-q tests coincides with the condition for undrained tests, so stress states in the p'–q plane corresponding to the vanishing of the volumetric strain increment during constant-q tests are the same as the stress states corresponding to the peaks of the stress-strain curves for the undrained tests. Figure 6 presents different results obtained on very loose Hostun sand $(D_r = 0\%)$. One can see that the condition of instability in the $p'-q$ plane is found to be the same for undrained and constant-q triaxial tests, in agreement with the theoretical developments presented above. This condition defines an instability line for a mobilized friction angle equal to 16°, much lower than the friction angle at critical state equal to 30°.

Instability conditions obtained from numerical simulations of undrained compression and constant-q tests are also plotted in the $p'-q$ plane together with the experimental results (Fig. 6). One can see that the model is capable of predicting very accurately the condition of instability associated with these two types of tests. The position of the instability line determined by model simulations is in very good agreement with the position obtained experimentally.

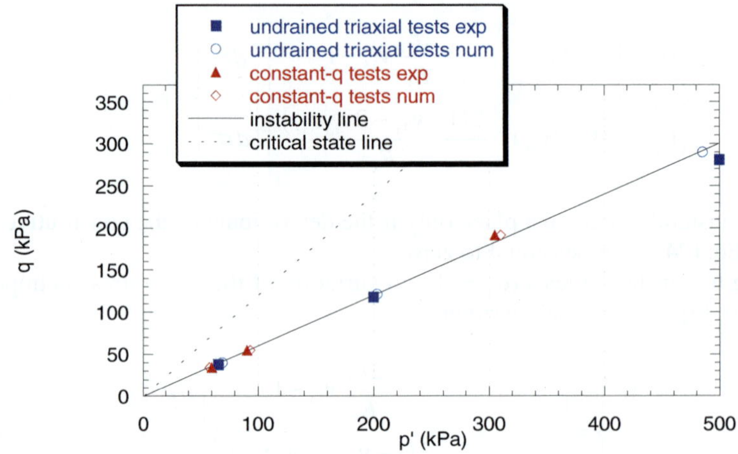

Fig. 6. Comparison of predicted and measured instability condition for loose Hostun sand determined from undrained triaxial and constant-q tests

3.5 Proportional Strain Paths

A series of two Proportional Strain Path (PSP) tests have been carried out. The applied strain path corresponds to

$$\begin{cases} d\varepsilon_1 > 0 \\ d\varepsilon_2 = d\varepsilon_3 \\ \varepsilon_1 + 2R\ \varepsilon_3 = 0 \end{cases} \tag{27}$$

where R is a strictly positive constant for a given loading path. R = 1 corresponds to the case of an undrained or isochoric test.

For the first PSP test, a constant strain ratio R = 0.68 is imposed on the specimen. This path corresponds to an imposed constant rate of dilatancy equal to $d\varepsilon_v / d\varepsilon_1 = -0.471$. For the second PSP test, a constant R = 1.36 is imposed on the specimen, which corresponds to an imposed contraction at a constant rate $d\varepsilon_v / d\varepsilon_1 = 0.265$. The test results are presented in Fig. 7.

Let us now define two control parameters $C'_1 = \sigma'_1 - \sigma'_3 / R$ (proportional stress path) and $C'_2 = \varepsilon_1 + 2R\varepsilon_3$ (the proportional strain path applied to the specimen.

The second order work can be written as (Darve et al. 2004):

$$d^2W = (d\sigma'_1 - d\sigma'_3 / R)d\varepsilon_1 + d\sigma'_3 / R(d\varepsilon_1 + 2Rd\varepsilon_3) \quad (28)$$

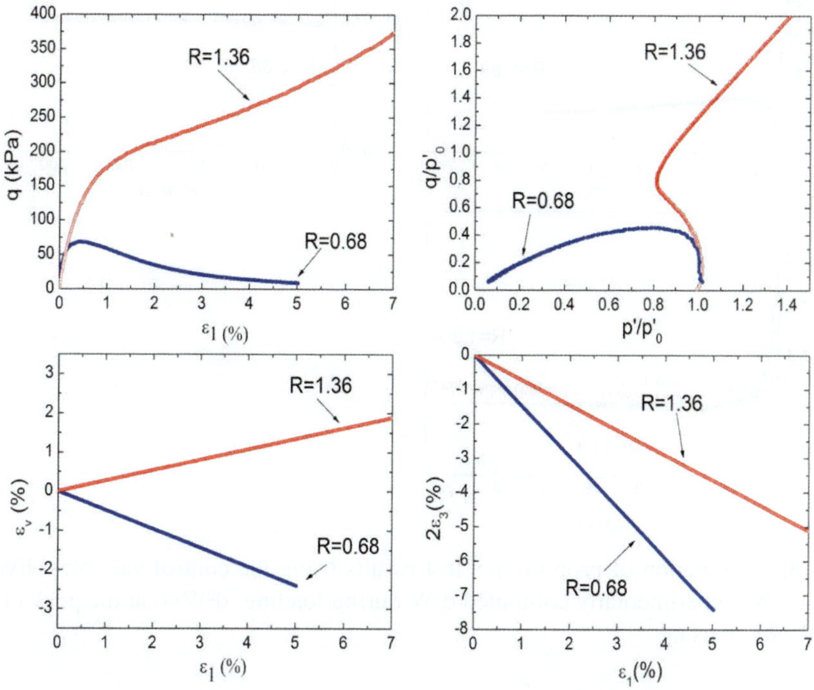

Fig. 7. Proportional strain tests with imposed constant dilatancy (R=0.68) and contractancy (R=1.36) rate. Classical plan are presented: q-ε_1, q-p' normalised by the initial effective mean pressure, ε_v-ε_1 and ε_1-2ε_3.

The proportional strain path is applied with the condition $(d\varepsilon_1 + 2R\,d\varepsilon_3) = 0$, hence $dC'_2 = 0$. Therefore, the second-order work, along this particular loading path, is reduced to

$$d^2W = (d\sigma'_1 - d\sigma'_3 / R)\,d\varepsilon_1 \tag{29}$$

The second order work vanishes whenever $dC'_1 = 0$, i.e. when $(d\sigma'_1 - d\sigma'_3 / R) = 0$ as $d\varepsilon_1 > 0$ was imposed during the test. The proper control variables C'_1 and C'_2 are used in Fig. 8.

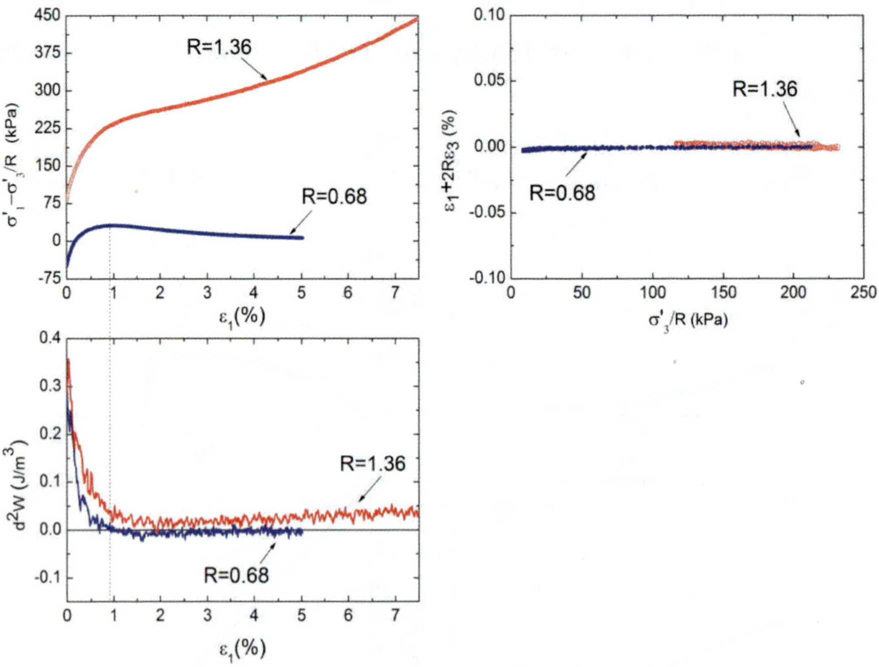

Fig. 8. Presentation of proportional test results using the control variables. Evolution of the experimentally computed d²W during loading: d²W=0 at the peak of σ'_1 - $d\sigma'_3/R$ versus ε_1.

One can see that the second control variable C'_2 is constant. The experimental second order work is given by Eq. (29). This quantity is computed using test data and is plotted against the axial deformation. For R = 0.68, it is noticeable that the second order work is positive (with a small magnitude) at the beginning of the loading and then becomes negative. As theoretically expected, the experimental point corresponding to the vanishing of the second order work coincides with the peak of the $(\sigma'_1 - d\sigma'_3 / R)$ versus ε_1 curve, but not with the peak of the $q - \varepsilon_1$ curve. The axial strain corresponding to the vanishing of the second order work is of the same magnitude for PSP test as for undrained or constant-q tests i.e. around 0.9%. Hence, it is not possible to deduce the collapse of the specimen by regarding the $q - \varepsilon_1$ curve, as peak occurs at an axial strain of 0.5 percent. The second order work vanishes after the $q - \varepsilon_1$ peak for this test.

However, for $R = 1.36$, the $(\sigma'_1 - d\sigma'_3 / R)$ versus ε_1 curve does not present any peak, so the second order work is expected to be positive during the whole loading. This point is experimentally verified. The specimen remains stable all along the loading path.

Proportional strain paths have been simulated by the microstructural model, using the set of parameters in Table 1. The values of R vary from 0.68 to 2.16. Results are presented in Fig. 9. One can see in the $(p' - q)$ plane that a decrease in the deviatoric stress is obtained in most of the tests, except for the highest values of R. However, the peak of the deviatoric stress does not correspond to the vanishing of the second order work. The condition of instability is reached after the peak for values of R smaller than 1 and before the peak for $R > 1$ (Fig. 10). We have seen in the previous section that the condition of instability coincides with the peak for undrained or isochoric tests $(R = 1)$. If we plot the results using the control variables, we can see that the second order work becomes nil at the peak of the $(\sigma'_1 - \sigma'_3 / R)$ versus ε_1 curve for each test (Fig. 11).

The numerical results for the test at $R = 0.68$ are in agreement with the experimental results presented above (Figs. 7 and 8). This is not the case for $R = 1.36$, since the numerical simulation shows that an unstable state exists for this test, contrary to the experiment. This is probably due to a difference in the initial void ratio. In the numerical simulations, the initial void ratio was the same for all tests, corresponding to a relative density $D_r = 0$ $(e = e_{\max})$, while it varies from $D_r = 0$ for $R = 0.68$ to $D_r = 20\%$ for $R = 1.36$.

3.6 *Analysis of Proportional Strain Paths in Terms of Instability Condition*

The constitutive relation (Eq. 22) is modified for proportional strain test using the proper control variables (Servant et *al.*, 2004) as

$$\begin{Bmatrix} d\sigma'_1 - d\sigma'_3 / R \\ d\varepsilon_1 + 2R d\varepsilon_3 \end{Bmatrix} = \begin{bmatrix} E_1 & (2R\dfrac{E_1 \nu_{13}}{E_3} - 1) \\ (1 - 2R\nu_{31}) & \dfrac{2R^2(1 - \nu_{33} - 2\nu_{13}\nu_{31})}{E_3} \end{bmatrix} \begin{Bmatrix} d\varepsilon_1 \\ d\sigma'_3 \end{Bmatrix} = [Q] \begin{Bmatrix} d\varepsilon_1 \\ d\sigma'_3 / R \end{Bmatrix} \quad (30)$$

For the PSP test, $d\varepsilon_1 + 2Rd\varepsilon_3 = 0$; therefore, Eq. (30) can be written as

$$\begin{Bmatrix} d\sigma'_1 - d\sigma'_3/R \\ 0 \end{Bmatrix} = \begin{bmatrix} E_1 & (2R\dfrac{E_1\nu_{13}}{E_3} - 1) \\ (1-2R\nu_{31}) & \dfrac{2R^2(1-\nu_{33}-2\nu_{13}\nu_{31})}{E_3} \end{bmatrix} \begin{Bmatrix} d\varepsilon_1 \\ d\sigma'_3 \end{Bmatrix} = [Q] \begin{Bmatrix} d\varepsilon_1 \\ d\sigma'_3/R \end{Bmatrix} \quad (31)$$

At the peak of $(\sigma'_1 - \sigma'_3/R)$, we get

$$\begin{Bmatrix} 0 \\ 0 \end{Bmatrix} = \begin{bmatrix} E_1 & (2R\dfrac{E_1\nu_{13}}{E_3} - 1) \\ (1-2R\nu_{31}) & \dfrac{2R^2(1-\nu_{33}-2\nu_{13}\nu_{31})}{E_3} \end{bmatrix} \begin{Bmatrix} d\varepsilon_1 \\ d\sigma'_3 \end{Bmatrix} = [Q] \begin{Bmatrix} d\varepsilon_1 \\ d\sigma'_3/R \end{Bmatrix} \quad (32)$$

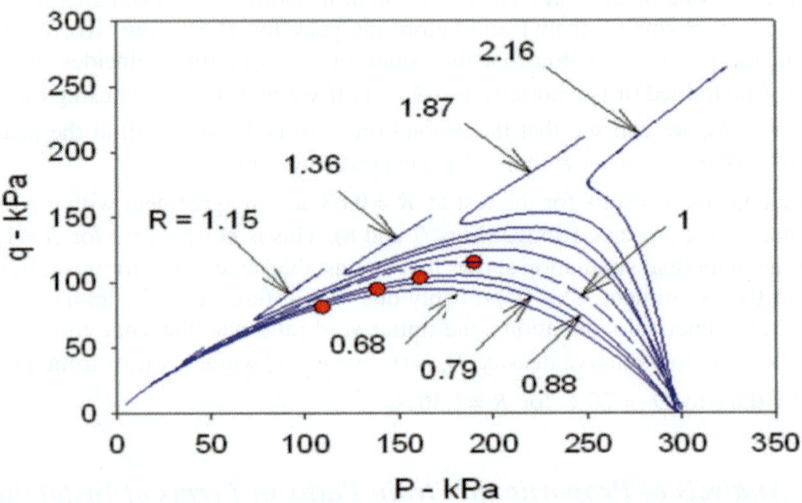

Fig. 9. Stress paths during proportional strain tests for various values of *R*

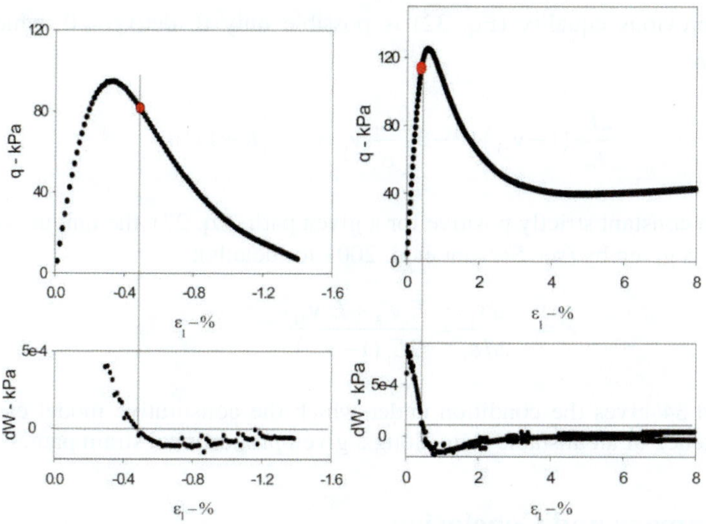

a) $R = 0.68$ b) $R = 1.15$

Fig. 10. Stress–strain behaviour along proportional strain paths for (a) R= 0.68 and (b) R = 1.15

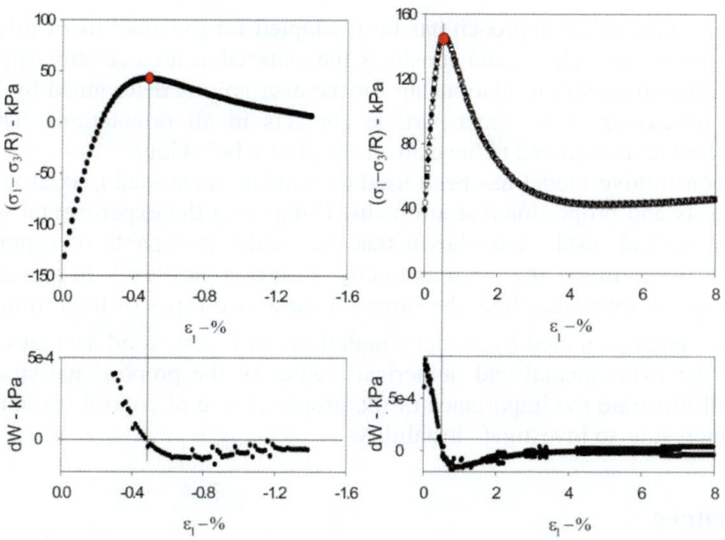

a) $R = 0.68$ b) $R = 1.15$

Fig. 11. Stress-strain behaviour using the control variables for (a) R = 0.68 and (b) R = 1.15

The previous equality (Eq. 32) is possible only if $\det(\mathbf{Q}) = 0$ which corresponds to:

$$\frac{2E_1}{E_3}(1-\nu_{33})R^2 - 2\left(\frac{E_1}{E_3}\nu_{13} + \nu_{31}\right)R + 1 = 0 \tag{33}$$

As R is a constant strictly positive for a given path (Eq. 27), the unique solution of Eq. (32) is given by (see Servant et al. 2004 for details):

$$R = -\frac{d\varepsilon_1}{2d\varepsilon_3} = \frac{E_1\nu_{13} + E_3\nu_{31}}{2E_1(1-\nu_{33})} \tag{34}$$

Equation 34 gives the condition under which the constitutive model can capture the existence of an unstable state along a given proportional strain path.

4 Summary and Conclusion

Under specific loading conditions, loose sand can succumb to instability at a shear stress level much lower than the critical state failure line. The instability condition can be studied by examining the sign of the second order work along a given loading path.

A micromechanical approach has been adopted for the analysis of this type of instability problem. The model considers the material as an assembly of particles. As such, the stress-strain relationship for the assembly is determined by integrating the behaviour of the inter-particle contacts in all orientations. The inter-particle contact is assumed to have an elasto-plastic behaviour.

The constitutive model has been used to simulate undrained triaxial tests, constant-q tests and proportional strain paths. Comparing the experimental with predicted numerical results has shown that the model is capable of capturing the modes of instability at the grain assembly (macroscopic) level. In particular, the position of the instability line, the same for these two types of tests in the $p'-q$ plane, has been predicted by model simulations with very good accuracy. In conclusion, the experimental and numerical studies of the proportional strain paths very well illustrate the importance of the proper choice of control variables when analyzing results to investigate instability.

References

Biarez, J., Hicher, P.Y.: Elementary Mechanics of Soil Behaviour, Balkema, p. 208 (1994)

Chang, C.S., Sundaram, S.S., Misra, A.: Initial Moduli of Particulate Mass with Frictional Contacts. Int. J. for Numerical & Analytical Methods in Geomechanics 13(6), 626–641 (1989)

Chang, C.S., Hicher, P.-Y.: An elastic-plastic model for granular materials with microstructural consideration. International Journal of Solids and Structures 42, 4258–4277 (2004)

Christofferson, J., Mehrabadi, M.M., Nemat-Nassar, S.: A micromechanical description on granular material behavior. ASME Journal of Applied Mechanics 48, 339–344 (1981)

Chu, J., Leong, W.K.: Recent progress in experimental studies on instability of granular soil. In: Labuz, J.F., Drescher, A. (eds.) International Workshop on Bifurcations and Instabilities in Geomechanics, pp. 175–191. Swets & Zeitlinger, Lisse (2003)

Darve, F., Servant, G., Laouafa, F., Khoa, H.D.V.: Failure in geomaterials: continuous and discrete analyses. Computer Methods in Applied Mechanics and Engineering 193 (27-29), 3057–3085 (2004)

Darve, F., Roguiez, X.: Homogeneous bifurcation in soils. In: Adachi, et al. (eds.) Localization and Bifurcation Theory for Soils and Rocks, pp. 43–50. Rotterdam, Balkema (1998)

Darve, F., Sibille, L., Daouadji, A., Nicot, F.: Bifurcations in granular media: macro-and micro-mechanics approaches. C. R. Mecanique 335, 496–515 (2007)

Gajo, A., Piffer, L., De Polo, F.: Analysis of certain factors affecting the unstable behaviour of saturated loose sand. Mechanics of Cohesive-Frictional Materials 5, 215–237 (2000)

Goddard, J.D.: Material instability in complex fluids. Annual Review of Fluid Mechanics 35, 113–133 (2003)

Hicher, P.-Y.: Experimental behavior of granular materials. In: Cambou, B. (ed.) Behavior of Granular Materials, pp. 1–97. Springer, Wien (1998)

Hicher, P.-Y., Chang, C.S.: An anisotropic non-linear elastic model for particulate materials. J. Geotechnical and Geoenvironmental Engrg., ASCE (8), 132 (2007)

Hill, R.: A general theory of uniqueness and stability in elasto-plastic solids. J. Mechanics and Physics of Solids 6, 236–249 (1956)

Lade, P.V.: Instability, shear banding, and failure in granular materials. International Journal of Solids and Structures 39(13-14), 3337–3357 (2002)

Lanier, J., Block, J.F.: Essais à volume constant réalisés sur presse tridimensionnelle. Greco Geomaterials Report, 240–243 (1989)

Laouafa, F., Darve, F.: Modelling of slope failure by a material instability mechanism. Computers and Geotechnics 29(4), 301–325 (2002)

Mindlin, R.D., Deresiewicz, H.: Elastic spheres in contact under varying oblique forces. ASME Trans. J. Appl. Mech. 20, 327–344 (1953)

Nova, R.: Controllability of the incremental response of soil specimens subjected to arbitrary loading programs. J. Mechanical Behaviour of Materials 5(2), 193–201 (1994)

Nova, R., Imposimato, S.: Non-uniqueness of the incremental response of soil specimens under true-triaxial stress paths. In: Pietruszczak, Pande (eds.) Numerical Models in Geomechanics, Balkema, pp. 193–198 (1997)

Oda, M.: Co-ordination Number and Its Relation to Shear Strength of Granular Material. Soils and Foundations 17(2), 29–42 (1977)

Rudnicki, J.W., Rice, J.: Conditions for the localization of deformation in pressure sensitive dilatant materials. International Journal of Solids and Structures 23, 371–394 (1975)

Rothenburg, L., Selvadurai, A.P.S.: Micromechanical definitions of the Cauchy stress tensor for particular media. In: Selvadurai, A.P.S. (ed.) Mechanics of Structured Media, pp. 469–486. Elsevier, Amsterdam (1981)

Sasitharan, S., Robertson, P.K., Sego, D.C., Morgenstern, N.R.: Collapse behavior of sand. Canadian Geotechnical Journal 30(4), 569–577 (1993)

Schofield, A.N., Wroth, C.P.: Critical State Soil Mechanics. McGraw-Hill, London (1968)

Failure in Granular Materials in Relation to Material Instability and Plastic Flow Issues

Richard Wan and Mauricio Pinheiro

University of Calgary, Calgary, Alberta, T2N 1N4, Canada
e-mail: wan@ucalgary.ca, MPinheiro@thurber.ca

Abstract. This work has its beginnings in the seminal works of Hill (1958) concerning a largely theoretical question as to the condition of instability in materials. The question has now taken on a more important and practical relevance over the intervening years, this in part motivated by various forms of failure such as strain localization and diffuse instability in geomaterials. According to classical theories, failure is largely seen as a condition of plastic limit with the implication of strain localization and surface discontinuities. However, other forms of failure such as of the diffuse type in the absence of any localization can be observed well before plastic limit conditions are met. Within this backdrop, we first examine Hill's stability criterion as a means to detect diffuse instability. Then, we discuss issues of controllability and sustainability in various load controlled conditions. Finally, we turn our attention to a related issue, i.e. unstable plastic flow in elastoplastic materials. Through discrete element simulations, we explore the nature of the plastic incremental response of a granular material and verify whether the plastic flow rule postulate holds under general three-dimensional conditions.

1 Introduction

The failure of geomaterials is a manifestation of the instability of an otherwise homogeneous state as observed in the form of rupture patterns with and without sharp discontinuities, e.g. shear bands and fractures. This instability is of the material or constitutive type as it arises from the interaction of particles down to the meso-scale in the absence of external boundary effects. Hence, it is important to understand and include micromechanical features such as Reynolds dilatancy when formulating constitutive models for granular materials. As such, material instability is a basic prerequisite for capturing the rich variety of unstable deformation modes commonly observed in boundary value problems. Among various manifestations of material instability, we are particularly interested in diffuse failure in the absence of any strain localization. As a prominent example, Fig. 1 shows the instability of homogeneous deformations resulting into the formation of unstable force chains to a sudden loss in inter-particle contacts to collapse in an analogue granular assembly of pentagonal photo-elastic disks subject to biaxial

R. Wan et al. (Eds.): Bifurcations, Instabilities and Degradations in Geomaterials, SSGG, pp. 33–52.
springerlink.com

shearing under load control. Apart from the evidence of a length scale larger than a single particle on material response, there is also an interesting analogy of this phenomenon with static liquefaction in loose sand under undrained shearing and load control. The effective stress ratio passes through a peak at which the material suddenly succumbs and loses strength. The deformations are diffuse with the important observation that the effective stress ratio at collapse is well below the usual plastic limit defined by Mohr-Coulomb.

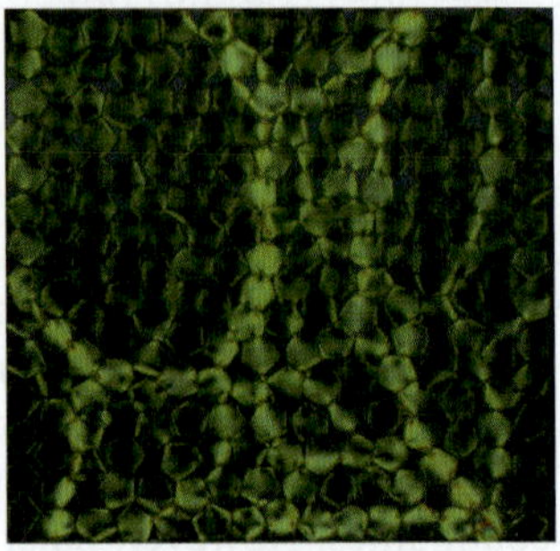

Fig. 1. Diffuse failure in an analogue granular material composed of photo-elastic material. Interference fringes reveal complex force chains in stable, metastable and unstable states as well as the micromechanical nature of material instability.

The diffuse type of instability has its origins in the loss of positive definiteness of the incremental constitutive relation (tensor **D**) under a certain loading program. The connection with the vanishing of the second-order work introduced by Hill (1958) is also well recognized. The non-symmetry of **D**, as is the case for non-associated plastic geomaterials, indicates the possibility of loss of determinacy in incremental material response well before peak conditions. At the boundary value problem level, this indeterminacy leads to a multiplicity of solutions for the underlying governing equations, and hence represents a bifurcation problem.

One important theoretical issue is that of the relation between the material instabilities mentioned in the above and the plastic flow rule. The postulate of flow rule has a direct link with plastic stability criteria as some authors have already shown within the last fifty years (Drucker 1956, 1959; Hill 1958; Mroz 1963; Rudnicki and Rice 1975; Raniecki 1979; Bigoni and Hueckel 1991; Bigoni 2000, to name a few). For example, Drucker (1956, 1959) has demonstrated that a hardening material obeying an associated flow rule is always stable except when the failure criterion is fulfilled. This holds true regardless of loading control conditions. On the other hand, non-associative materials may present unstable response even during hardening stage of the loading process (Bigoni and Hueckel, 1991; Bigoni 2000).

In this contribution, we first discuss about diffuse instability based on compelling examples of lab experimental observations and a few salient mathematical aspects. These instabilities are then mathematically captured using a multi-surface elasto-plastic constitutive model which encapsulates all the basic prerequisites of material instability. We note the importance of the plasticity flow rule and establish a connection between material instability and the regularity of the flow rule. One important question is whether the incremental plastic strains are a sole function of current stress, or does it also depend on the direction of loading as described by incrementally nonlinear relationships (Darve and Labanieh, 1982). Against this last theoretical issue, we use discrete element modelling for verifying this conjecture in general three dimensional conditions. We also attempt to establish any relationship that may exist between the loading history and the nature of the incremental plastic response.

2 Experimental Evidence of Diffuse Instability

Conventional undrained tests have received most of the attention when discussing diffuse instability, especially because of their relation to the well-known liquefaction phenomenon (Castro 1969; Kramer and Seed 1988; Lade 1992). In these conventional tests, the specimen is first consolidated to an initial isotropic stress state under drained conditions, and then is axially loaded through a servo-controlled device while confining stresses are kept constant. During this second phase, fluid is prevented to move in or out of the specimen; therefore, no drainage is allowed. It has been experimentally observed (*loc. cit.*) and theoretically shown (Darve et al. 2004) that the peak of the stress curve in the p-q plane corresponds to a point of energetic instability according to Hill's stability criterion. At this peak, where the second-order work is first null, stress states are yet far away from plastic limit condition described by conventional failure criteria such as Coulomb (1776) and von Mises (1913) or more elaborate ones such as Matsuoka and Nakai (1974) and Lade and Duncan (1975).

A recent more compelling example of material instability in the form of diffuse failure is provided by the so-called constant shear test (Chu et al. 2003; Gajo 2004; Darve et al. 2007) consisting of shearing a sand sample at constant deviatoric stress with decreasing mean effective stress. Like the conventional undrained test, the constant shear test has been found to be very susceptible to premature failure (before plastic limit) with the characteristic of diffuse deformations and instability depending on the mode of control of the test. The connection with an engineering application of such tests is in a slope problem undergoing water infiltration at dead (constant) loads, which have been apparently found to fail in a catastrophic manner rather than in a plastic localized fashion.

As for illustrative purposes, Fig. 1 shows a series of constant shear tests on Hostun S28 sand under axi-symmetric conditions reported in Darve et al. (2007).

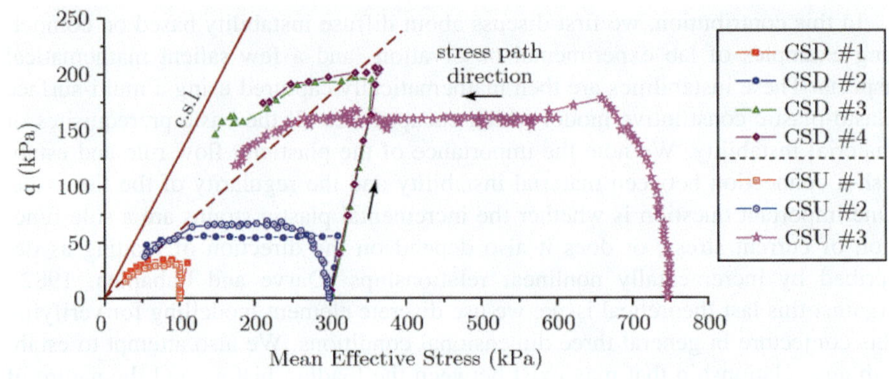

Fig. 2. Constant shear drained tests and constant shear undrained tests performed on Hostun S28 sand at various stress levels (data from Darve et al. 2007)

Both drained (CSD) and undrained (CSU) conditions were explored at various stress levels. Curiously, all specimens showed an unstable response and a loss of test controllability roughly at a same reference defined by a line in the p-q plane. Unstable response of the test was marked by the inability of the operator to control the test according to the loading program. For instance, referring to test CD#1 in Fig. 2, maintenance of a constant q at zero excess pore pressure for realizing drained conditions could not be satisfied anymore at some point during loading history when the sample suddenly succumbed to escalating axial strains. The deformation pattern involved a collapsing structure (probably the same as the one shown in Fig. 1) with no visible localization pattern within the kinematic field. Also, this loss of controllability of the test depends apparently on the loading program, i.e. loading direction. As a generalization of the latter, the locus of points for which instability occurs for a given loading direction defines a boundary in the p-q plane whose existence can be explained within Hill's second order work framework and related bifurcation issues.

3 Diffuse Instability, Controllolability and Sustainability Concepts

The notion of diffuse instability is linked to the early works of Hill (1958) which associate instability to the loss of the positivity of the so-called second-order work as the product of the incremental stress and strain during a loading increment, i.e.

$$W_2 = d\boldsymbol{\sigma} \cdot d\boldsymbol{\varepsilon} \tag{1}$$

where $d\boldsymbol{\sigma}$ is the stress increment vector and $d\boldsymbol{\varepsilon}$ its strain work-conjugate related through the tangent constitutive matrix $\mathbf{D}$. This result arises from the study of equilibrium states subject to small disturbances which indicates that the internal energy minus the work done by external forces must be strictly positive for stability to prevail. Furthermore, recalling the tangent constitutive matrix, Eq. (1) can be recast into:

$$W_2 = d\boldsymbol{\varepsilon}^T \cdot \mathbf{D} \cdot d\boldsymbol{\varepsilon} \tag{2}$$

which points to the positive-definiteness of **D** for $W_2 > 0$. Furthermore, viewed more generally, the positive definiteness of **D** is equivalent to the positive definiteness of the symmetric part of **D**, denoted by $\mathbf{D}^{sym}$. Therefore, the second order work criterion has the alternate form of:

$$\det(\mathbf{D}^{sym}) > 0 \Rightarrow stability \tag{3}$$

Advocating the theorem of Ostrowski and Taussky (1951) as introduced by Nova (1994) leads to a mathematically attractive result stating that for **D** to be positive definite, $\det(\mathbf{D}) \geq \det(\mathbf{D}^{sym}) > 0$. Hence, the conclusion is that the second-order work criterion provides a lower bound for the plastic limit condition given by $\det(\mathbf{D}) = 0$.

Yet, a more subtle question surrounds the issue of loss of uniqueness of material response and its relationship to the second-order work. In a lab experiment, a soil specimen may be subjected to mixed loading programs whereby either pressure, or force, or displacement is being controlled simultaneously. It may happen that during the course of loading, one or a combination of these control parameters can no longer be controlled. Nova (1994) referred to that phenomenon as loss of controllability, a concept synonymous to non-uniqueness of the incremental solution of the underlying constitutive equations. In other words, controllability signifies that a loading programme can only be implemented if a unique incremental response is produced at every incremental loading step (Nicot and Darve 2009).

As such, consider the following generic constitutive relation:

$$\mathbf{A} \cdot d\mathbf{r} = d\mathbf{c} \tag{4}$$

where the non-zero vector d**c** contains the controlled parameters and d**r**, the measured or response variables for a given load program (or test). The 2×2 matrix **A** is related to the constitutive tensor **D** as a re-arrangement of rows and columns. Control of the loading program is lost whenever det (**A**) is zero. As such, this evokes the following corollary: controllability of load parameters and uniqueness of incremental response are guaranteed if **A** is positive definite. In turn, whenever **A** is positive definite, it is straightforward to show that the second-order is also positive:

$$W_2 = d\mathbf{r} \cdot d\mathbf{c} = d\mathbf{r}^T \cdot \mathbf{A} \cdot d\mathbf{r} > 0 \text{ for } \forall d\mathbf{r} \neq 0 \tag{5}$$

Therefore, the above establishes the connection between second-order work and loss of controllability or uniqueness in material response. No diffuse instability takes place when load parameters are controllable. Nova (2003) has also showed that the occurrence of localized instabilities can be treated within this framework.

More recently, Nicot et al. (2007) have introduced the concept of sustainability of equilibrium states as a means to provide a physical meaning to the loss of uniqueness of incremental material response. It is proposed that, when the sustainability of a mechanical state is lost, a rapid burst of kinetic energy might take place if a proper infinitesimal perturbation is imposed (Nicot and Darve 2009). As such, there is a sudden transition from a quasi-static regime to a dynamic one so that equations used before are no longer valid.

In order to mathematically demonstrate the aforementioned relation between the burst of kinetic energy, E_c, and the second-order work, W_2, Nicot et al. (2007) have first shown that for a given time increment δt at a time t:

$$2E_c(t + \delta t) = P - W_2 \tag{6}$$

where P represents the energy due to the external control parameters applied to the system. Then, they further examined the particular case where P vanishes, i.e. controllability is lost. For this case, Eq. (6) simplifies into:

$$2E_c(t + \delta t) = -W_2 \tag{7}$$

Thus, Eq. (7) provides the direct connection between kinetic energy and second-order work. Nicot and Darve (2009) proceed to comment that the collapse of the system, characterized by rapid growth in kinetic energy, corresponds to a negative second-order work. In summary, a loss of uniqueness in material response is synonymous to a loss of controllability and hence a negative second-order work accompanied by a burst of kinetic energy as a loss of sustainability of an equilibrium state. This is the missing physical link in the original Hill's work on the second order work.

4 Material Instability Analysis in Element Tests

In the following, we present numerical explorations of diffuse instability in the constant shear test described in Fig. 2 within the framework of the second-order work. A rather theoretical investigation is performed in order to understand the roles played by diffuse instability and control parameters. It is recalled that no attempt is made to quantitatively match numerical simulations with experimental results. The soil (sand) is described by an existing elastoplastic model earlier proposed by Wan and Guo (2001a, b, 2004) and recently modified in Wan et al. (2011). In the next subsection we briefly recall the essential ingredients of this model, here referred to as WG-model. Then, we turn to the stability analysis of constant shear tests under various sets of control parameters including undrained conditions.

4.1 Elasto-plastic Constitutive Model

The WG-model is an elastoplastic model not only based on the theory of multi-surface plasticity but also on recent concepts of micromechanics embedded in a stress-dilatancy law used as a plasticity flow rule. This model is an outgrowth of a double yield surface constitutive law originally developed by Wan and Guo (1999) and founded on two solid frameworks: (a) Rowe's stress-dilatancy theory, which establishes a linear relationship between stress ratio and strain increment ratio (i.e. dilatancy) through energetic principles (Rowe 1962) and (b) critical state soil mechanics which defines a theoretical state where the material is continuously distorted under no change in volume and stress ratio (Roscoe and Burland 1968; Schofield and Wroth 1968). These two frameworks were enriched in order to incorporate pyknotropy (density), barotropy (stress level) and anisotropy (fabric) dependencies, as well as cyclic loading regime conditions (Wan and Guo 2001a, b, 2004).

The mathematical structure of WG-model is summarized in Table 1 for the particular case of axisymmetric stress-strain states. The incremental elastic response is nonlinear, arising from the assumption of an increasing shear modulus, G, under compressive loading. Although the model is characterized by two yield surfaces: one that treats deviatoric loading dominated by dilatancy and another that accounts solely for isotropic loading producing plastic volumetric compressive strains, we restrict our simulations of the constant shear tests to the deviatoric (shear) yield surface only due to the nature of the loading direction. Compressive stresses and strains are taken as positive as is customary in soil mechanics.

In summary, the WG-model comprises of a plastic potential, flow rule and hardening law, in addition to a yield surface. The resulting incremental plastic response is described by a non-associated flow rule derived from the enriched stress-dilatancy theory proposed in Wan and Guo (1999). It is reminded that the non-associativity of the plastic flow rule is one of the prerequisites for triggering material instability as mentioned in the beginning of the paper. Finally, the updating of the shear-yield surface is governed by the mobilized friction angle that acts as a main hardening variable, which in turn is directly controlled by the plastic strains and density state as shown in Table 1.

Table 1. Main equations of WG-model for loading involving the shear-yield surface under axisymmetric (Wan and Guo 2004; Wan et al. 2009)

Definition	Formula	Variables and parameters
Elastic	$d\varepsilon_v^e = dp/K;\ \ d\gamma^e = dq/3G$ $K = 2G(1+\nu)/3(1-2\nu)$ $G = \frac{(2-e)^2}{1+e}\left(\frac{p}{p_0}\right)^{1/2} G_0 p_0$	$p = (\sigma_1 + 2\sigma_3)/3$: mean effective stress; $q = \sigma_1 - \sigma_3$: deviatoric stress; $\varepsilon_v = \varepsilon_1 + 2\varepsilon_3$: volumetric strain;
Yield surface	$f = q - M_\varphi p$ $M_\varphi = \frac{6\sin\varphi_m}{3-\sin\varphi_m}$	$\gamma = 2/3\cdot(\varepsilon_1 - \varepsilon_3)$: deviatoric strain; Superscripts "e" and "p" stand for elastic and plastic, respectively.
Potential function	$g = q - \sin\psi_m p$, $\sin\psi_m = \frac{\sin\varphi_m - \sin\varphi_f}{1-\sin\varphi_m \sin\varphi_f}$ $\sin\varphi_f = \frac{\alpha_F + \gamma^p}{\alpha_0 + \gamma^p}\left(\frac{e}{e_{cs}}\right)^{n_f} \sin\varphi_{cs}$	$\varphi_m, \varphi_f, \varphi_{cs}$: friction angles mobilized at failure and at critical state, respectively; ψ_m: mobilized dilatancy angle; e, e_{cs}: void ratio at current and at critical states, respectively;
Hardening law	$\sin\varphi_m = \frac{\gamma^p}{\alpha_0 + \gamma^p}\left(\frac{e}{e_{cs}}\right)^{-n_m} \sin\varphi_{cs}$	G_0, ν, e_{cs0}, α_F, α_0, n_f, n_m, n_{cs} and h_{cs} are material parameters; $p_0 = 1$ kPa: reference stress.
Evolution law	$e_{cs} = e_{cs0}\exp\left[-h_{cs}(p/p_0)^{n_{cs}}\right]$	

4.2 Analysis of Diffuse Instability

In the following, we perform a series of numerical simulations of constant shear tests at the material point level in a homogeneous test. These tests were earlier idealized as a means to simulate soil response during water infiltration process of slopes led to failure (Brand 1981; Sasitharan et al. 1993). A straightforward simulation of the test ideally proceeds in three steps. First, the sample, here taken as homogeneous, is isotropically consolidated to a given stress level ($d\sigma_1 = d\sigma_3 > 0$). Then, shear loading proceeds along a triaxial compression path under drained condition up to a prescribed deviatoric stress level ($d\sigma_1 > 0$, $d\sigma_3 = 0$). Finally, the mean effective stress is either directly or indirectly decreased while the deviatoric stress is being kept constant ($d\sigma_1 = d\sigma_3 < 0$). From a numerical standpoint, a direct method of simulating the tests consists in decreasing the effective stress components under drained conditions. However, in lab testing, there are two ways to indirectly decrease the mean effective stress, namely, by either increasing the pore water pressure at constant total stress or forcing the sample to dilate ($d\varepsilon_v < 0$) by water injection. In our simulations, we have applied both the direct and the forced dilation (injection) methods.

The initial mean stress level in all tests is 300 kPa, whereas three distinct levels of deviatoric stresses are investigated, i.e. 50, 100 and 200 kPa. The material studied is a loose sand with an initial void ratio equal to 0.80. The parameters listed in Table 2 are used to describe the mechanical behaviour of the sand with the WG-model. Fig. 3 shows the numerical simulations of the above tests by integrating the constitutive equations using a sub-stepping stress point algorithm of the Forward-Euler type with over 100,000 time steps.

Table 2. Material parameters for WG-model

G_0	ν	e_{cs0}	α_F	α_0	n_f,	n_m	n_{cs}	h_{cs}	φ_{cs}
200	0.30	0.75	0.0	0.005	1.50	1.50	0.5	0.005	30°

The results for the constant shear tests under stress-controlled mode (direct method) are presented in Fig. 3a. In Fig. 3b, we display similar plots for the very same constant shear tests simulated under mixed-controlled mode (forced dilation).

The second-order work is computed all along the complete loading paths. Since $dq = 0$ and $dp < 0$ throughout the constant shear drained test, a vanishing second-order work is signalled whenever the volumetric strain reaches a peak ($d\varepsilon_v = 0$). This condition is indeed verified in the numerical simulations as indicated by small arrows pointing down in Fig. 3a. The locus of points for which the second-order work first vanishes defines a bifurcation boundary (line) reminiscent of Lade's (1992) instability line which arbitrarily connects the peaks of effective stress paths of loose sand responses under undrained conditions. However, here in stark contrast with Lade's instability line, the treatment of instability as the vanishing of the second-order work carries both a mathematical and a physical meaning. Also, Lade's instability line does not coincide with the bifurcation limit, but is found inside it.

Fig. 3a also shows that the test can proceed past the bifurcation line due to the nature of the loading program, here stress-controlled. The test eventually stops whenever the effective stress path reaches the plastic limit surface giving way to a different operating failure mode than the one (diffuse failure) at the bifurcation point. On the other hand, we reveal in Fig. 3b that when the same test is conducted in mixed loading mode, the solution breaks down as soon as effective stresses reach the bifurcation line previously defined under stress-controlled mode. It is also interesting to notice that this occurs well below the plastic limit surface.

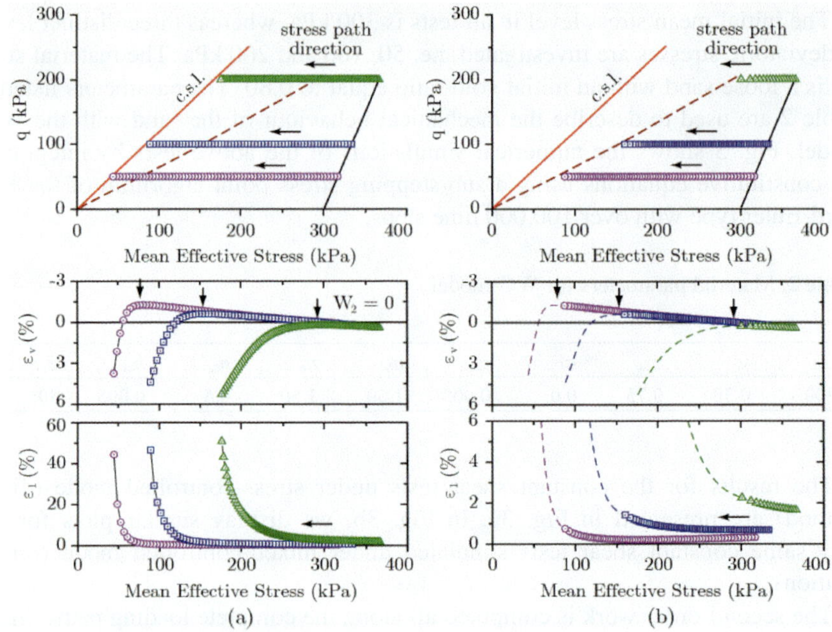

Fig. 3. Numerical results of constant shear tests using: (a) direct method through stress-controlled loading program and (b) forced dilation method through mixed-controlled loading program. Note that at the end of simulations presented in part (a), we are no longer under the realm of small strain assumption ($\varepsilon_1 > 10\%$). Nonetheless, the conclusions arrived at here refer to stress levels yet away from the plastic limit failure, and thus we are still within domain of small strains.

Fig. 4 helps to clarify the findings revealed in the previous paragraphs as it shows the results of the stability analysis for the constant shear tests performed at a deviatoric stress level of 50 kPa. Stability analyses for the other deviatoric stress levels are not presented here as they led to similar conclusions. In Fig. 4a, we show that, although the second-order work becomes null, the loading program is still controllable because the determinant of the constitutive matrix, **D**, is nonzero. Also notice that the determinant of the symmetric part of **D**, det $\mathbf{D}_{sym}$, and the second-order work, W_2, become non-positive at the same point in time. At that moment, the constitutive matrix has lost positive-definiteness but that does not suffice to lead to a non-unique material response. As pointed in Wan et al. (2009) and Nicot and Darve (2009), the directional character of the loading program needs to be taken into account. Curiously, in Fig. 4b we show that the second-order work never becomes zero even when controllability is lost as effective stresses reach the bifurcation line. It actually increases asymptotically to an infinite value, which can be interpreted as the collapse of the specimen marking failure. Nicot and Darve (2009), through the notion of sustainability, established a physical link between the growth of second-order work and the outburst of kinetic energy as mentioned in the previous section. However, we notice here that the second-order work does not necessarily need to pass through a zero value in order for this to happen as apparently stated by these authors. Another observation here is that one of the control parameters is $d\varepsilon_v$ and set to be a non-zero negative value (dilation) and hence the condition of $d\varepsilon_v = 0$ cannot be fulfilled for the second-order work to become zero.

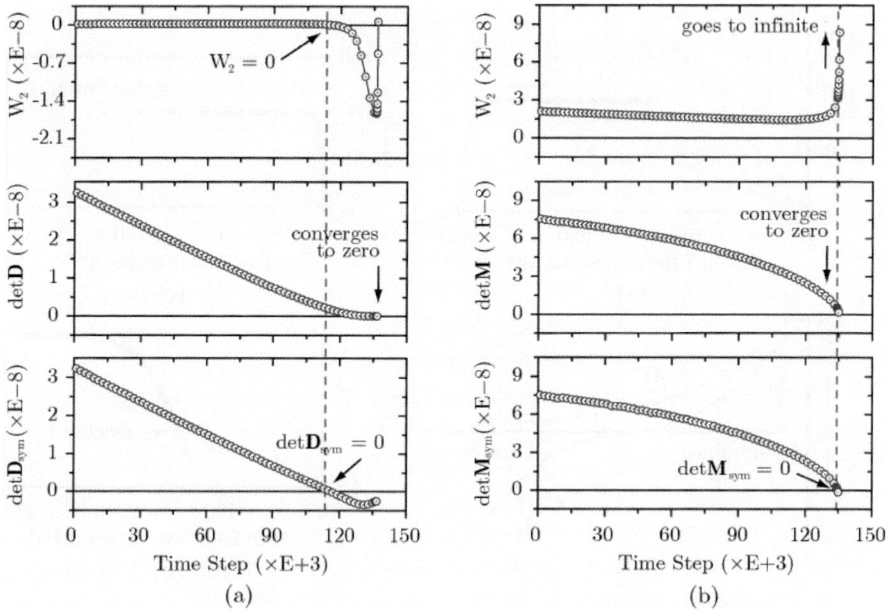

Fig. 4. Stability analysis: (a) direct method through stress-controlled loading program and (b) forced dilation method through mixed-controlled loading program

For further exploration of instability, we performed another series of numerical experiments similar to the constant shear tests described earlier, except that step 2 which corresponds to the drained conventional triaxial shearing is now replaced with a conventional undrained loading until the peak deviatoric stress is reached. At this peak, the constant shear drained path is pursued as before. The same material parameters listed in Table 2 are used. Fig. 5 summarizes a few findings. With regards to the constant shear drained phase, we point out that the simulation could be controlled throughout the entire test despite the non-positiveness of second-order work from the start of the simulation. This is essentially because det **D** is never zero. However, we see in Fig. 5a that det (**D**) approaches zero as the effective stress path reaches the plastic limit line. On the other hand, when the forced dilation procedure was attempted to purse the constant drained phase after the conventional undrained phase, the numerical simulations broke down right away. No solution for the constant shear path could be obtained regardless of the size of the control parameter increment. These numerical exercises are useful in that they reinforce the main conclusions arrived at in Darve et al. (2007). In other words, there certainly exist other failure surfaces within the plastic limit one when examining non-associated materials like geomaterials. In fact, there is a whole domain in axisymmetric stress-strain space where bifurcation, non-uniqueness, instabilities and failure appear and that depends on a series of factors, in special, the stress-strain history, current loading direction and loading mode.

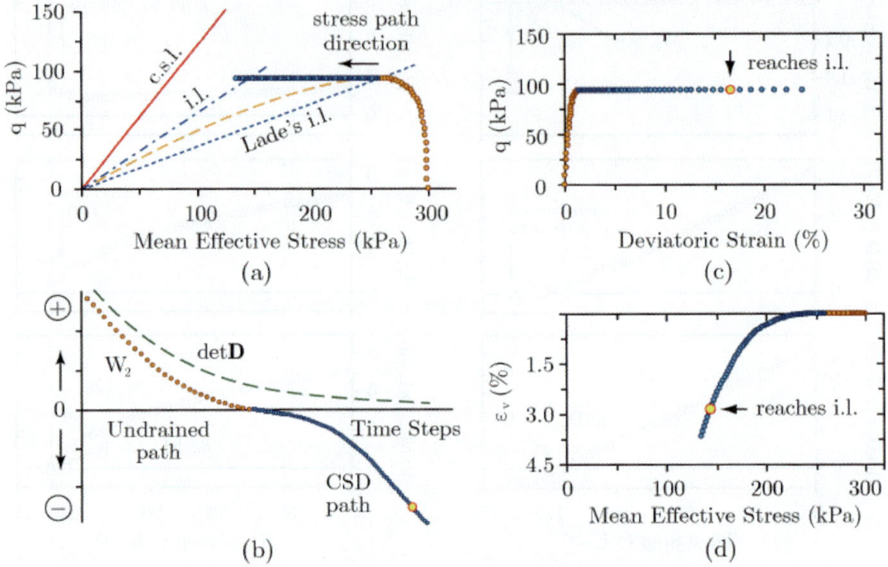

Fig. 5. Stress-controlled constant shear test preceded by conventional undrained test up to peak point

5 Plastic Flow Rule

In plasticity, it is common to assume that the direction of plastic flow does not depend on the direction of stress increments (Hill 1950). Depending on how this postulate is regarded, such plastic flow rule can either be associated or non-associated with a yield locus that encloses an elastic domain. In general, the plastic flow rule is expressed as follows:

$$d\boldsymbol{\varepsilon} = d\lambda \frac{\partial g}{\partial \boldsymbol{\sigma}} \tag{8}$$

where dλ is a plastic multiplier and g is the plastic potential function. When g is chosen to be identical to the yield function f, the flow rule is said to be associative;

otherwise it is called non-associative. In either case, it is important to note that a link between stability and flow rule type has already been established more than fifty years ago. For example, Drucker (1956, 1959) has shown that a hardening material obeying an associated flow rule is always stable but when the failure criterion is fulfilled. This holds true regardless of the loading conditions. On the other hand, non-associative materials may present unstable response even during the hardening stage of loading process (Bigoni and Hueckel, 1992). In this section, we explore the validity of the flow rule postulate and as a consequence its link to instability.

Kishino (2003) has recently revealed through numerical analysis via granular element method (GEM) that the direction of plastic strain increment depends on the direction of the applied stress increment, at least, under 'true' triaxial stress conditions. From another perspective, Darve and Nicot (2005) have also arrived to similar conclusions using a multidirectional model. In this section, we perform a series of numerical simulations using the discrete element method (DEM) as a means to investigate these recent findings under a much broader loading and stress history conditions. The commercial software PFC-3D from Itasca (1999) is employed in all DEM simulations.

The discrete element model used in the simulations consists of a 50 mm cubic assembly with 5,305 unbonded polydispersed spherical particles (see Fig. 6) following a uniform particle size distribution generated from specified values of minimum and maximum radii (1-2 mm). The resulting initial porosity of the assembly is 0.40. The parameters used in all simulations are: Young's modulus equal to 1 GPa, stiffness ratio (k_n/k_s) equal to 2 and particles density equal to 2,650 kg/m^3. It is noted that we have also used a larger number (about 10,000) of particles to test the validity of the representative elementary volume (REV) assumption. However, this led to essentially the same results so that for practical and numerical efficiency purposes the smaller system of particles was retained in all subsequent simulations.

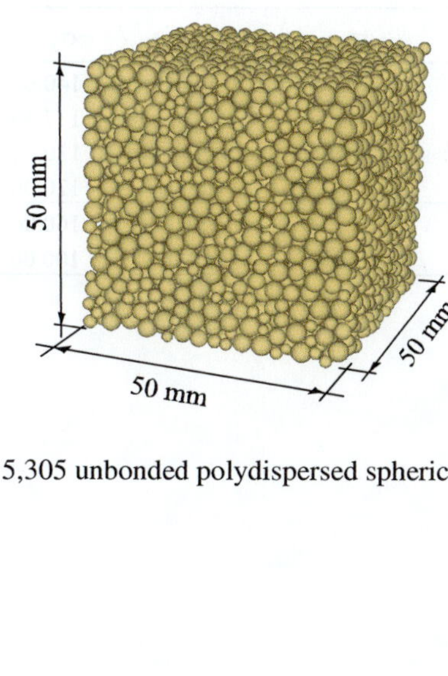

Fig. 6. Cubic assembly of 5,305 unbonded polydispersed spherical particles

The directional analysis earlier introduced by Gudehus (1979) was extended to a three-dimensional setting in order to examine the incremental character of the granular material stress-strain response and therefore, the nature of plastic flow rule. As such, spherical stress probe tests are here carried out in the sense that equal magnitude stress probes are applied in all directions of the principal stress space. This type of numerical experiments is a practical alternative to similar lab experiments on real sand which are very difficult and tedious to perform under true triaxial conditions. A set of 652 spherical stress probes are realized at different stress states reached after a series of loading histories as illustrated in Fig. 7 with respect to the Rendulic plane and the π-plane. The probes performed here have a much smaller magnitude, 0.1 kPa, as compared to the one used in the work of Calvetti et al. (2003), 10 kPa, even though similar stress levels (mean stress around 100 kPa) are dealt with. The increment norm is defined as $\|\Delta\sigma\| = \sqrt{(\Delta\sigma_x^2 + \Delta\sigma_y^2 + \Delta\sigma_z^2)}$. Such strictly smaller increment norm value was chosen so as to guarantee that incremental response is linear.

One series of tests consists of moving along the hydrostatic axis to a mean stress equal to 100 kPa, and then deviating from it at various angles in the deviatoric plane at a constant value q = 60 kPa for comparison purposes. Hence, working within a sextant of the deviatoric plane, various radial paths can be obtained starting from triaxial compression (TC) to triaxial extension (TE) passing through various Lode's angle θ values. Another series of tests refer to the classical conventional triaxial compression (CTC) and conventional triaxial extension (CTE) tests. In the former, the confining pressure is maintained constant with increasing axial stresses, whereas in the latter, the confining pressure is increased with constant axial stresses. The final stress values prior to the stress probing phase are listed in Table 3.

Table 3. Target stress states prior to probe tests

Paths	σ_x	σ_y	σ_z
TC	80.00	80.00	140.00
θ = 20°	93.05	69.36	137.59
θ = 40°	106.95	62.41	130.64
TE	120.00	60.00	120.00
CTC	100.00	100.00	160.00
CTE	160.00	160.00	100.00

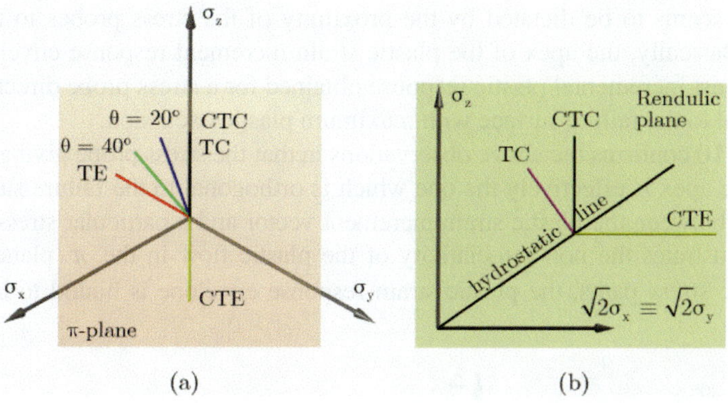

Fig. 7. Response envelopes for various stress histories in the space of plastic strain increments (a) π-plane view and (b) Rendulic plane view

Figure 8 shows the material response under various stress probes plotted as points in the incremental plastic strain space and revealing a number of interesting constitutive features. First, we recall that if a plastic flow rule were to exist as postulated by the theory of plasticity, the incremental plastic strain response would be independent of loading direction, and hence would plot as a straight line in the incremental plastic strain space. Instead, the incremental plastic strain response plots as a series of points clearly defining an oval shaped envelope as shown in Fig. 7a with a projection on the deviatoric (π) plane, and thus destroying the plastic flow rule postulate.

The second noteworthy remark is that, regardless of the initial stress level and stress history, all incremental plastic strain envelopes plot on a plane roughly coincident with the deviatoric plane as illustrated in Fig. 8b. However, having these two planes to coincide would mean zero plastic incremental volume change, which is at any rate not the case for granular materials. Viewing the material response through the incremental relation $d\boldsymbol{\varepsilon}^p = \mathbf{C} \cdot d\boldsymbol{\sigma}$ subject to a constant stress probe norm $\|d\boldsymbol{\sigma}\|$, the resulting incremental plastic strain increments should normally plot as an ellipsoid in the plastic strain space. However, it turns out that the ellipsoid collapses into a planar surface consisting of nested oval shaped plastic strain contours probably due to the vanishing of the determinant of plastic tangent compliance matrix **C**. As such, the order of multiplicity of the null eigenvalue is one with the associated null eigenvector being the normal to the plane containing the incremental plastic strains. This curious result needs to be explored further.

Finally, the last remark relates to the symmetry of the plastic strain envelope about the direction of the stress path prior to stress probing as soon as the previous history involves two equal stress components, thus pointing to axi-symmetry conditions. This occurs for axi-symmetric stress histories such as in TE, CTC, TC and CTE as shown in Fig. 8. For stress paths corresponding to Lode's angle equal to 20° and 40° though, there is a deviation of the plastic strain increment envelope with respect to the direction of the previous stress history (see Fig. 8a and Fig. 9). This deviation seems to be dictated by the proximity of the stress probes to the failure surface. Basically, the apex of the plastic strain increment response envelope refers to the largest incremental plastic response obtained for a stress probe direction that is orthogonal to the failure surface with maximum plastic excursion.

Figure 10 confirms the above observations in that the stress probe giving the direction of the apex is effectively the one which is orthogonal to the failure surface. The deviation between the plastic strain increment vector and a particular stress probe direction illustrates the non-associativity of the plastic flow in the π plane. For axisymmetric stress paths, the plastic strain response envelope is bound to be (nearly)

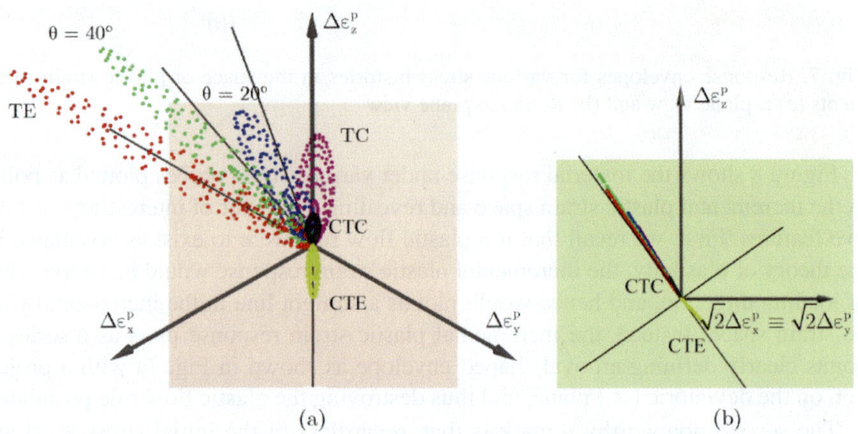

Fig. 8. Response envelopes for various stress histories in the space of plastic strain increments (a) π-plane view and (b) Rendulic plane view

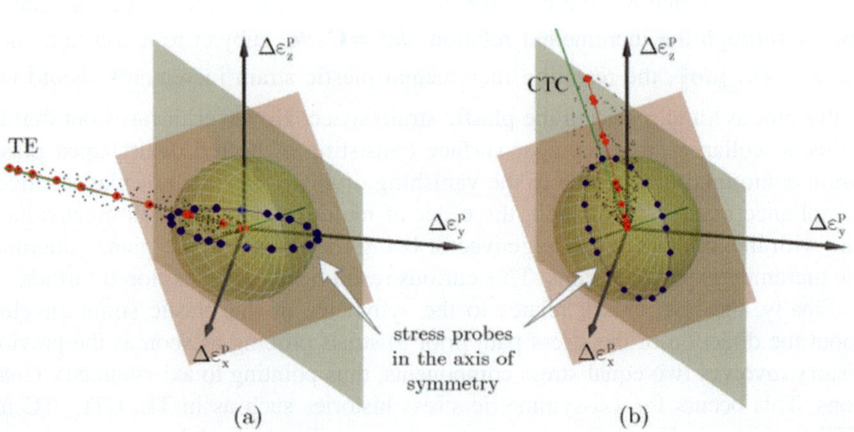

Fig. 9. Plastic strain incremental responses for stress probes with axisymmetric loading history: (a) Triaxial extension – TE and (b) Conventional triaxial compression – CTC cases. The globe represents the spherical stress probes

symmetrical about the previous stress path as the latter coincides with the direction of closest stress probe to the failure surface. Under these conditions, the stress probe and plastic strain directions are almost associated due to the close proximity of the current stress to failure. It is a known fact as deduced from true triaxial tests on sand (Nakai 1989) that close to failure, the plastic flow tends to be associated.

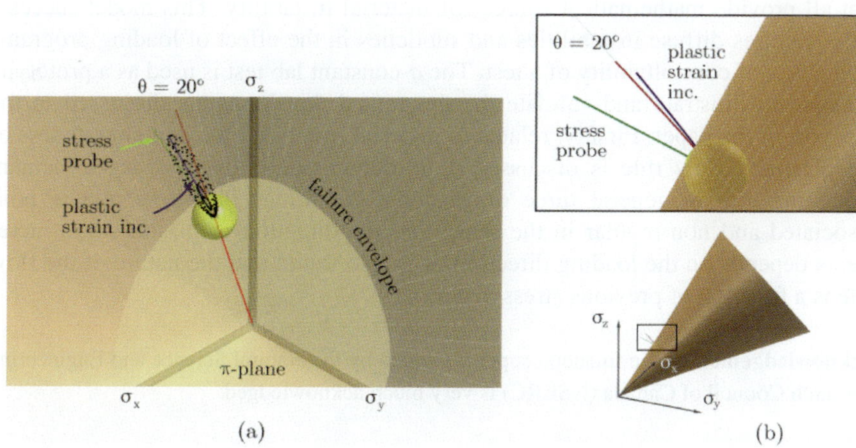

Fig. 10. Plastic strain response for Lode's angle equal to 20°: (a) relation between strain response and proximity to failure envelope; (b) isometric view of failure surface. The inset highlights three things: the stress path direction; the stress probe direction closest to the failure surface; and the largest value of incremental plastic strain.

6 Conclusions

This paper provides a synthesis of some recent theoretical, experimental and numerical studies of material (constitutive) instability in elastoplastic solids such as geomaterials. The condition of a plastic limit as defined by a Mohr-Coulomb criterion has virtually dominated the mathematical and computational analysis of failure. However, it is now fully recognized by experimental evidence and mathematical developments that there exist subcases of failure where material instability can be manifested without any apparent discontinuity in kinematic field and the corresponding localized deformation. It is important to consider this so-called diffuse type of failure since it presents a lower bound to the plastic limit condition. We show that notions of loss of positivity of the second-order work, uniqueness in material response and controllability during a loading program are all related in spirit. The famous Hill's second-order work criterion lacks physical insight and the missing physical link is provided by the notion of loss of sustainability of equilibrium states as put forward by Nicot & Darve (2009). A physical interpretation of the vanishing of the second-order work is the loss of uniqueness in material response marked by an outburst in kinetic energy as an equilibrium state is perturbed. There is then a transition from static to dynamic regime. We also show that for the second-order work to vanish, one of the pre-requisites is to have a non-symmetric tangent constitutive tensor for describing the behaviour of the geomaterial. As such, we introduce a multi-surface elastoplastic model with a non-associated plastic flow rule as well as stress, void ratio and fabric dependencies that all provide mathematical sources of material instability. This model successfully captures diffuse instabilities and subtleties in the effect of loading programs on the loss of controllability of a test. The q-constant lab test is used as a prototype example to illustrate and validate the theoretical developments discussed in the first part of the paper. Finally, related to material instability the non-smoothness of the plasticity flow rule is discussed. It is demonstrated using discrete element modelling that in general three dimensional conditions, the flow rule is non-associated and non-regular in the sense that the direction of plastic strain increments depends on the loading direction. It is also found that the nature of the flow rule is a function of previous stress history.

Acknowledgements. The financial support granted by the Natural Science and Engineering Research Council of Canada (NSERC) is very much acknowledged.

References

Bigoni, D.: Bifurcation and instability of non-associative elastoplastic solids. In: Petryk, H. (ed.) Materials Instabilities in Elastic and Plastic Solids. CISM Courses and Lectures, vol. 414, pp. 1–52 (2000)

Bigoni, D., Hueckel, T.: Uniqueness and localization - I. Associative and non-associative elastoplasticity. Intl. J. Solids Structures 28(2), 197–213 (1991)

Brand, E.W.: Some thoughts on rain-induced slope failures. In: Proc. 10th Intl. Conf. Soil Mech. Fnd. Eng., Stockholm, Balkema, Rotterdam, The Netherlands, vol. 1, pp. 373–376 (1981)

Calvetti, F., Viggiani, G., Tamagnini, C.: A numerical investigation of the incremental behavior of granular soils. Rivista Italiana di Geotecnica 37(3), 11–29 (2003)

Castro, G.: Liquefaction of sand. Harvard Soil Mechanics Series, Cambridge, vol. 81 (1969)

Chu, J., Leroueil, S., Leong, W.K.: Unstable behaviour of sand and its implication for slope instability. Can. Geotechnical J. 40, 873–885 (2003)

Coulomb, C.A.: Essai sur une application des regles des maximis et minimis a quelquels problemes de statique, relatifs a l'architecture. Mem. Acad. Roy. Div. Sav. 7, 343–387 (1776)

Darve, F., Servant, G., Laouafa, F., Khoa, H.D.V.: Failure in geomaterials, continuous and discrete analyses. Comp. Methods Appl. Mech. Eng. 193(27-29), 3057–3085 (2004)

Darve, F., Sibille, L., Daouadji, A., Nicot, F.: Bifurcations in granular media: macro- and micro-mechanics approaches. C. R. Mecanique 335, 496–515 (2007)

Drucker, D.C.: On uniqueness in the theory of plasticity. Q Appl. Math. 14, 35–42 (1956)

Drucker, D.C.: A definition of stable inelastic material. J. Appl. Mech. 26, 101–106 (1959)

Gajo, A.: The influence of system compliance on collapse of triaxial sand samples. Can. Geotechnical J. 41, 257–273 (2004)

Gudehus, G.: A comparison of some constitutive laws for soils under radially symmetric loading and unloading. In: Proc. 3rd Num. Meth. Geomech., Balkema, vol. 4, pp. 1309–1323 (1979)

Hill, R.: The mathematical theory of plasticity. Oxford University Press, London (1950)

Hill, R.: A general theory of uniqueness and stability in elastic-plastic solids. J. Mech. Phys. Solids 6, 236–249 (1958)

Itasca, C.G.: PFC 3D - User manual. Itasca Consulting Group, Minneapolis (1999)

Kramer, S.L., Seed, H.B.: Initiation of soil liquefaction under static loading conditions. Geotech. Eng. 114(4), 412–430 (1988)

Kishino, Y.: On the incremental non-linearity observed in a numerical model for granular media. Italian Geotechnical J. 3, 30–38 (2003)

Lade, P.V.: Static instability and liquefaction of loose fine sandy slopes. J. Geotech. Eng., ASCE 118(1), 51–71 (1992)

Lade, P.V., Duncan, J.M.: Elastoplastic stress-strain theory for cohesionless soil. J. Geotechnical Eng. ASCE 101, 1037–1053 (1975)

Matsuoka, H., Nakai, T.: Stress deformation and strength characteristics of soil under three different principal stresses. In: Proc. Jap. Soc. Civil Engineering, vol. 232, pp. 59–70 (1974)

Mroz, Z.: Nonassociated flow laws in plasticity. J. de Mécanique 2, 21–42 (1963)

Nakai, T.: An isotropic hardening elastoplastic model for sand considering the stress path dependency in three-dimensional stresses. Soils and Foundations 29(1), 119–137 (1989)

Nicot, F., Darve, F.: A micro-mechanical investigation of bifurcation in granular materials. Intl. J. Solids Structures 44, 6630–6652 (2007)

Nicot, F., Darve, F.: A unified framework for failure in geomaterials? In: Proceedings of the 1st International Symposium on Computational Geomechanics (ComGeo I), London, pp. 158–168 (2009)

Nova, R.: Controllability of the incremental response of soil specimens subjected to arbitrary loading programs. J. Mech. Behavior Materials 5(2), 193–201 (1994)

Nova, R.: The failure concept in soil mechanics revisited. In: Labuz, J.F., Drescher, A. (eds.) Bifurcations and Instabilities in Geomechanics, Balkema, Lisse, pp. 3–16 (2003)

Raniecki, B.: Uniqueness criteria in solids with non-associate plastic flow laws at finite deformations. Bul. Acad. Pol. Sciences. Serie Sciences Techniques 27, 391–399 (1979)

Roscoe, K.H., Burland, J.B.: On the generalized stress-strain behavior of 'wet' clay. In: Engineering Plasticity, pp. 535–609. Cambridge University Press, Cambridge (1968)

Rowe, P.W.: The stress-dilatancy relation for static equilibrium of an assembly of particles in contact. Proc. Royal Soc. London A, Math. Phys. Sciences 269(1339), 500–527 (1962)

Rudnicki, J.W., Rice, J.R.: Conditions for the localization of deformation in pressure sensitive dilatant material. J. Mech. Phys. Solids 23, 371–394 (1975)

Sasitharan, S., Robertson, P.K., Sego, D.C., Morgenstern, N.R.: Collapse behavior of sand. Canadian Geotechnical J. 30, 569–577 (1993)

Schofield, A.N., Wroth, C.P.: Critical state soil mechanics. McGraw-Hill, London (1968)

von Mises, R.: Mechanik der festen Korper im plastisch deformablen Zustand. Gottingen Nachrichten, Math. Phys. 4(1), 582–592 (1913)

Wan, R.G., Guo, P.J.: A pressure and density dependent dilatancy model for granular materials. Soils and Foundations 39(6), 1–12 (1999)

Wan, R.G., Guo, P.J.: Effect of microstructure on undrained behaviour of sands. Can. Geotechnical J. 38, 16–28 (2001a)

Wan, R.G., Guo, P.J.: Drained cyclic behavior of sand with fabric dependence. J. Eng. Mech. 127(11), 1106–1116 (2001b)

Wan, R.G., Guo, P.J.: Stress dilatancy and fabric dependencies on sand behavior. J. Eng. Mech. 130(6), 635–645 (2004)

Wan, R.G., Pinheiro, M., Guo, P.J.: Elastoplastic modelling of diffuse instability response of geomaterials. Intl. J. Num. Anal. Meth. Geomech. 35(2), 140–160 (2011)

Loss of Controllability in Partially Saturated Soils

Giuseppe Buscarnera and Roberto Nova

Politecnico di Milano, Dipartimento di Ingegneria Strutturale, Italy

Abstract. A study on saturation induced soil instability is presented. A constitutive model for unsaturated soils is linked to a theoretical approach able to deal with mechanical instability of fully saturated geomaterials. The theoretical approach is therefore extended to the more general case of partially saturated soils. The controllability of generalised loading tests is then considered, focusing on oedometric tests at varying water content. It is shown that instability phenomena, such as collapse at constant axial loading, can be described. In addition, it is discussed how oedometric instability modes can be interpreted as possible compaction band bifurcations. Finally, similar results are reported for the simpler case of isotropic state of stress, defining the most relevant material parameters in determining such instability modes.

1 Introduction

The concept of failure and the identification of instability conditions for a solid material have always represented major topics for engineering and applied science. The mechanical behaviour of soils, and more in general of geomaterials, is affected by the occurrence of instability phenomena, either localized (shear or compaction bands) or diffused (e.g. static liquefaction). The possibility of predicting under which conditions an instability phenomenon takes place can have important engineering and even social relevance, especially in those cases in which a possible failure is related to extremely catastrophic consequences.

The purpose of the present work is to address the general problem of soil instability extending this concept to unsaturated states. Homogeneous bifurcations of the material response are considered (Darve and Chau 1987, Darve 1994, Nova 1989, 1994). Localized instability modes can be considered as a particular case of homogeneous bifurcations, however, the localization of the unstable response being a consequence of a boundary value problem. In particular, Nova and Imposimato (1997) showed that loss of controllability of geotechnical tests was possible at different stress states depending on the loading programme and the control variables. Nicot and Darve (2007) showed that the loss of controllability is equivalent to the loss of sustainability of a given stress and strain state, i.e. it is a proper instability condition.

Under partially saturated conditions surface tension effects provide additional stiffness and strength to geomaterials, making less likely the attainment of an

R. Wan et al. (Eds.): Bifurcations, Instabilities and Degradations in Geomaterials, SSGG, pp. 53–68.
springerlink.com

instability under load perturbations. Nevertheless, since wetting processes produce a sort of material degradation, a less stable condition is progressively approached. Instabilities are therefore possible due to hydro-mechanical perturbations.

In the following it is shown that the occurrence of soil instabilities upon soaking can be easily put into the framework of the loss of controllability theory. In order to achieve this goal, it is first necessary to recall the basic principles on which such a theory is based and to explicit what type of constitutive model has been used to deal with partially saturated soils.

Several examples of application of the theory are then shown, focusing on oedometric and isotropic stress conditions. It is shown that the so called phenomenon of *wetting collapse* can be considered in some cases as a proper unstable process, provided that certain condition are fulfilled by material and test conditions. Finally a brief discussion on the possible localized nature of these oedometric instabilities (compaction banding) is given on the basis of further model predictions.

2 Fundamentals of the Loss of Controllability Theory

The goal of an element test on a material specimen is that of determining the material response to a given loading programme, that can be either statically or kinematically controlled. In order to retrieve such information from the actual test on a finite specimen, the state of stress and strain must be assumed to be uniform so that the relationship between forces and displacements we can measure in the test can be substituted by a stress strain relationship. For instance, a load controlled test, under the hypothesis of small strains, can be considered as an element test in which stresses are imposed and the corresponding strains are determined by means of the constitutive law, i.e.:

$$\dot{\boldsymbol{\sigma}}' = \mathbf{D}\dot{\boldsymbol{\varepsilon}} \tag{1}$$

where a superposed dot stands for increment and a dash indicates effective stresses.

Since stress rates are imposed, they can be considered as known. The response of the material in terms of strain rates is therefore given by:

$$\dot{\boldsymbol{\varepsilon}} = \mathbf{D}^{-1}\dot{\boldsymbol{\sigma}}' \tag{2}$$

A finite response to the stress perturbation is obtained whenever the determinant of the stiffness matrix is not zero. On the contrary, when the stiffness matrix is singular, it is not possible to assign arbitrarily the stress increment. Furthermore, even under zero stress increment an infinity of solutions exist:

$$\dot{\boldsymbol{\varepsilon}} = \alpha\dot{\boldsymbol{\varepsilon}}^{*} \tag{3}$$

where $\dot{\boldsymbol{\varepsilon}}^{*}$ is the eigen-vector of the stiffness matrix and α is an undetermined scalar. The controllability of the loading programme is lost, therefore, and an infinity of homogeneous solutions are possible under constant stress state, i.e. the condition:

$$\det \mathbf{D} = 0 \tag{4}$$

is the condition for the occurrence of homogeneous bifurcations under load control.

Geotechnical tests are usually partly stress and partly strain controlled (e.g. a drained triaxial test). The constitutive law (1) can be therefore rewritten grouping at the l.h.s. the control terms, either static or kinematic, and partitioning the stiffness matrix:

$$\begin{Bmatrix} \dot{\sigma}'_\alpha \\ \dot{\varepsilon}_\beta \end{Bmatrix} = \begin{bmatrix} \boldsymbol{D}_{\alpha\alpha} - \boldsymbol{D}_{\alpha\beta}\boldsymbol{D}_{\beta\beta}^{-1}\boldsymbol{D}_{\beta\alpha} & \boldsymbol{D}_{\alpha\beta}\boldsymbol{D}_{\beta\beta}^{-1} \\ -\boldsymbol{D}_{\beta\beta}^{-1}\boldsymbol{D}_{\beta\alpha} & \boldsymbol{D}_{\beta\beta}^{-1} \end{bmatrix} \begin{Bmatrix} \dot{\varepsilon}_\alpha \\ \dot{\sigma}'_\beta \end{Bmatrix} \tag{5}$$

It can be shown (Nova 1989) that loss of control, corresponding to a vanishing determinant of the matrix of Eq. (5), occurs when:

$$\det \boldsymbol{D}_{\alpha\alpha} = 0 \tag{6}$$

If the stiffness matrix is symmetric, what implies an associated flow rule for elastoplastic constitutive laws, condition (6) is satisfied 'after' condition (1), i.e. for a lower hardening modulus. Since condition (6) is associated to ordinary failure, i.e. to zero hardening modulus, it means that loss of controllability cannot occur in the hardening regime. On the contrary if the flow rule is non-associated condition (6) can be met when the determinant of the stiffness matrix is still postive, i.e. in the hardening regime (Nova 1989).

An example is given by static liquefaction in undrained triaxial tests on loose sand specimens. Assuming that the material is isotropic, in axisymmetric conditions, the constitutive law can be written in terms of two independent stress variables (e.g. the isotropic effective pressure p' and the deviator stress q) and two independent strain variables, conjugated to the stresses by the work density equation (i.e., the volumetric strain ε_v and the deviatoric strain ε_d). In an undrained load controlled test, at small strains, the relationship between control variables and the soil element response can be then written as:

$$\begin{Bmatrix} 0 \\ \dot{q} \end{Bmatrix} = \begin{bmatrix} C_{pp} - C_{pq}C_{qq}^{-1}C_{qp} & C_{pq}C_{qq}^{-1} \\ -C_{qq}^{-1}C_{qp} & C_{qq}^{-1} \end{bmatrix} \begin{Bmatrix} \dot{p}' \\ \dot{\varepsilon}_d \end{Bmatrix} \tag{7}$$

where C_{ij} are the elements of the compliance matrix. Loss of control related to the nullity of the determinant of Eq. (7) occurs at the peak of the deviatoric stress, taking place when:

$$C_{pp} = 0 \tag{8}$$

It can be shown (di Prisco et al. 1995) that Eq. (8) is fulfilled when the state of stress is on a locus given by a straight line passing through the origin of the stress plane p', q, that is known as the Lade's instability line (Lade 1992). For loose sands this condition is met for a mobilized friction angle usually lower than 20°, considerably less than the drained failure value, that is of the order of 30°.

As shown in Imposimato and Nova (1998) many other examples of loss of control can be envisaged in true triaxial testing of soils. In particular it can be shown that instabilities can occur in loading programmes where linear combinations of stresses or strains are controlled, as it is actually the case for the test considered above.

3 Generalized Stress Variables and Constitutive Modelling for Unsaturated Soils

To tackle the problem of test controllability in unsaturated soils it is first necessary to define an appropriate constitutive model for these materials. The complete formulation of the model is available in Buscarnera and Nova (2009a). For this reason only a general description is here given.

The first approach for modelling the constitutive behaviour of unsaturated soils was based on the adoption of two independent stress variables, namely net stress and suction, and on the introduction of the loading collapse yield curve concept (Alonso et al., 1990). Such approach was the first one to provide a common interpretation framework to the general understanding of unsaturated soil behaviour. Nevertheless, some issues were not fully accounted for and several works after it have attempted to provide a more complete description of the hydro-mechanical response of partially saturated soils.

In the development of constitutive laws, in fact, a point of capital importance is represented by the choice of conjugate stress and strain measures. In the particular case of partially saturated soils, the energy input is not only due to the mechanical work done by the external forces. Since the soil is unsaturated, in fact, the possible change in water content can alter the volume occupied by fluids without causing any strain of the soil skeleton, and this water volume change is associated to an energy exchange between water and air at the water menisci. This is in turn related to the surface tension of water that must be proportional to the difference between the air and water pressures, i.e. to suction. Houlsby (1997) has proven on a micromechanical basis that the total specific energy is given by:

$$\begin{aligned}\dot{E}_s &= (\sigma_{ij} - S_r \cdot u_w - (1 - S_r) \cdot u_a) \cdot \dot{\varepsilon}_{ij} - n(u_a - u_w) \cdot \dot{S}_r \\ &= \sigma_{ij}'' \cdot \dot{\varepsilon}_{ij} - ns \cdot \dot{S}_r\end{aligned} \tag{9}$$

where σ_{ij} is the total stress, u_w the pore water pressure, u_a the air pore pressure, s the suction, S_r the degree of saturation and n the porosity of the soil.

This result suggests that a possible alternative as stress measure for unsaturated soils would be the so called average soil skeleton stress (Jommi 2000) defined by the following equation:

$$\sigma_{ij}'' = \sigma_{ij} - S_r \cdot u_w - (1 - S_r) \cdot u_a \tag{10}$$

as it appears to be the intensive measure work conjugate to the strain rate. A similar choice was suggested for the very first time by Jommi and di Prisco (1994), and appears now to be a widely accepted approach (Gens et al. 2006, Nuth and Laloui 2007). Moreover, Eq. (9) confirms the experimental evidence that a single effective stress theory is not able to model unsaturated soil behaviour since this is also affected by a further energy term, i.e. by a smeared suction, defined as:

$$s^* \equiv n \cdot s \tag{11}$$

having the opposite of the rate of the degree of saturation as conjugate strain measure. For brevity sake, the stress defined in Eq. (10) will be henceforth referred to as skeleton stress.

Starting from this framework, a coupled hydro-mechanical constitutive law for unsaturated soils has been developed. The model is conceptually similar to other models already available, but is formulated in a way that is convenient for the analysis of the possible occurrence of instabilities, which constitutes the major goal of the paper. The constitutive law is based on an enlarged form including the modified suction and the degree of saturation as further stress and strain variables, as follows:

$$\dot{\boldsymbol{\Sigma}} = \begin{Bmatrix} \dot{\sigma}''_{ij} \\ \dot{s}^* \end{Bmatrix} = \begin{bmatrix} \mathbf{D}_{\boldsymbol{\sigma\sigma}} & \mathbf{D}_{\boldsymbol{\sigma}\mathbf{w}} \\ \mathbf{D}_{\mathbf{w}\boldsymbol{\sigma}} & \mathbf{D}_{\mathbf{ww}} \end{bmatrix} \cdot \begin{Bmatrix} \dot{\varepsilon}_{ij} \\ -\dot{S}_r \end{Bmatrix} = \mathbf{D}_{\mathbf{ext}} \cdot \dot{\mathbf{E}} \tag{12}$$

In Eq. (12), $\mathbf{D}_{\mathbf{ext}}$ is the extended stiffness matrix for a partially saturated soil, while the vectors $\mathbf{D}_{\boldsymbol{\sigma}\mathbf{w}}$ and $\mathbf{D}_{\mathbf{w}\boldsymbol{\sigma}}$ represent the coupling terms expressing hydraulic and mechanical coupled contributions. $\mathbf{D}_{\boldsymbol{\sigma\sigma}}$ is the constitutive matrix governing the mechanical behaviour of the soil skeleton under constant degree of saturation, while $\mathbf{D}_{\mathbf{ww}}$ is actually a scalar function related to the water retention curve of the soil. For the sake of simplicity, the model is developed assuming that the material behaviour is isotropic and that hydraulic hysteresis effects can be neglected. However, the introduction of these aspects within the above described framework is in principle straightforward.

The hardening law controlling the size of the yield domain (Fig. 1) in the skeleton space is assumed to be governed by two separate contributions, a mechanical one and a hydraulic one. The analytical expression of the hardening law in a rate form is:

$$\dot{p}_S = \rho_S p_S \left(\dot{\varepsilon}_V^p + \xi_S \dot{\varepsilon}_S^p \right) - r_{sw} p_S \dot{S}_r \tag{13}$$

where ρ_S, ξ_S and r_{sw} are hardening constitutive parameters. The first term is the usual hardening relationship for granular materials, that is conveniently expressed in terms of both volumetric and deviatoric plastic strains, to account for dilation effects on soil strength (see e.g. Nova 1977).

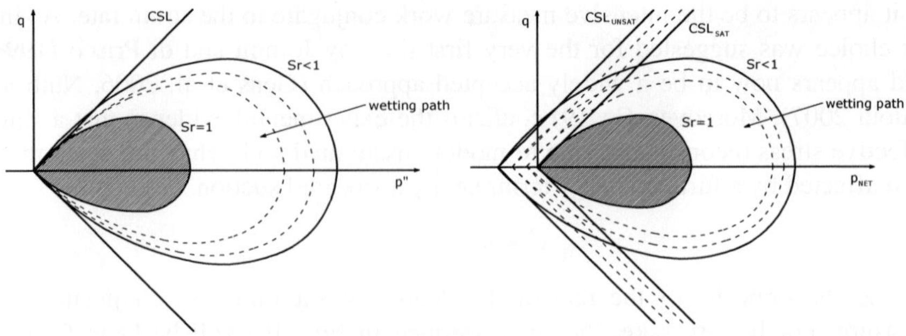

Fig. 1. Contraction of the yield surface due to wetting processes depicted in both the skeleton stress space and the net stress space

4 Loss of Controllability in Partially Saturated Soils

In this section the theory of controllability developed for dry and fully saturated soils (Nova 1989, Imposimato and Nova 1997) is extended to unsaturated states. In order to achieve this goal, reference is made to the extended stiffness matrix of Eq. (12). The extension of the theory here presented is mainly based on a previous work by the authors (Buscarnera and Nova 2009b). Only the most relevant features of the theory will then be recalled.

In order to tackle the problem of test controllability in unsaturated soils a generalized concept of loading programme must be used, which includes also hydraulic variables. Loading programmes in which either the smeared suction or the saturation index varies, together with appropriate combinations of stresses and strains have to be considered. A similar approach has been used for the first time by Vaunat et al. (2002).

If static and kinematic control variables are defined as a simple linear combination of the generalised stresses and generalised strains, respectively, it is possible to define:

$$\dot{\xi} = T_\sigma \dot{\Sigma} \tag{14}$$

$$\dot{\eta} = T_\varepsilon \dot{E} \tag{15}$$

Since ξ and η must be work conjugate:

$$T_\varepsilon = \tilde{T}_\sigma^{-1} \tag{16}$$

From Eq. (12) it follows that:

$$\dot{\xi} = T_\sigma D_{ext} \tilde{T}_\sigma \dot{\eta} = \Delta \dot{\eta} \tag{17}$$

Eqs. (14)-(15) represent the most simple approach to the loss of controllability, when control variables are constituted by linear combinations of either stresses or strains. Since, in fact, control variables are often partly generalised strains and partly generalised stresses, Eq. (17) can be rearranged as in Eq. (5):

$$\begin{Bmatrix} \dot{\xi}_\alpha \\ \dot{\eta}_\beta \end{Bmatrix} = \begin{bmatrix} \Delta_{\alpha\alpha} - \Delta_{\alpha\beta}\Delta_{\beta\beta}^{-1}\Delta_{\beta\alpha} & \Delta_{\alpha\beta}\Delta_{\beta\beta}^{-1} \\ -\Delta_{\beta\beta}^{-1}\Delta_{\beta\alpha} & \Delta_{\beta\beta}^{-1} \end{bmatrix} \begin{Bmatrix} \dot{\eta}_\alpha \\ \dot{\xi}_\beta \end{Bmatrix} \tag{18}$$

The loading programme will reach a condition of instability whenever the determinant of the matrix in Eq. (18) vanishes. This occurs when:

$$\det \Delta_{\alpha\alpha} = 0 \tag{19}$$

The generalised loading programmes for unsaturated soils are more complex than those usually considered for dry or fully saturated soils, however. If classic variables used in common laboratory tests are used (e.g suction or net stresses), in fact, the definition of the loading programme to be used will be a mixed function of stress and strain rates.

Formally it would be then possible to obtain again Eqs. (16)-(17) and follow the same procedure as in Eq. (5), but there would be some shortcomings with this approach (i.e mixed nature of governed stress variables) related to the somewhat ambiguous definition of intensive and extensive variables.

A more convenient approach, based on the definition of controlling variables ϕ, has therefore been adopted. The vector ϕ collects the variables actually governed during the test, whose increments are assumed to be known as a function of stress and strain rates. In this case, a formal representation of both controlling variables and controlled variables can be expressed as follows:

$$\begin{Bmatrix} \dot{\phi} \\ \dot{\psi} \end{Bmatrix} = \begin{bmatrix} \Omega_{\phi\sigma} & \Omega_{\phi\varepsilon} \\ \Omega_{\psi\sigma} & \Omega_{\psi\varepsilon} \end{bmatrix} \begin{Bmatrix} \dot{\Sigma} \\ \dot{E} \end{Bmatrix} \tag{20}$$

In Eq. (23) the controlled variable rate $\dot{\psi}$ is linked to stress and strain rates by a formal relationship similar to that of controlling variables.

The two sets of variables are connected by the following expression:

$$\dot{\phi} = \mathbf{X}\dot{\psi} \tag{21}$$

where the control matrix $\mathbf{X}$, can be derived from Eqs. (20) and (12) to be equal to:

$$\begin{aligned} \mathbf{X} &= \left(\Omega_{\phi\sigma} D_{ext} + \Omega_{\phi\varepsilon}\right)\left(\Omega_{\psi\sigma} D_{ext} + \Omega_{\psi\varepsilon}\right)^{-1} \\ &= \left(\Omega_{\phi\sigma} + \Omega_{\phi\varepsilon} C_{ext}\right)\left(\Omega_{\psi\sigma} + \Omega_{\psi\varepsilon} C_{ext}\right)^{-1} \end{aligned} \tag{22}$$

Where C_{ext} is the extended constitutive compliance matrix.

For a general laboratory test, the control variables rate $\dot{\phi}$ can be explicitly defined in tems of generalised stresses and strains, being matrices $\Omega_{\phi\sigma}$ and $\Omega_{\phi\varepsilon}$

always known. It is then possible to obtain the loss of controllability condition for a given set of mixed stress-strain control variables, as it is the case for partially saturated soils modelled within this framework.

Assuming, in fact, that the definition of controlled variables ψ is not singular (i.e, $\det\left(\Omega_{\psi\sigma}D_{ext}+\Omega_{\psi\varepsilon}\right)^{-1}\neq 0$ or equivalently $\det\left(\Omega_{\psi\sigma}+\Omega_{\psi\varepsilon}C_{ext}\right)^{-1}\neq 0$) the control of variables ϕ is lost when matrix $\mathbf{X}$ for that given control becomes singular, i.e when:

$$\det\left(\Xi_1\right)=\det\left(\Omega_{\phi\sigma}+\Omega_{\phi\varepsilon}C_{ext}\right)=0 \tag{23}$$

or equivalently when:

$$\det\left(\Xi_2\right)=\det\left(\Omega_{\phi\sigma}D_{ext}+\Omega_{\phi\varepsilon}\right)=0 \tag{24}$$

The two quantities defined by Eqs. (23)-(24) can be therefore considered as *instability indices*, being their vanishing related to the onset of a loss of controllability of the loading programme for which the two quantities have been evaluated. Eqs. (23)-(24) show that this specific form of instability condition only depends on the two matrices $\Omega_{\phi\sigma}$ and $\Omega_{\phi\varepsilon}$ (and therefore on the specific control considered), and either on the extended constitutive stiffness matrix D_{ext} or on the extended constitutive compliance matrix C_{ext} (and therefore on the current hydro-mechanical stress level).

5 Some Applications of the Loss of Controllability Theory for Unsaturated Soils

5.1 Oedometric Wetting Test

The first example presented in this section refers to a classical typology of laboratory tests used in unsaturated soil mechanics: an oedometric test with a final wetting stage under constant vertical net stress. This test is largely adopted to characterize the so called *collapse potential* of unsaturated soils, i.e. the tendency exhibited by some soils to develop volumetric compaction when saturated. Even though the the term collapse implicitely recalls an unstable response of the specimen, the mechanical nature of the phenomenon is still debated.

In this example the value of $\det\left(\mathbf{X}\right)$ is monitored for a specific loading programme during the test, checking if a transition from a stable to an unstable response can occur in terms of the condition given by Eqs. (23)-(24).

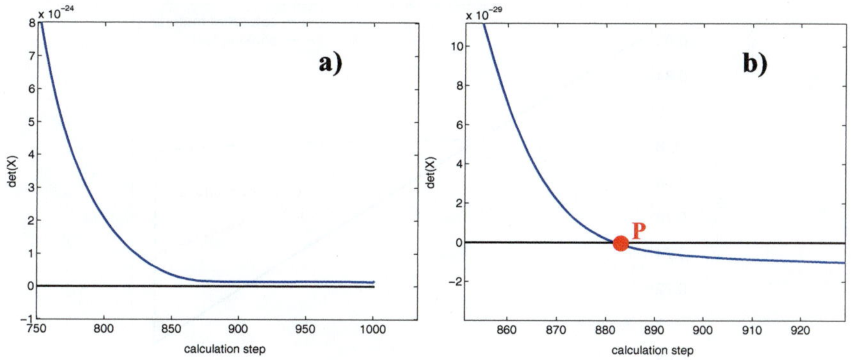

Fig. 2. (a) Sign of det(X) calculated for vertical net stress and suction control; (b) sign of det(X) calculated for vertical net stress and water content control

Figure 2 shows that, while $\det(\mathbf{X})$ is positive throughout the entire test for suction and vertical net stress control (Fig. 2a), the same quantity vanishes and then becomes negative during the simulation if it is calculated in terms of a different set of control variables (Point P, Fig. 2b). As a consequence, the test is perfectly controllable in the first case, but there is the possibility of an unstable compaction if a non controllable loading programme is followed (e.g. the control monitored in Fig. 2b, representing the response if vertical net stress and water content would be governed).

The material therefore comes across an unstable state of equilibrium during the saturation process (Point P in Fig. 3). This possible instability can be actually activated only provided that the control is accordingly changed, i.e. that a switch from suction control to water content control is imposed after the loss of controllability condition is achieved.

In a loading programme with water content control rather than suction control a switch in test control conditions may represent a wetting stage in which saturation is obtained injecting a water flux in the unsaturated soil specimen, rather than through suction reduction. Under these conditions suction and degree of saturation are not directly governed, being their value a consequence of the overall constitutive response. In this way the evolution of the hydraulic stress-strain variables can in principle become uncontrollable, producing a general material instability.

In order to show an example of the possibility above described, the evolution of the water ratio e_w during the simulation is given in Fig. 4.

It is remarkable that e_w reaches a peak during wetting, and therefore an upper water content threshold can be identified. It can be shown that the peak in water ratio coincides with the loss of controllability condition given by Eqs. (23)-(24). Therefore, if water content would be directly controlled at that point, e.g. injecting a water flux in the specimen, instability under constant deviatoric stress would be predicted by the model (Fig. 4b).

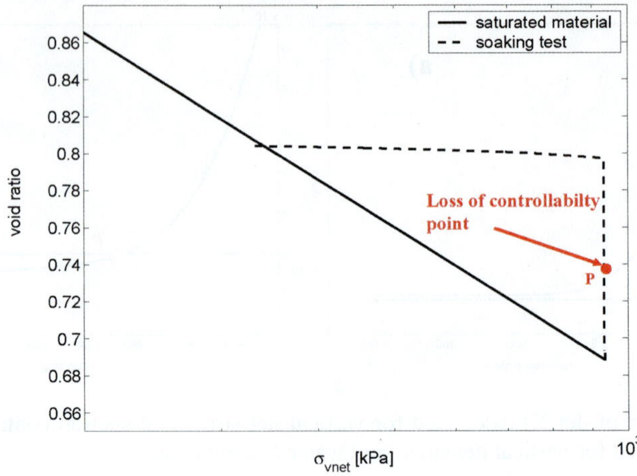

Fig. 3. Achievement of a loss of controllability point upon saturation

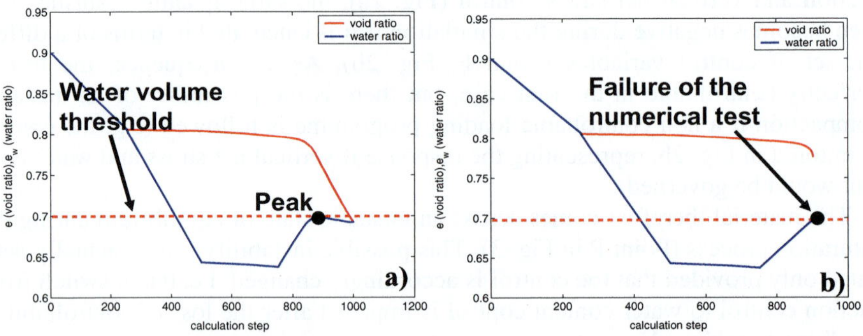

Fig. 4. (a) Peak of water content attained during soaking; (b) Loss of controllability for water control under constant net stress

It is worth noting that the physical meaning of this result is that in a test in which water is added to the specimen, after the peak shown in Fig. 4, the specimen expels more water than that injected, which implies a collapse of the soil skeleton, i.e. an uncontrollable soil response.

This numerical result can be a possible explanation of the uncontrolled compaction often suffered by unsaturated soil samples wetted in oedometric conditions. If the soil is in fact characterised by a loose deformable structure, which is actively sustained by suction effects, the loss of capillary contributions during wetting can become uncontrollable when a certain hydraulic threshold is reached, with the soil rapidly evolving towards a more stable condition.

5.2 Compaction Banding in Unsaturated Soils

All the instabilities described in the framework of the theory of controllability have to be interpreted as homogeneous bifurcations. Nevertheless, given the particular case considered and the fact that the onset of an instability is strongly affected by small imperfections of the real specimen, it is reasonable to assume that the phenomena so far described can in principle be related to an unstable localized compaction (compaction bands, Fig. 5).

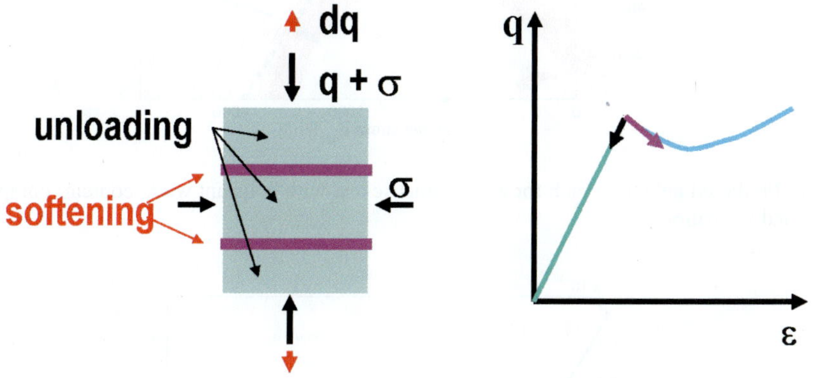

Fig. 5. Simplified scheme of compaction banding instabilities

In order to better explain this concept other simulations have been performed, in which a mechanical loading is applied under a hydraulic constraint imposing an increase of the degree of saturation. The resulting hydraulic degradation can produce a form of mechanical instability characterized by the reduction of the vertical stress carried by the sample upon oedometric straining. The results of a numerical simulation of an oedometric test with constant water content are shown in Fig. 6 in terms of predicted net stress path.

A similar behaviour has been theoretically obtained, and then experimentally observed, in bonded geomaterials, where grain bond destructuration is the responsible factor of the material degradation and of the instability (Arroyo et al. 2005, Castellanza et al. 2007). In the case here presented the phenomenon is the result of the vanishing of a hydraulic bonding effect due to the reduction of suction during loading.

Also this instability mode obtained with the model can be reinterpreted using the theoretical approach presented in the previous sections. This is evident in Fig. 7, where the evolution of the instability index for vertical stress control under the constraint of constant water content is shown.

It is worth noting that this simulation has to be seen only as an example aimed at showing the possibility of such phenomena in the light of a theoretical model. The oedometric test at constant water content therefore represents only one of the

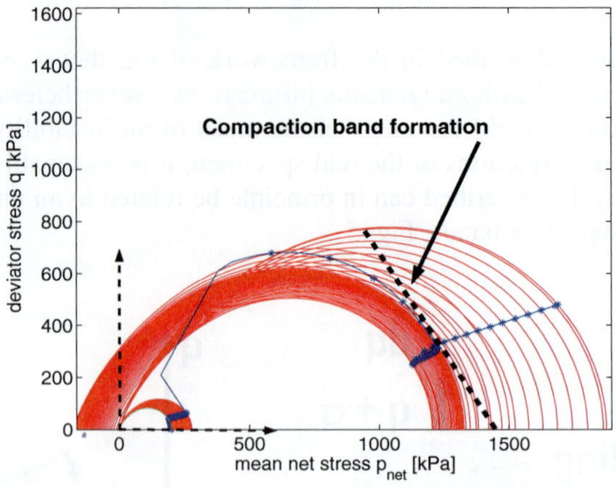

Fig. 6. Predicted net stress path for an oedometric test with constant water content: compaction band formation

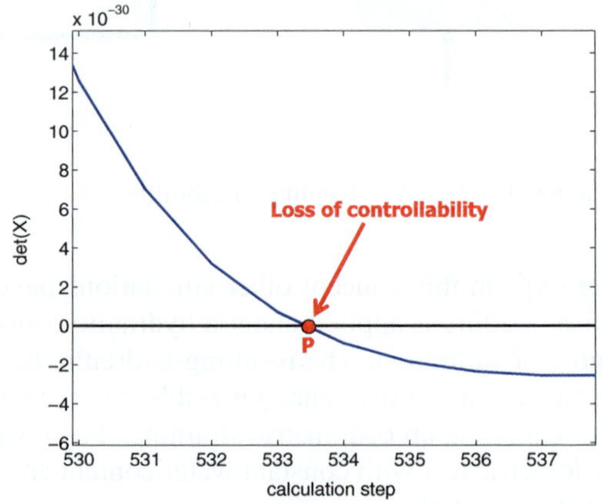

Fig. 7. Evolution of the instability index during the oedometric test at constant water content

possible hydro-mechanical constraints for which the onset of such instabilities can occur. More specifically, a compaction band instability can occur in this type of test only provided that the loss of suction during loading and the resulting soil weakening are extremely relevant.

Any other test in which the applied load produces a relevant reduction of degree of saturation can produce similar results, however. This is shown with reference to an oedomeric test simulation in which a mixed hydro-mechanical perturbation is imposed, i.e. in which the material is wetted and deformed at the same time (Fig. 8).

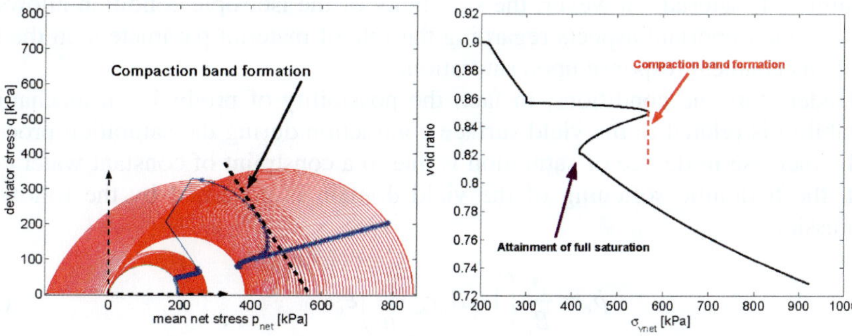

Fig. 8. Results of the simulation of an oedometric test with straining and wetting

5.2 *Volume Implosion and Influence of Material Parameters*

The examples shown in this last section refer to the case of saturation induced instabilities predicted in isotropic conditions. The typology of phenomena predicted under such state of stress are similar to those observed under oedometric conditions and can be defined as a sort of *volume implosion*. The simplicity of the isotropic case, however, allows a better insight in the role played on the onset of mechanical instabilities by some constitutive parameters describing the material behaviour.

Two examples of instabilities predicted by the model under isotropic conditions are shown in Fig. 9, with reference to a wetting test under constant mean net pressure and a isotropic loading test at constant water content.

An instability under isotropic conditions is a very particular and relatively uncommon phenomenon, and as such these results have to be considered as the evidence of extreme possibilities induced by significant hydraulic softening of an

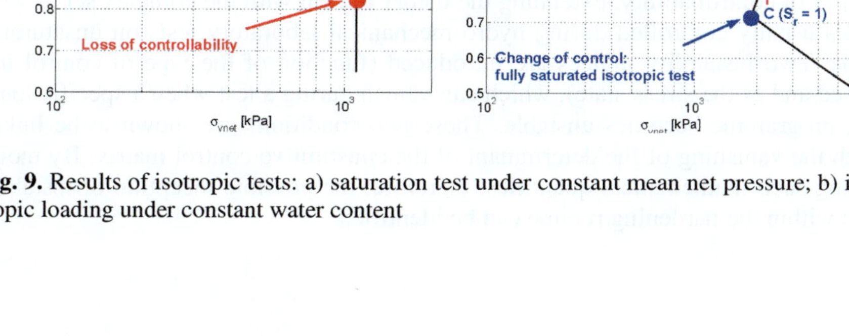

Fig. 9. Results of isotropic tests: a) saturation test under constant mean net pressure; b) isotropic loading under constant water content

unsaturated material. However, the simplicity of the isotropic condition allows to grasp some important aspects regarding the role of material parameters on the unstable mechanical response upon saturation.

Under isotropic conditions, in fact, the possibility of producing a mechanical instability is related to the yield surface contraction during the saturation process. If the increase in degree of saturation is due to a constraint of constant water content, the hydraulic softening of the yield domain is governed by the following expression:

$$\dot{p}_s = \frac{p_s}{B_p}\left(1 - B_p r_{sw} \frac{S_r}{n}\right)\dot{\varepsilon}_v \tag{25}$$

which gives the evolution of the internal variable governing the size of the yield surface during the test.

The simplicity of the isotropic loading therefore allows to point out that the most important parameters influencing the onset of a mechanical instability under saturation are the plastic compressibility B_p, i.e. the amount of plastic deformation developed during wetting, and the hydraulic hardening parameter r_{sw}, related to the increase of the apparent over-consolidation effect with suction.

Within the present modelling framework, the evolution of the yield surface is basically treated as an isotropic process. As a result, these remarks can be considered to be general for any other test typology, even characterized by saturation processes performed under relevant shear stresses. Therefore, instability conditions that can for instance rule the onset of a landslide can be dependent also by material properties not directly related with shear strength, such as volumetric compressibility and hydraulic hardening phenomena.

6 Conclusions

A study aimed at predicting the onset of instability in partially saturated soils has been presented. A constitutive model for unsaturated soil behaviour has been used for the purpose. The model is formulated in terms of work conjugate stress variables, adopting a stress measure (average soil skeleton stress) allowing for the study of mechanical instabilities both within the partially saturated and the fully saturated regime. This allowed us to easily investigate stress paths characterized by suction and saturation variations due to a mechanical loading.

The constitutive law for unsaturated soils has then been linked to the so called theory of controllability, extending the theory to deal with the complex set of variables usually controlled during hydro-mechanical laboratory tests on unsaturated soils. Two instability indices are introduced (function of the type of control imposed and of the stress state), which can vanish during a test when a specific loading programme becomes unstable. These two conditions are shown to be linked with the vanishing of the determinant of the constitutive control matrix. By monitoring such mathematical quantities, therefore, the possible inception of instabilities within the hardening regime can be identified.

Some examples of applications have finally been discussed, focusing on oedometric and isotropic test conditions. Concerning the oedometric case, the so called wetting collapse has been discussed, trying to suggest an alternative mechanical explanation of the process. The model shows the possibility of describing the phenomenon of wetting collapse as a proper unstable process. It is worth noting, however, that this possibility is related to the fact that the saturation process is not suction controlled and the material parameters allow the achievement of a loss of controllability condition. If these two conditions are not simultaneously fulfilled the collapse phenomenon has to be interpreted as a simple plastic compression of the specimen driven by a suction reduction process. In addition some considerations and examples aimed at discussing the possible localized nature of these phenomena in form of compaction banding have been given.

Finally isotropic test conditions have been considered, with the main purpose to discuss the role of material parameters in the onset of a saturation instability. It is shown that general instability conditions, possibly ruling the onset of rainfall induced landslides, can be dependent also by material properties not directly related with shear strength, such as plastic compressibility and hydraulic hardening.

References

Alonso, E.E., Gens, A., Josa, A.: A constitutive model for partially saturated soils. Géotechnique 40, 405–430 (1990)

Arroyo, M., Castellanza, R., Nova, R.: Compaction bands and oedometric tests in cemented soils. Soils Found. 45, 181–195 (2005)

Buscarnera, G., Nova, R.: An elastoplastic strainhardening model for soil allowing for hydraulic bonding-debonding effects. Int. J. Numer. Anal. Meth. Geomech. 33(8), 1055–1086 (2009a)

Buscarnera, G., Nova, R.: Loss of controllability in unsaturated soils. European Journal of Environmental and Civil Engineering 13(2), 235–250 (2009b)

Castellanza, R., Gerolymatou, E., Nova, R.: Compaction Bands in Oedoemetric Tests on High Porosity Soft Rock. In: Proceedings of the 13th International Conference on Experimental Mechanics, Alexandroupolis, Greece, July 1-6 (2007)

Darve, F.: Liquefaction phenomenon: modelling, stability and uniqueness. In: Arulanandan, Scott (eds.) Verification of Numerical Procedures for the Analysis of Soil Liquefaction Problems, pp. 1305–1319 (1994)

Darve, F., Chau, B.: Constitutive instabilities in incrementally non linear modelling. In: Desai, C.S., Gallagher, G.H. (eds.) Constitutive Laws for Engineering Materials, New York, pp. 301–310 (1987)

di Prisco, C., Matiotti, R., Nova, R.: Theoretical investigation of the undrained stability of shallow submerged slopes. Géotechnique 45(3), 479–496 (1995)

Gens, A., Sanchez, M., Sheng, D.: On constitutive modelling of unsaturated soils. Acta Geotechnica 1, 137–147 (2006)

Jommi, C.: Remarks on the constitutive modelling of unsaturated soils. In: Proceedings of the International Workshop on Unsaturated Soils, Trento, pp. 139–153 (2000)

Jommi, C., di Prisco, C.: Un semplice approccio teorico per la modellazione del comportamento meccanico dei terreni granulari parzialmente saturi. In: Atti Convegno sul Tema: Il Ruolo deiFluidi nei Problemi di Ingegneria Geotecnica, Mondovi', pp. 167–188 (1994)

Houlsby, G.T.: The work input to an unsaturated granular material. Géotechnique 47, 193–196 (1997)

Imposimato, S., Nova, R.: An investigation on the uniqueness of the incremental response of elastoplastic models for virgin sand. Mechanics of Cohesive-Frictional Material 3, 65–87 (1998)

Lade, P.V.: Static instability and liquefaction of loose fine sandy slopes. J. Geotech. Engin. ASCE 118, 51–71 (1992)

Nicot, F., Darve, F.: A micro-mechanical investigation of bifurcation in granular materials. Int. J. Solids Struct. 44, 6630–6652 (2007)

Nova, R.: On the hardening of soils. Arch. Mech. Stos. 29, 445–458 (1977)

Nova, R.: Liquefaction, stability, bifurcations of soil via strain-hardening plasticity. In: Gdansk, Dembicki, E., Gudehus, G., Sikora, Z. (eds.) Proc. Int. Works, Numerical Methods for the Localisation and Bifurcation of granular bodies, pp. 117–132. TUG (1989)

Nova, R.: Controllability of the incremental response of soil specimens subjected to arbitrary loading programmes. J. Mech. Behav. Mater 5, 193–201 (1994)

Nova, R., Imposimato, S.: Non-uniqueness of the incremental response of soil specimens under true-triaxial stress paths. In: Pande, Pietruszczak (eds.) Proc. VI NUMOG, Montreal, Balkema, pp. 193–197 (1997)

Nova, R., Castellanza, R., Tamagnini, C.: A constitutive model for bonded geomaterials subject to mechanical and/or chemical degradation. Int. J. Numer. and Anal. Meth. Geomech. 27, 705–732 (2003)

Nuth, M., Laloui, L.: Effective stress concept in unsaturated soils: Clarification and validation of a unified framework. Int. J. Numer. and Anal. Meth. Geomech. 32, 771–801 (2008)

Vaunat, J., Gens, A., Pontes Filho, I.D.S.: Aplication of Localization Concepts to Discontinuos Water Content Patterns in Unsaturated Media. In: Eighth International Symposium on Numerical Models in Geomechanics - NUMOG VIII, vol. 01, pp. 179–184. Swets & Zeitlinger, Roma (2002)

Critical Plane Approach to Analysis of Failure Criteria for Anisotropic Geomaterials

Zenon Mróz[1] and Jan Maciejewski[1,2]

[1] Institute of Fundamental Technological Research, ul. Świętorzyska 21, 00- 049 Warsaw, Poland
e-mail: zmroz@ippt.gov.pl
[2] Institute of Construction Machinery Engineering, Warsaw University of Technology, Narbutta Str. 84, 02-524 Warsaw
e-mail: jmaciej@ippt.gov.pl

Abstract. Numerous geomaterials such as rock and soil exhibit structural anisotropy related to material fabric elements such as crack pattern, bedding, layering, contact arrangements, among others. The fundamental problem is associated with the specification of effective properties of the representative material element, accounting for microstructure and defect distribution. The present work is aimed at the derivation of failure criteria for materials with anisotropic microstructure, such as crack pattern, microlaminate structure, or grain contact arrangement. The assumed density distribution function specifies the microstructure used in deriving the failure criteria and damage evolution rules for specified deformation histories. The state of a material is described by the damage density distribution on the physical planes. The critical plane approach is used with account for a damaged and an intact area fraction. The maximum of failure function is specified for all potential failure planes and critical plane orientation is determined. The derived failure condition is applied to study strength evolution for triaxially compressed specimens with varying orientation of principal stress and damage tensor axes. Also a general stress state is considered and the representative failure condition is derived. The application of failure criteria to particular cases is discussed. In particular, the limit states are specified for engineering problems, such as embedded anchor plate pull-out and rigid tool penetration into the material.

1 Introduction

Many geomaterials, such as rocks and soils exhibit structural anisotropy related to material fabric elements such as crack pattern, bedding, layering, contact arrangements, among others. Similarly, metals under large plastic strains exhibit crystallographic texture and morphological texture manifested macroscopically by anisotropic response. The major problem is associated with the specification of effective properties of the representative element, so that the constitutive relations can be formulated in terms of the effective parameters.

R. Wan et al. (Eds.): Bifurcations, Instabilities and Degradations in Geomaterials, SSGG, pp. 69–89.
springerlink.com

The formulation of yield or failure criteria for such materials constitutes an important part of their description. The analytical formulations usually employ linear and quadratic terms in stress components referred to principal anisotropy axes. An example is the orthotropic criterion formulated by Hill [1] for metals, subsequently extended by Tsai and Wu [2] for composites and by Pariseau [3] for rocks. The representation in terms of invariants of stress and structure tensors was also used by some writers, cf. Boehler and Sawczuk [4], Nova [5], Mróz and Jemioło [6]. The other group of failure criteria was derived by applying the critical plane approach requiring the failure condition to be satisfied on a potential failure plane corresponding to a maximum of failure condition, cf. Walsh and Brace [7], Hoek and Brown [8], Hoek [9], etc. A simplified approach to formulation of anisotropic failure criteria using the microstructure tensor was recently discussed by Pietruszczak and Mróz [10, 11]. In these papers the critical plane approach incorporating spatial distribution of microcracks has been presented and applied in quantitative description of compressive strength variation with orientation of principal stress axis relative to anisotropy axes. In the subsequent work by Mróz and Maciejewski [12,13] the critical plane approach was applied in quantitative analysis of directional strength variation with relative orientation of stress and anisotropy axes. The nonlocal critical plane approach to specify crack propagation and fatigue damage prediction was analyzed extensively by Seweryn and Mróz [14,15]. The present paper provides formulation of a nonlinear failure condition using critical plane approach and presents its application to limit analysis problems.

2 Critical Plane Approach to Formulation of Yield or Failure Criteria

The present approach follows the framework initiated by Coulomb in which the failure function is defined in terms of traction components acting on physical plane. The critical plane is specified by maximizing the failure function with respect to the plane orientation.

Consider a physical plane Π specified by a unit normal vector $\boldsymbol{n}$. Denoting the stress state by $\boldsymbol{\sigma}$ and the traction vector $\boldsymbol{t} = \boldsymbol{\sigma n}$, the normal and tangential traction components are

$$\boldsymbol{t}^n = (\boldsymbol{n} \cdot \boldsymbol{\sigma n})\mathbf{n}\,,\ \boldsymbol{t}^s = (\mathbf{1} - \boldsymbol{n} \otimes \boldsymbol{n})\boldsymbol{\sigma n} \tag{1}$$

where $\mathbf{1}$ is a unit tensor, dot between symbols denotes the scalar product and tensor product is denoted by $\otimes$, so that the $\boldsymbol{\sigma n} = \sigma_{ij} n_j$, $\boldsymbol{n} \cdot \boldsymbol{\sigma n} = n_i \sigma_{ij} n_j$, $\boldsymbol{n} \otimes \boldsymbol{n} = n_i n_j$. The tangential traction $\boldsymbol{t}^s$ can also be specified by introducing a unit vector $\boldsymbol{s}$ in the plane Π following the orientation of $\boldsymbol{t}^s$, so that

$$\boldsymbol{t}^s = (\boldsymbol{s} \cdot \boldsymbol{\sigma n})\boldsymbol{s}\ ,\ \boldsymbol{s} = \frac{\boldsymbol{t} - \boldsymbol{t}^n}{\left\| \boldsymbol{t} - \boldsymbol{t}^n \right\|} \tag{2}$$

Consider now a failure function with a single strength parameter c

$$F = f\left(t^n, t^s\right) - c(\boldsymbol{n}) = 0 \tag{3}$$

Here the parameter $c(\boldsymbol{n})$ is assumed to vary with the orientation of the plane. At the same time, the material response within the plane Π is assumed to be isotropic, so that the failure function depends on the moduli of the $\boldsymbol{t}^n$ and $\boldsymbol{t}^s$, thus

$$\begin{aligned} &F = f(\sigma_n, \tau_n) - c(\boldsymbol{n}) = 0, \\ &\sigma_n = \boldsymbol{n}\cdot\boldsymbol{\sigma n},\ \tau_n = \left\|\boldsymbol{t}^s\right\| \end{aligned} \tag{4}$$

The variation of the strength parameter can also be described by introducing a microstructure tensor whose principal axes specify the material anisotropy. A simple distribution function is obtained in a form

$$c(\boldsymbol{n}) = c_0\left(1 + \Omega_{ij} n_i n_j\right) \tag{5}$$

where c_0 is the orientation average of c and Ω_{ij} is a deviatoric symmetric tensor describing the variation in the spatial distribution with respect to the mean value.

A more general distribution function can be obtained by introducing higher order expansion, thus

$$c(\boldsymbol{n}) = c_0\left(1 + \Omega_{ij} n_i n_j + \Gamma_{ijkl} n_i n_j n_k n_l + \ldots\right) \tag{6}$$

where Γ_{ijkl} is a fourth order tensor distribution function. In particular it can be assumed as a tensor product of the second order tensors; $\Gamma_{ijkl}=\Gamma_{ij}\Gamma_{kl}$.

The orientation of the critical plane can now be specified from the maximization procedure

$$\max_{\boldsymbol{n}} F = \max_{\boldsymbol{n}}\left[f(\sigma_n, \tau_n) - c(\boldsymbol{n})\right] = 0 \tag{7}$$

A more general case can be considered in which the strength parameter c exhibits anisotropic properties within the plane Π. The distribution function can then be assumed in the form

$$c(\boldsymbol{n},\boldsymbol{s}) = c_0\left(1 + \Omega_{ij} n_i n_j + \Lambda_{ij} s_i s_j\right) \tag{8}$$

where Λ_{ij} specifies the strength variation within the plane Π. The value of τ_n can now be specified by the maximization with respect to an arbitrary $\boldsymbol{s}$-vector

$$\max_{\boldsymbol{n},\boldsymbol{s}} F = \max_{\boldsymbol{n},\boldsymbol{s}}\left[f(\sigma_n, \tau_n) - c(\boldsymbol{n},\boldsymbol{s})\right] = 0 \tag{9}$$

subject to constraints

$$\boldsymbol{n}\cdot\boldsymbol{n} = 1,\quad \boldsymbol{s}\cdot\boldsymbol{s} = 1,\quad \boldsymbol{s}\cdot\boldsymbol{n} = 0 \tag{10}$$

The critical plane orientation can also be specified by a *kinematic approach*. Assuming the failure mode by shear and dilatancy along the critical plane (or a set of planes), the velocity gradient and the strain rate components can be written in the form satisfying velocity continuity in the tangential direction, thus

$$\begin{aligned} \boldsymbol{L} &= \boldsymbol{b} \otimes \boldsymbol{n} \\ \dot{\boldsymbol{\varepsilon}} &= \tfrac{1}{2}[\boldsymbol{b} \otimes \boldsymbol{n} + \boldsymbol{n} \otimes \boldsymbol{b}], \quad \dot{\varepsilon}_n = b_n, \ \dot{\gamma}_n = b_s \end{aligned} \tag{11}$$

where b_n, b_s are the normal and tangential components of the vector $\boldsymbol{b}$. Consider a representative material element under stress $\boldsymbol{\sigma}$ and strain rate $\dot{\boldsymbol{\varepsilon}}$, so we have the dissipation rate in the form

$$D = \boldsymbol{\sigma}\dot{\boldsymbol{\varepsilon}} = \sigma_n \dot{\varepsilon}_n + \tau_n \dot{\gamma}_n \tag{12}$$

Now, for the specified strain rate $\dot{\boldsymbol{\varepsilon}}$ the dissipation rate D is minimized with respect to the plane orientation, thus

$$\min_{\boldsymbol{n}} D[c(\boldsymbol{n}), \dot{\varepsilon}_n, \dot{\gamma}_n] \tag{13}$$

In particular, when the associated flow rule occurs, we have

$$\dot{\varepsilon}_n = \dot{\lambda}\frac{\partial f}{\partial \sigma_n}, \quad \dot{\gamma}_n = \dot{\lambda}\frac{\partial f}{\partial \tau_n} \tag{14}$$

and when the function $f(\sigma_n, \tau_n)$ is homogenous of degree one, there is

$$D = \dot{\lambda}\left(\sigma_n \frac{\partial f}{\partial \sigma_n} + \tau_n \frac{\partial f}{\partial \tau_n}\right) = \dot{\lambda}(\dot{\varepsilon}_n, \dot{\gamma}_n) c(\boldsymbol{n}) \tag{15}$$

3 Model Formulation

Consider a plane or axisymmetric material element shown in Fig. 1, subjected to the principal stresses σ_1 and σ_2, where the compressive stresses are assumed as positive. Consider a physical plane Π inclined at the angle β to the x-axis. The normal and shear stresses acting on the plane are

$$\sigma_n = \sigma_1 \sin^2\beta + \sigma_2 \cos^2\beta, \ \tau_n = \tfrac{1}{2}(\sigma_2 - \sigma_1)\sin 2\beta \tag{16}$$

It is assumed that the failure will be attained when the slip on the critical plane occurs. We introduce failure condition in the form $F(\sigma_n, \tau_n, \omega_i(\boldsymbol{n})) \le 0$, where ω_i are material parameters for the particular plane with normal $\boldsymbol{n}$. For example when Coulomb failure condition is applied, the parameters ω_i corresponds to the effective friction angle and cohesion, $\phi(\boldsymbol{n})$, $c(\boldsymbol{n})$.

Assume that the portions l_t and l_d of the segment AB represent the intact and damaged surfaces. In fact, for the plane strain problem, the values l_t and l_d are proportional to damaged and intact areas A_d and A_t of the plane. Introducing the non-dimensional parameters η_t and η_d, specifying the relative areas of intact and damaged portions on Π, we can write :

$$\begin{aligned}\tau_n &= \tau_n^t \eta_t + \tau_n^d \eta_d \\ \sigma_n &= \sigma_n^t \eta_t + \sigma_n^d \eta_d\end{aligned}, \quad \eta_t + \eta_d = 1 \tag{17}$$

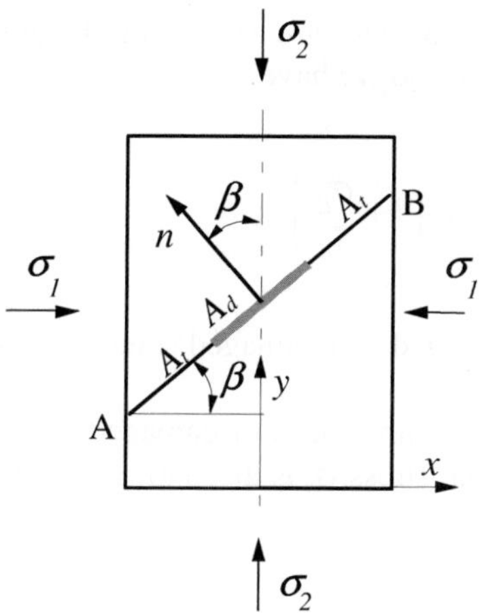

Fig. 1. Damage distribution on the physical plane Π.

where σ_{nt}, τ_{nt} and σ_{nd}, τ_{nd} are normal and tangential stress components on the plane Π for intact and damaged phases.

We assume a non-linear failure condition in the form:

$$F = |\tau_n| - a(\boldsymbol{n})\left(1 + \frac{\sigma_n}{S(\boldsymbol{n})}\right)^m = 0, \qquad \sigma_n \geq -S, \quad S > 0 \tag{18}$$

where τ_n and σ_n denote the shear and the normal stresses on the failure plane and a, m, S are the material parameters, Fig. 2. Here S denotes the tensile strength, a is the shear strength and $0<m<1$ is the non-dimensional parameter. Both the tensile strength and shear strength parameters are assumed to depend on the damage distribution function, $S = S(\eta_d(\boldsymbol{n})), a = a(\eta_d(\boldsymbol{n}))$. Introducing the dilatancy angle δ specified by the associated flow rule is:

$$\tan\delta = -\left(\frac{\frac{\partial F}{\partial \sigma_n}}{\frac{\partial F}{\partial \tau_n}}\right) = \frac{ma}{S}\left(1 + \frac{\sigma_n}{S}\right)^{m-1}, \tag{19}$$

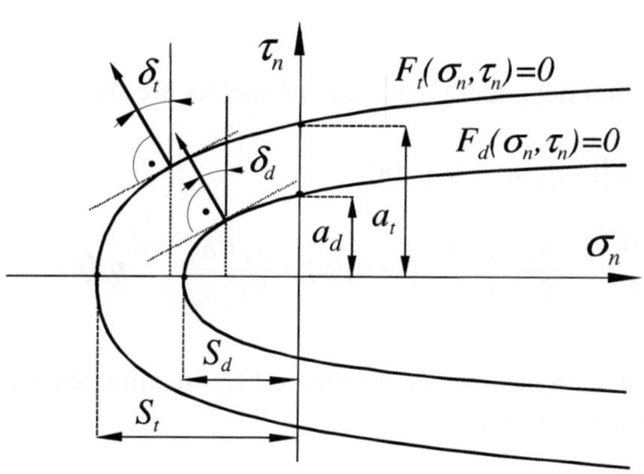

Fig. 2. Nonlinear failure condition on τ_n, σ_n plane

The non-linear condition can now be presented in the form:

$$\sigma_n = S\left[\left(\frac{\tan\delta}{m}\frac{S}{a}\right)^{\frac{1}{m-1}} - 1\right], \quad \tau_n = a\left(\frac{S}{a}\frac{\tan\delta}{m}\right)^{\frac{m}{m-1}} \tag{20}$$

The limit surfaces for the intact and fully damaged states are assumed to be of the same form (18) with different material parameters, so we have:

$$F_t = |\tau_n| - a_t\left(1 + \frac{\sigma_n}{S_t}\right)^m, \quad F_d = |\tau_n| - a_d\left(1 + \frac{\sigma_n}{S_d}\right)^m, \tag{21}$$

where a_t, S_t and a_d, S_d corresponds to the undamaged and damaged states, and parameter m is the same in both limit surfaces.

Assume now that the dilatancy angle δ is the same in both damaged and undamaged states on the physical plane. The normal stress σ_n is discontinuous on the surface portions A_t and A_d, namely:

$$\sigma_{nt} = S_t\left[\left(\frac{\tan\delta}{m}\frac{S_t}{a_t}\right)^{\frac{1}{m-1}} - 1\right], \quad \sigma_{nd} = S_d\left[\left(\frac{\tan\delta}{m}\frac{S_d}{a_d}\right)^{\frac{1}{m-1}} - 1\right]. \tag{22}$$

Specifying the averaged stresses σ_n, τ_n on the physical plane by means of Eqs. (17), the failure condition can be expressed in the form analogous to (21), namely

$$F(\eta_d) = \|\tau_n\| - a_\eta\left(1 + \frac{\sigma_n}{S_\eta}\right)^m = 0, \tag{23}$$

where the orientation dependent material parameters are specified as follows:

$$\begin{gathered} S_\eta = S_t[1-\eta_d(\boldsymbol{n})] + S_d\eta_d(\boldsymbol{n}) \\ \left(\frac{a_\eta}{S_\eta}\right)^{\frac{1}{1-m}} = \frac{S_t}{S_\eta}\left(\frac{a_t}{S_t}\right)^{\frac{1}{1-m}}[1-\eta_d(\boldsymbol{n})] + \frac{S_d}{S_\eta}\left(\frac{a_d}{S_d}\right)^{\frac{1}{1-m}}\eta_d(\boldsymbol{n}) \end{gathered} \tag{24}$$

The orientation of critical plane is determined by maximization of F with respect to orientation normal vector $\boldsymbol{n}$.

The specific dissipation function $D=\sigma_n\dot{\varepsilon}_n+\tau_n\dot{\gamma}_n$ associated with the yield condition (18) is expressed as follows

$$D=D(\dot{\varepsilon}_n,\dot{\gamma}_n,a(\boldsymbol{n}),S(\boldsymbol{n}))=a(\boldsymbol{n})\left(\frac{S}{ma}\frac{\dot{\varepsilon}_n}{|\dot{\gamma}_n|}\right)^{\frac{m}{m-1}}\dot{\gamma}_n+S(\boldsymbol{n})\left[\left(\frac{S}{ma}\frac{\dot{\varepsilon}_n}{|\dot{\gamma}_n|}\right)^{\frac{1}{m-1}}-1\right]\dot{\varepsilon}_n=$$

$$=a(\boldsymbol{n})\left(\frac{S}{ma}\tan\delta\right)^{\frac{m}{m-1}}\dot{\gamma}_n+S(\boldsymbol{n})\left[\left(\frac{S}{ma}\tan\delta\right)^{\frac{1}{m-1}}-1\right]\dot{\varepsilon}_n \tag{25}$$

and it should be minimized in order to specify the critical plane orientation.

Consider a general case when the material orthotropy axes S_1, S_2, S_3 are arbitrarily oriented with respect to the principal stress axes, coinciding with the global reference system x, y, z, Fig. 3. The transformation of any vector from the global system to the structural system is specified by the relation:

$$s_i=Q_{ij}^T x_i\,,\quad \boldsymbol{n}^s=Q^T\boldsymbol{n} \tag{26}$$

where Q_{ij} is the rotation matrix which can be expressed in terms of the Euler angles $\nu, \psi\ \phi$ defining the relative orientation of two systems. Referring to Fig. 3, we have

$$Q=\begin{bmatrix} l_1 & l_2 & l_3 \\ m_1 & m_2 & m_3 \\ n_1 & n_2 & n_3 \end{bmatrix} \tag{27}$$

where

$$\begin{aligned}
&l_1=c_2c_3-c_1s_2s_3,\ \ l_2=-c_2s_3-c_1s_2c_3,\ \ l_3=s_1s_2,\\
&m_1=s_2c_3+c_1c_2s_3,\ \ m_2=-s_2c_3+c_1c_2c_3,\ \ m_3=-s_1c_2,\\
&n_1=s_1s_3\,,\ \ n_2=s_1c_3\ \ ,\ \ n_3=c_1
\end{aligned} \tag{28}$$

and

$$c_1=\cos\nu,\quad c_2=\cos\psi,\quad c_3=\cos\phi,\quad s_1=\sin\nu,\quad s_2=\sin\psi,\quad s_3=\sin\phi. \tag{29}$$

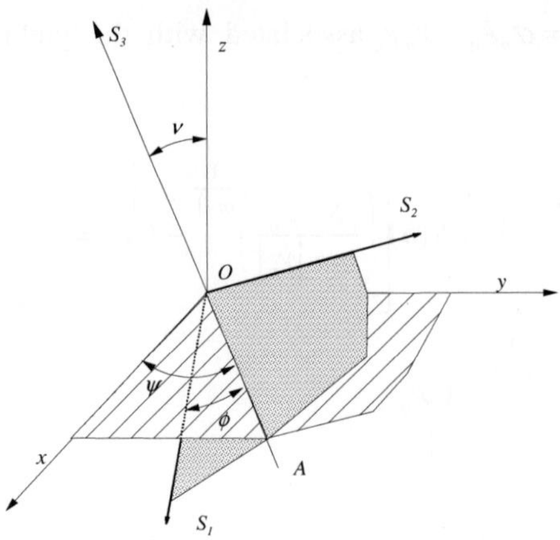

Fig. 3. Euler angles characterizing relative orientation of the reference frame S_1,S_2,S_3 relative to the x,y,z axes.

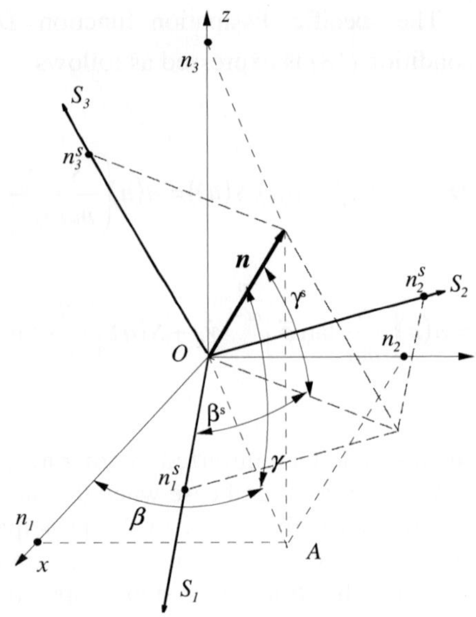

Fig. 4. Definition of angles β,γ specifying the unit vector *n*.

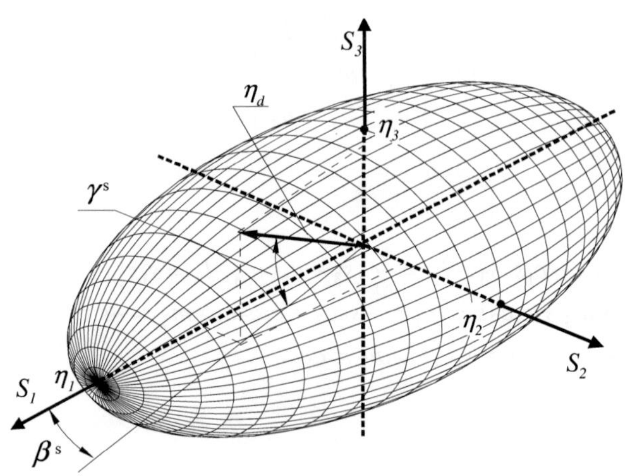

Fig. 5. Ellipsoidal distribution damage density factor η_d

Consider first the reference to the principal orthotropy axes. The normal unit vector $\boldsymbol{n}$ specifies the orientation of the physical plane on which the normal and shear stresses are specified as follows:

$$\sigma_n = \sigma_{ij} n_i n_j = \sigma_{11} n_1^2 + \sigma_{22} n_2^2 + \sigma_{33} n_3^2 + 2\sigma_{12} n_1 n_2 + 2\sigma_{23} n_2 n_3 + 2\sigma_{13} n_1 n_3 ,$$
$$\tau_n = \sqrt{\sigma_{ij} n_i \sigma_{ik} n_k - \left(\sigma_{ij} n_i n_j\right)^2} \tag{30}$$

Using the spherical coordinate system r, β^s, γ^s, the Cartesian coordinates of any radius vector r in the reference system S_1, S_2, S_3 are

$$s_1 = r \cos\beta^s \cos\gamma^s , \quad s_2 = r \sin\beta^s \cos\gamma^s , \quad s_3 = r \sin\gamma^s \tag{31}$$

Assuming that the principal values η_1, η_2, η_3 of the damage tensor correspond to the axes S_1, S_2, S_3 the damage distribution of any physical plane is specified by the relation:

$$\eta_d = \eta_{ij} n_i n_j = \eta_1 n_1^2 + \eta_2 n_2^2 + \eta_3 n_3^2$$
or
$$\eta_d = \eta_1 \cos^2\beta^s \cos^2\gamma^s + \eta_2 \sin^2\beta^s \cos^2\gamma^s + \eta_3 \sin^2\gamma^s \tag{32}$$

A more general description of damage distribution can be obtained by assuming the higher order powers with respect to the orientation angles γ^s, β^s, for instance in the form

$$\eta_d = \eta_1 \cos^2\beta^s \cos^l\gamma^s + \eta_2 \sin^2\beta^s \cos^l\gamma^s + \eta_3 \sin^l\gamma^s \tag{33}$$

where l is a natural even number (l=2, 4, 6,...).

An alternative distribution is obtained by postulating the damage η_d to be specified by an ellipsoidal distribution, Fig. 5, thus

$$\left(\frac{s_1}{\eta_1}\right)^2 + \left(\frac{s_2}{\eta_2}\right)^2 + \left(\frac{s_3}{\eta_3}\right)^2 = 1 \tag{34}$$

Hence,

$$\eta_d = \left[\frac{\cos^2\beta^s \cos^2\gamma^s}{\eta_1^2} + \frac{\sin^2\beta^s \cos^2\gamma^s}{\eta_2^2} + \frac{\sin^2\gamma^s}{\eta_3^2}\right]^{-\frac{1}{2}} \tag{35}$$

This form will be applied in the limit analysis problems. Obviously the damage distribution function should be specified from experimental identification of anisotropy response.

In the global system x,y,z, the critical plane orientation is now specified by the condition

$$F^C(\sigma_n, \tau_n, \omega_j(\beta,\gamma)) = \max_{\beta,\gamma} F\left(\sigma_n, \tau_n, \omega_j(\beta,\gamma)\right) = 0 \tag{36}$$

4 Application Examples

In this section several specific examples will be provided for the case of anisotropic rocks for which the directional strength variation is observed. The uniaxial or triaxial tests are carried out for specimens cut at different angles with respect to of the axes orthotropy.

The test results for the Angers schist are reported in [15]. The specimens were tested for different values of transverse stresses $\sigma_2=\sigma_3$ (0,5,10,20,30,40 MPa) and for varying orientation angles θ *(for* $\theta=\nu$, $\psi=\phi=0$*)* of the principal orthotropy axes.

Fig. 6 presents the variation of critical stress and of the failure plane angle β_c with the orientation θ. It is seen that for some values of θ the critical plane orientation follows the orientation of the principal orthotropy axis, $\beta_C=\theta$. However for θ close to 0° and 90° this orientation essentially differs from θ. The experimental data for $\sigma_2=\sigma_3=5$ MPa and 40 MPa are also marked in the figure. It is seen that the agreement is satisfactory. Table 1 presents the model parameters.

Let us now investigate the shape of the failure surface in the octahedral plane. Introduce the stress invariants σ_m, ρ, ξ, where

$$\sigma_m = \frac{1}{3}\sigma_{kk}, \quad \rho = \sqrt{2J_2}, \quad \xi = \frac{1}{3}\arccos\left(3\sqrt{6}\frac{J_3}{\rho^3}\right) \tag{37}$$

and J_2, J_3 are the second and third invariants of the stress deviator. The principal stress components are now expressed as follows:

$$\begin{bmatrix}\sigma_1\\ \sigma_2\\ \sigma_3\end{bmatrix} = \sigma_{\mathrm{m}}\begin{bmatrix}1\\ 1\\ 1\end{bmatrix} + \sqrt{\frac{2}{3}}\rho\begin{bmatrix}\cos(\xi)\\ \cos(\xi+\frac{2}{3}\pi)\\ \cos(\xi-\frac{2}{3}\pi)\end{bmatrix} \tag{38}$$

Figure 7 presents the failure surfaces on the octahedral plane for different orientations of the orthotropy and stress axes. It was assumed that σ_m=100 MPa, $\eta_1=\eta_2=0.14$, $\eta_3=1.0$ and the model parameters selected for Angers schist were assumed, cf. Table 1. The damage distribution function (33) was assumed.

Fig. 8 presents the failure surfaces for the plane stress case, and different orientations of the anisotropy axes. Here again, a strong dependence of failure surface shape on anisotropy orientation relative to principal stress axes is exhibited.

Figure 9 presents the result of sensitivity analysis of model predictions for varying damage distributions specified by the generalized tensorial rule (32) with damage distribution parameters $\eta_1=\eta_2=0.05$, $\eta_3=0.97$ and the values of exponent l=2,4,20. The diagrams illustrate the effect of damage distribution on the variation of strength on the octahedral plane. It was assumed that σ_m=100 MPa and the orthotropy orientation is specified by the angles $\nu=0°$, $\phi=0°$, $\psi=0°$. The simulation was performed by applying model parameters selected for Angers schist.

It is seen that model provide sufficiently accurate simulation of compressive strength variation and the orientation of critical plane.

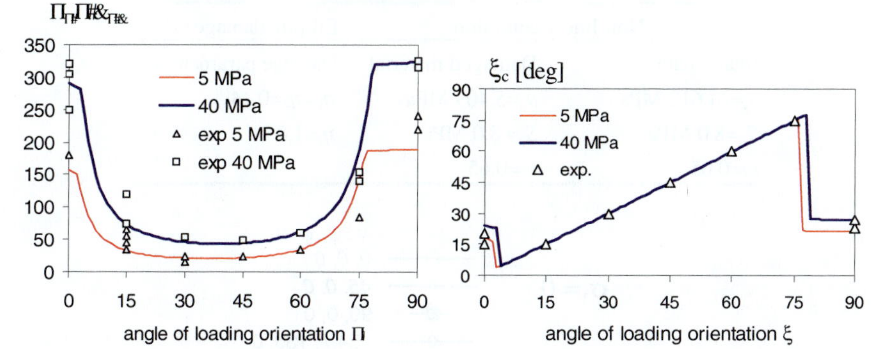

Fig. 6. Strength and failure angle variation on Angers schist for non-linear failure condition

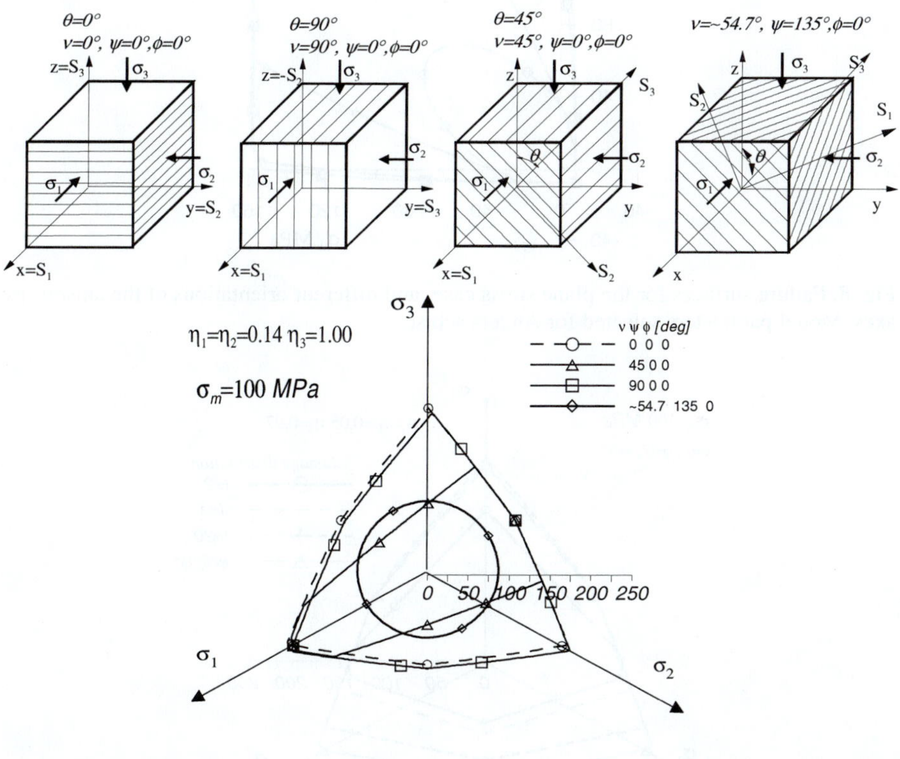

Fig. 7. Failure surfaces in the octahedral plane for different orientations of the orthotropy axes with model parameters selected for Angers schist

Table 1. Non-linear model parameters for Angers schist.

Non-linear condition		Elliptic damage
Intact material	Damaged material	Damage parameters
a_t=24.612 MPa	a_d=5.409 MPa	$\eta_1=\eta_2$=0.14,
S_t =8.0 MPa	S_d= 8.0 MPa	η_3=1.0
m=0.65	m=0.65	

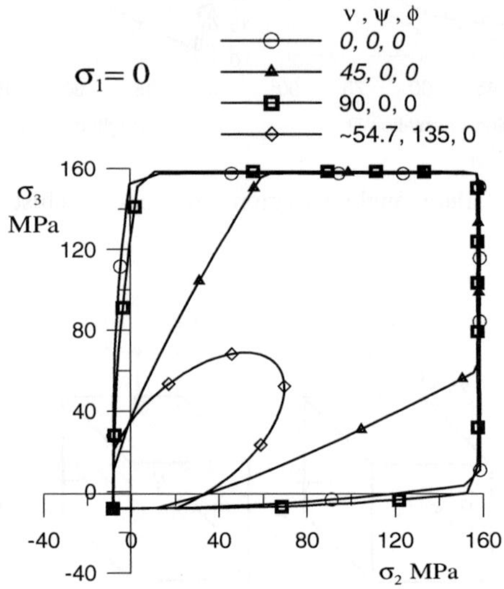

Fig. 8. Failure surfaces for the plane stress case, and different orientations of the anisotropy axes. Model parameters selected for Angers schist.

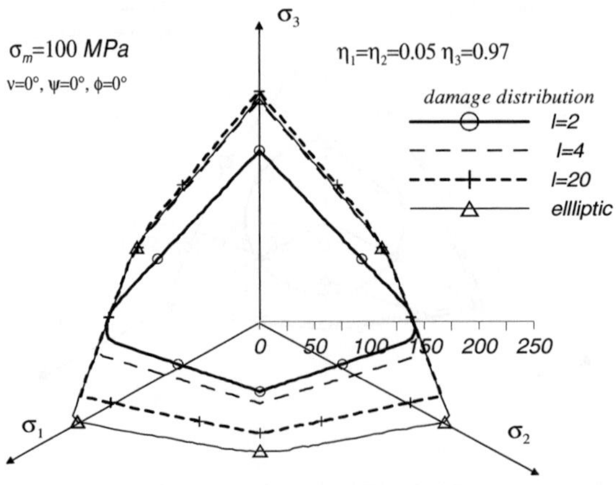

Fig. 9. Variation of failure surfaces on the octahedral plane for various tensor damage distributions (σ_m=100 MPa,)

4.1 Active Pressure of a Rigid Wall on Anisotropic Material

The methods of limit analysis for geomaterials satisfying Coulomb condition were exposed in numerous books, cf. Derski et al [17], Chen [18]. However, there are few applications of nonlinear conditions accounting for anisotropic response. The present analysis follows the previous works [12,13,19] and is aimed to expose the methodology to anisotropic strength conditions indicating importance of orientational distribution of strength parameters.

Consider first the case of a rigid vertical plate penetrating into an anisotropic geomaterial. This is typical case for soil or rock cutting problem, Fig. 10. The non-linear failure condition (23) in now applied with the following parameters: γ=20 kN/m^3, m=0.7, S_d=1.0 kN/m^2, a_d=1, S_t=5.0 kN/m^2, a_t=9.255. The Coulomb friction at the interface between plate and material is characterized by the cohesion c_s and the friction angle δ_s. The material anisotropy is specified by the ellipsoidal damage distribution (35) and the parameters η_1=0.2, η_2=0.2, η_3=1.0, representing the transverse isotropy for which the weakest material plane orientation corresponds to the axis η_3. The Cartesian reference system x,y,z, is introduced with the x-axis normal to the plane of motion and *z-axis* to the semi-plane boundary surface.

The vertical plate is assumed to move along the y-axis with the velocity V_0. The failure mechanism shown in Fig. 10 is composed of two rigid blocks OAB and OBC moving along the discontinuity lines AB and BC with velocities $[V_1]$ inclined at the angle δ_1 to AB and $[V_2]$ inclined at the angle δ_2 to BC. The relative sliding velocity $[V_{21}]$ along the discontinuity line OB is inclined at the angle δ_{21} to OB. The rigid block OAB translates relatively to the plate with velocity $[V_n]$ inclined at the angle δ_s to plate OA. The velocity hodograph is presented in Fig. 10b. The failure mechanism geometry is specified by three angles α_1, α_2, α_3, shown in Fig. 10a. Assuming the associated flow rule, the discontinuity vectors are inclined at the dilatancy angles δ_1, δ_2, δ_{21} to lines AB, BC and OB. However, the values of dilatancy angles are related to the stress components σ_n, τ_n on these lines and constitute the elements of the solution. In the following, the block equilibrium method will be used in order to specify the mean normal and shear stresses on the discontinuity lines. The effective strength parameters on the discontinuity lines are specified by determining the values η_d from (35) on these lines. For instance, the unit normal vector to OB has components $n_1=0$, $n_2=-\sin\alpha_3$ i $n_3=\cos\alpha_3$ and the relations (24) and (35) provide the effective strength parameters on the discontinuity line OB.

The diagram of force acting on the block OBC is presented in Figs. 10 c,d. The block is acted on by the gravity force Q_2 and the reactions on the discontinuity lines. On the line BC the reaction $\boldsymbol{R}_{z2}$ has the normal and tangential components S_{z2}, T_{z2}, the reaction $\boldsymbol{R}_{12}$ =- $\boldsymbol{R}_{21}$ on OB has the normal and tangential components S_{12}, T_{12}. The equilibrium conditions of OBC are expressed by the equation

$$\boldsymbol{Q}_2 + \boldsymbol{R}_{z2} = \boldsymbol{R}_{21} \tag{39}$$

presented graphically in Fig. 10d. Plotting the normals to BC and OB in Fig. 10d and the parabolic limit state diagrams for contact force components, the graphical

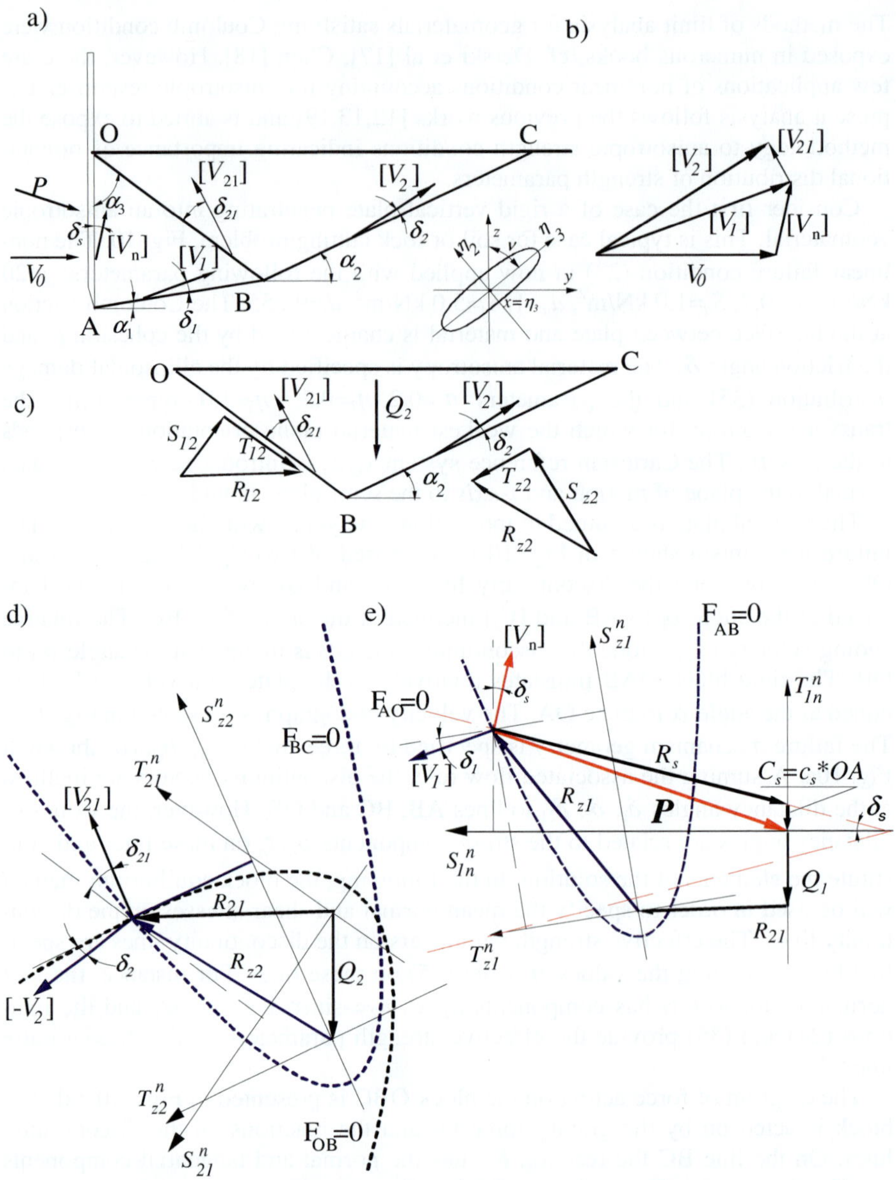

Fig. 10. The failure mechanism for the active plate pressure on the nonlinear anisotropic material a) kinematically admissible failure mechanism b) velocity hodograph for the associated flow rule, c) force acting on the block OBC; d) force diagram for the block OBC at the limit equilibrium , e) force diagram for the block OAB at the limit equilibrium.

solution of the Eq. (39) specifying forces $\boldsymbol{R}_{z2}$ and $\boldsymbol{R}_{21}$ is obtained by the intersection point of limit diagrams. The velocity discontinuity vectors are then normal to the limit state lines at the intersection point. The analytical solution of Eq. (39) is obtained by expressing first the components of the contact interaction forces in terms of the dilatancy angles δ_2 and δ_{21}, In view of Eq. (20), there is

$$
\begin{gathered}
S_{z2} = l_{BC} S_{BC} \left[\left(\frac{\tan\delta_2}{m} \frac{S_{BC}}{a_{BC}} \right)^{\frac{1}{m-1}} - 1 \right],\ T_{z2} = l_{BC} a_{BC} \left(\frac{S_{BC}}{a_{BC}} \frac{\tan\delta_2}{m} \right)^{\frac{m}{m-1}} \\
S_{21} = l_{OB} S_{OB} \left[\left(\frac{\tan\delta_{21}}{a_{OB} m} \frac{S_{OB}}{a_{OB}} \right)^{\frac{1}{m-1}} - 1 \right],\ T_{21} = -l_{OB}\, a_{OB} \left(\frac{S_{OB}}{a_{OB}} \frac{\tan\delta_{21}}{m} \right)^{\frac{m}{m-1}}
\end{gathered}
\tag{40}
$$

where a_{BC}, S_{BC} , a_{OB}, S_{OB} are the effective strength parameters on the discontinuity lines BC and OB , whose lengths are denoted by l_{BC} and l_{OB}. The components of $\boldsymbol{R}_{z2}$ and $\boldsymbol{R}_{21}$ along the y,z axes are

$$
\begin{gathered}
R_{z2}^{y} = S_{z2}\cos(\tfrac{\pi}{2}+\alpha_2) - T_{z2}\sin(\tfrac{\pi}{2}+\alpha_2),\ R_{z2}^{z} = S_{z2}\sin(\tfrac{\pi}{2}+\alpha_2) + T_{z2}\cos(\tfrac{\pi}{2}+\alpha_2) \\
R_{21}^{y} = S_{21}\cos(\pi+\alpha_3) - T_{21}\sin(\pi+\alpha_3),\ R_{21}^{z} = S_{21}\sin(\pi+\alpha_3) + T_{21}\cos(\pi+\alpha_3)
\end{gathered}
\tag{41}
$$

The equilibrium equation (40) now has the form

$$
\begin{cases}
R_{z2}^{y} = R_{21}^{y} \\
-Q_2 + R_{z2}^{z} = R_{21}^{z}
\end{cases}
\tag{42}
$$

and can be solved numerically with respect to δ_2, δ_{21} by applying the Newton-Raphson procedure. The force diagram for the block OAB is presented in Fig. 10e. It is acted on by the gravity force Q_1 contact reaction $\boldsymbol{R}_{21}$ on AB and the plate force $\boldsymbol{P}$. On the contact interface OA between the plate and the material the Coulomb friction condition is assumed, so the interaction force is composed of the cohesion force $C_s = c_s l_{OA}$ acting along OA oppositely to the tangent velocity discontinuity and the frictional force R_s inclined at the angle δ_s to the normal vector to OA. The equilibrium equation now is:

$$
\boldsymbol{R}_{21} + \boldsymbol{Q}_1 + \boldsymbol{R}_{z1} + \boldsymbol{P} = 0, \quad \boldsymbol{P} = \boldsymbol{C}_s + \boldsymbol{R}_s \tag{43}
$$

The intersection of limit state diagrams $F_{AB}=0$ and $F_{AO}=0$ provides the graphical solution of the equilibrium equation (43), Fig 10e. The velocity discontinuity vectors are normal to limit loci of the contact forces at the intersection point.

The optimal failure mechanism is specified by minimizing the plate force component P_y with respect to α_1, α_2, α_3.

Figure 11 presents the dependence of the plate force P_y on the orientation of anisotropy axes, specified by the Euler angles ν, ϕ, ψ. Setting $\phi=0°$, $\psi=0°$, only the effect of varying values of the inclination angle $0° \le \nu \le \pi$, is analyzed.

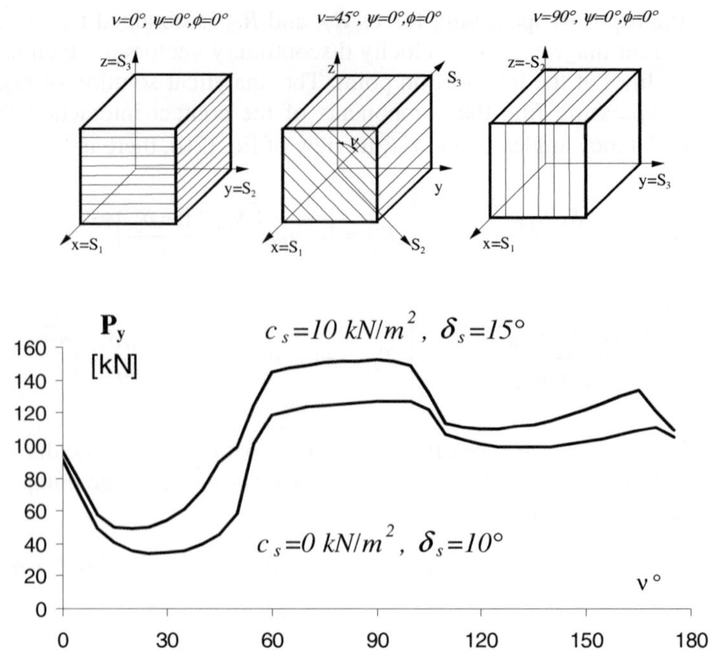

Fig. 11. The limit active force for different anisotropy orientation ν

For ν=0° the weakest material plane is parallel to the *x,y*-plane, for $\nu = \frac{1}{2}\pi$, the weakest plane is parallel to *x,z*-plane. The evolution of the plane force was specified for two cases: one with the cohesive friction on OB, c_s=10 kN/m^2 and δ_s=15°, the other with cohesive friction c_s=0 kN/m^2 and δ_s=10°. It is seen that P_y=$P_y(\nu)$ diagram exhibits two minima and two maxima for $0° \leq \nu \leq \pi$.

4.2 Load Carrying-Capacity of Embedded Plate Anchors

Consider now the problem of limit pulling load of a plate anchor placed within the material parallel to the boundary plane at the distance *h*, Fig.12. Let us analyze first the plane strain case. The assumed failure mechanism is shown in Fig.12. At the limit state the material block *A'ABB'* is pulled with the plate and the velocity vector on AA' and BB' equals the plate velocity V_0. The dilatancy angles on AA' and BB' are δ_1=90°-α_1 and δ_2=90°-α_2. The upper bound on the limit load is determined from the balance of work rate and plastic dissipation power on the discontinuity lines. The normal vectors to these lines are now: on AA' n_1=0, n_2=-sinα_1 , n_3=cosα_1, on BB' n_1=0, n_2=sinα_2 , n_3=cosα_2. The strength distribution parameters $\eta_d(n_i)$ and the stress components satisfying the limit state condition are specified from (21) and (35), thus:

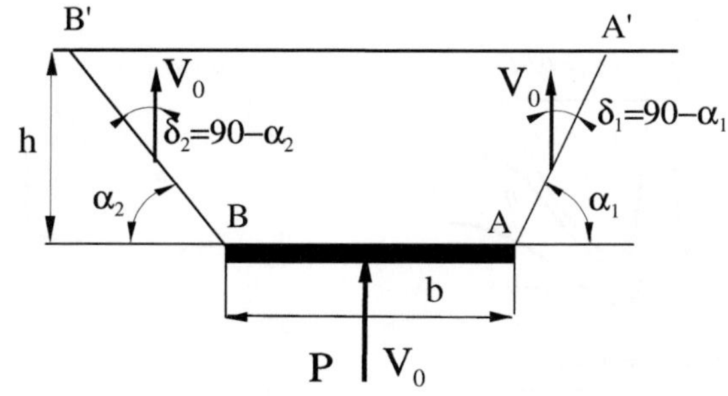

Fig. 12. Kinematically admissible failure mechanism for the plane strain case

$$\sigma_n^{AA'} = S_1\left[\left(\frac{\tan\left(\frac{\pi}{2}-\alpha_1\right)}{m}\frac{S_1}{a_1}\right)^{\frac{1}{m-1}}-1\right],\ \tau_n^{AA'} = a_1\left(\frac{S_1}{a_1}\frac{\tan\left(\frac{\pi}{2}-\alpha_1\right)}{m}\right)^{\frac{m}{m-1}} \tag{44}$$

where a_1 and S_1 are the effective strength parameters on AA'. Similar relations are derived for stress the components on BB'.

The dissipation power on the discontinuity line with a finite thickness d is expressed as

$$D = \left(\sigma_n\dot{\varepsilon}_n + \tau_n\dot{\gamma}_n\right)d = \tau_n[V_t] - \sigma_n[V_n] = [V]\cos\delta\left(\tau_n - \sigma_n\tan\delta\right), \tag{45}$$

Thus, the dissipation power on the discontinuity line AA' equals

$$D^{AA'} = AA'\left(\tau_n^{AA'}[V_t] - \sigma_n^{AA'}[V_n]\right) = hV_0\left(\tau_n^{AA'} - \sigma_n^{AA'}\tan\left(\tfrac{\pi}{2}-\alpha_1\right)\right) \tag{46}$$

and similar expression is obtained for $D^{BB'}$. Equating the total plastic dissipation power $D^{AA'}+D^{BB'}$ to the rate of work, the limit force P is expressed as follows

$$P = \frac{1}{V_0}\left(D^{AA'}\left(\alpha_1,Q_{ij}\right) + D^{BB'}\left(\alpha_2,Q_{ij}\right)\right) + \gamma h\left(\frac{h}{2}\left(\frac{1}{\tan\alpha_1}+\frac{1}{\tan\alpha_2}\right)+b\right) \tag{47}$$

The optimal failure mechanism is obtained by minimizing the value of P with respect to α_1, α_2. It will be shown that the optimal values of α_1 and α_2 depend essentially on the orientation of anisotropy axes.

Fig. 13 presents the dependence of the limit pulling force on the nutation angle ν for $\psi=\phi=0$ and different values of the anisotropy distribution parameters, assuming $\eta_1=\eta_2=1$ and varying η_3. The limit loads for two isotropic cases: $\eta_1=\eta_2=\eta_3=1$ and $\eta_1=\eta_2=\eta_3=0.01$ are lower und upper bounds for the anisotropy cases. Fig. 14 presents the evolution of α_1 and α_2 with the variation of the anisotropy axis orientation ν. It is seen that the optimal values of α_1 and α_2 are very sensitive with respect to the values of the orientation angle ν and may vary discontinuously.

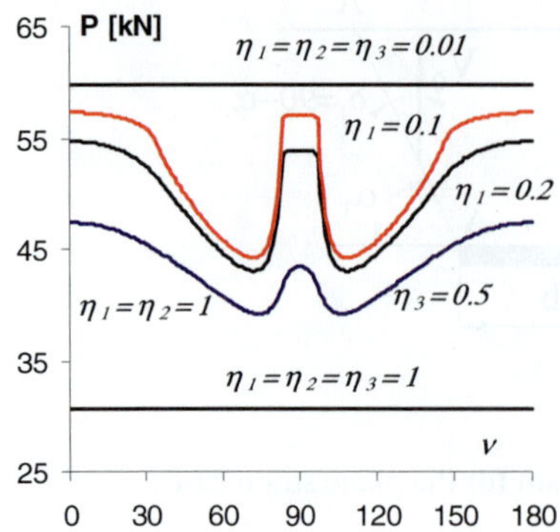

Fig. 13. Variation of the limit load of anchor plate versus the anisotropy orientation angle ν for varying values of η_3, $\eta_1=\eta_2=1$

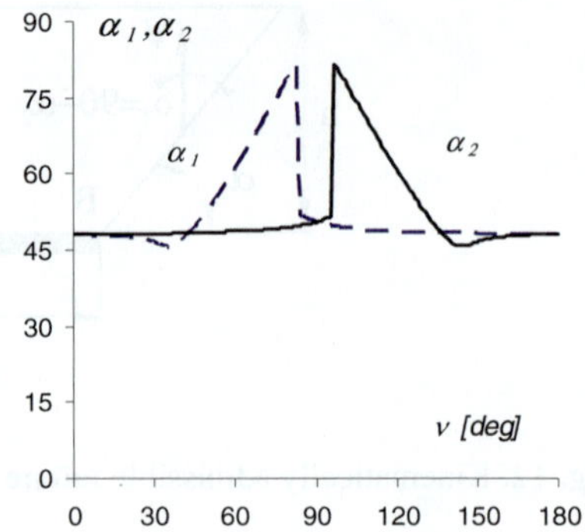

Fig. 14. The evolution of orientation of two discontinuity planes versus orientation of the anisotropy axes

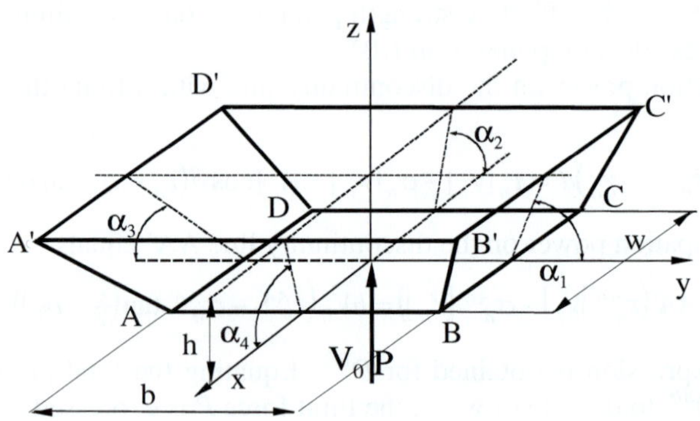

Fig. 15. Three dimensional failure mechanism for the case of pulled anchor plate

Consider now the case of a plate of finite dimensions inducing three dimensional failure mechanism, Fig. 15. The plate of dimensions *b, w* is placed within the material at the depth h=1m. The failure mechanism is assumed in a form allowing the rigid block to move vertically with the plate velocity V_0 with four discontinuity planes inclined at the angles α_1, α_2, α_3, α_4 to the plate.

The limit load is calculated from the balance of work rate and the plastic dissipation power. It is obtained as

$$P = \frac{1}{V_0}\left[D^1(\alpha_1, Q_{ij}) + D^2(\alpha_2, Q_{ij}) + D^3(\alpha_4, Q_{ij}) + D^4(\alpha_5, Q_{ij})\right] + \gamma V \tag{48}$$

where V is the volume of block and D^1, D^2, D^3, D^4 denote the dissipation powers on the discontinuity planes. For the plane BB'C'C the dissipation power is expressed as follows:

$$D^{AA'} = F^{BB'C'C}\left(\tau_n^{BB'C'C}[V_t] - \sigma_n^{BB'C'C}[V_n]\right) = \\ = F^{BB'C'C}V_0\left(\tau_n^{BB'C'C}\sin\alpha_1 - \sigma_n^{BB'C'C}\cos\alpha_1\right) \tag{49}$$

where

$$F^{BB'C'C} = \tfrac{1}{2}\left(2w + h\left(\frac{1}{\tan\alpha_2} + \frac{1}{\tan\alpha_4}\right)\right)\frac{h}{\sin\alpha_1} \tag{50}$$

is the area of the discontinuity plane.

The contact stress values are specified from (20) with account for the effective strength parameters on the discontinuity planes. The optimal failure mechanism is specified by minimizing the value of pulling force P with respect to inclination angles α_i, i=1,2,3,4.

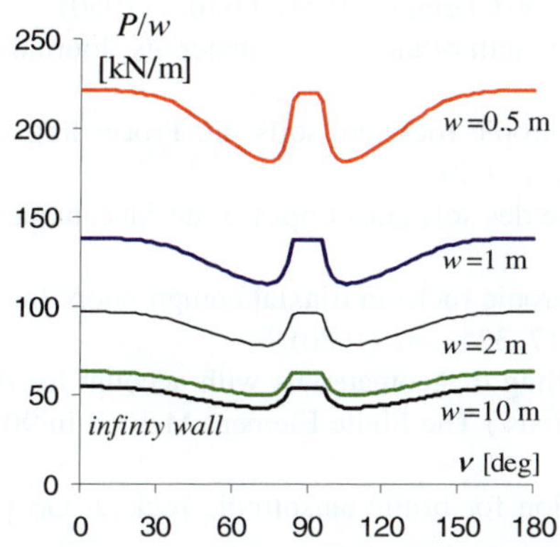

Fig. 16. Vertical force variation versus angles ν orientation of the anisotropy axes for different plate width w and $b=1$, $h=1$m. The plane strain case occurring for $w \to \infty$.

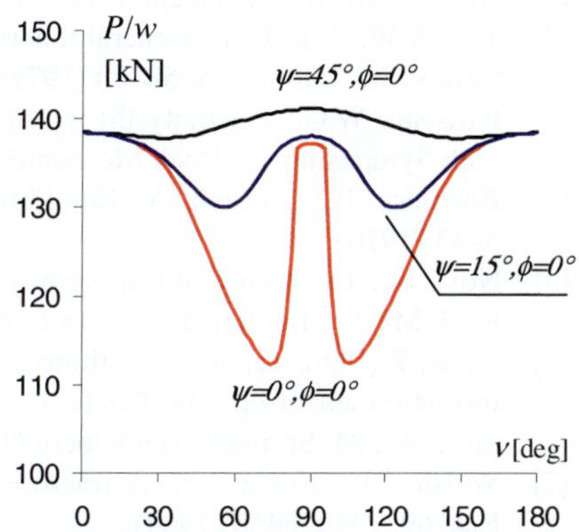

Fig. 17. Vertical force variation versus different angles of the orientation anisotropy axes ($w=h=b=1m$, ϕ=0°)

Fig. 16 presents the ratio of limit load P and plate length w versus the anisotropy orientation angle $0° \le \nu \le \pi$ and $\psi=\phi=0°$ for several values of w and $b=h$=1m. The materials parameters are the same as in the previous example of active pressure. The effect of plate length is significant. In fact, for a square plate, $w=b$=1m, the ratio P/w is about 2.5 times larger than that for the case of plane strain $w \to \infty$. Fig. 17. presents the dependence of P/w on ν for the square plate and several values of the orientation angle ψ=45°, 15°, 0° and ϕ=0°. For ψ=45°,ϕ=0° the variation of P/w versus ν is not significant but it increases for ψ=15° and ψ=0°. In fact, the strongest variation occurs when the orthotropy axis η_3 lies in the z, x-plane, $\psi=\phi$=0.

5 Concluding Remarks

The present paper provides the description of nonlinear failure condition with account for material anisotropy. In particular, the anisotropic distribution can be accounted for by introducing the damage or strength distribution function. Two examples illustrate the applicability of the proposed condition to specification of passive pressure exerted by a moving plate and of limit pulling force of the anchor plate. Both limit equilibrium method and method of balance of work and dissipation rates were applied. They provide upper bounds to limit loads provided the associated flow rule is used. It was shown that the anisotropy effects are very significant and should be introduced in the usual assessment of limit loads of geotechnical structures.

References

[1] Hill, R.: The mathematical theory of plasticity. Clarendon Press, Oxford (1950)

[2] Tsai, S.W., Wu, E.A.: General theory of strength of anisotropic materials. Journal of Composite Materials 5, 58–80 (1971)

[3] Pariseau, W.G.: Plasticity theory for anisotropic rock and soils. In: Proceedings of 10th Symposium on Rock Mechanics AIME (1972)

[4] Boehler, J.P., Sawczuk, A.: Equilibre limite des sols anisotropes. J. de Mecanique 3, 5–33 (1970)

[5] Nova, R.: The failure of transverally anisotropic rocks in triaxial compression. Int. J. Rock Mech. Min. Sci. & Geomech. Abstr. 17, 325–332 (1980)

[6] Mróz, Z., Jemioło, A.: Constitutive modelling of geomaterials with account for deformation anisotropy. In: Onate, E., et al. (eds.) The Finite Element Method in 90's, pp. 274–284. Springer, Heidelberg (1991)

[7] Walsh, J.B., Brace, J.F.: A fracture criterion for brittle anisotropic rock. J. Geoph. Res. 69, 3449–3456 (1964)

[8] Hoek, E., Brown, E.T.: Empirical strength criterion for rock masses. J. Geotech. Eng. Div. ASCE 106, 1013–1035 (1980)

[9] Hoek, E.: Strength of jointed rock masses. Geotechnique 33, 187–205 (1983)

[10] Pietruszczak, S., Mróz, Z.: Formulation of anisotropic failure criteria incorporating a microstructure tensor. Comp. & Geotechnics 24, 105–112 (2000)

[11] Pietruszczak, S., Mróz, Z.: Formulation of failure criteria for anisotropic frictional materials. Int. J. Num. Anal. Meth. Geomech. 25, 509–524 (2001)

[12] Mróz, Z., Maciejewski, J.: Failure criteria of anisotropically damaged materials based on the critical plane concept. Int. J. Numer. Anal. Meth. Geomech. 26, 407–431 (2002)

[13] Mróz, Z., Maciejewski, J.: Failure Criteria and Compliance Variation of Anisotropically Damaged Materials. In: Skrzypek, J.J., Ganczarski, A. (eds.) Lecture Notes in Applied and Computational Mechanics, vol. 9, pp. 75–112. Springer, Heidelberg (2003)

[14] Seweryn, A., Mróz, Z.: A non-local stress failure condition for structural elements under multiaxial loading. Engineering Fracture Mechanics 51, 499–512 (1995)

[15] Seweryn, A., Mróz, Z.: On the criterion of damage evolution for variable multiaxial stress states. Int. J. Solids Structures 35(14), 1589–1616 (1998)
[16] Duveau, G., Shao, J.F., Henry, J.P.: Assessment of some failure criteria for strongly anisotropic geomaterials. Mechanics of Cohesive-Frictional Materials 3, 1–26 (1998)
[17] Derski, W., Izbicki, R., Kisiel, I., Mróz, Z.: Rock and Soil Mechanics. Elsevier, PWN, Warszawa (1989)
[18] Chen, W.F.: Limit analysis and soil plasticity. Developments in Geotechnical Engineering, vol. 7. Elsevier Scientific Publishing Co., Amsterdam (1975)
[19] Maciejewski, J., Jarzębowski, A.: Application of kinematically admissible solutions to passive earth pressure problems. Int. J. of Geomechanics 4(2), 127–136 (2004)

A Simple Method to Consider Density and Bonding Effects in Modeling of Geomaterials

Teruo Nakai, Mamoru Kikumoto, Hiroyuki Kyokawa,
Hassain M. Shahin, and Feng Zhang

Nagoya Institute of Technology, Nagoya, Japan

Abstract. A simple method to describe stress-strain behavior of structured soils under normally and over-consolidated states in one-dimensional stress condition is first presented by introducing a state variable to represent the influence of density. To describe the one-dimensional stress-strain behavior of structured soils, attention is focused on the density and the bonding as the main factors that affect a structured soil, because it can be considered that the soil skeleton structure in a state which is looser than that of a normally consolidated soil is formed by bonding effects. The extension from one-dimensional model to three-dimensional model can be done only by defining the yield function using the invariants of modified stress 't_{ij}' instead of one-dimensional stress 'σ' and assuming the flow rule in modified stress space t_{ij}.

1 Introduction

It is known that under one-dimensional consolidation (or isotropic consolidation) remolded normally consolidated clay shows typical strain-hardening elastoplastic behavior, so that clay is assumed to be non-linear elastic in the region where the current stress is smaller than the yield stress (over-consolidation region). However, real clay shows elastoplastic behavior even in over-consolidation region. Furthermore, natural clay behaves intricately compared with remolded clay which is used in laboratory tests, because natural clay develops a complex structure in its deposition process. Such structured clay can exist in a region where its void ratio is greater than that of non-structured normally consolidated clay under the same stress condition. This type of structured clay shows more brittle and more compressive behavior than non-structured clay.

Most of the constitutive models for soils have been developed to describe mainly the behaviors of non-structured soil, even though their density (influence of over-consolidation ratio) is more or less considered. The influence of soil structure is sometimes interpreted as the result of aging effect and is intended to be described using elasto-viscoplastic theories and others. However, such modeling does not seem to be essential. Asaoka et al. (2002) developed a unique elastoplastic model for structured soils, employing subloading and superloading surfaces together with normal yield surface (three surfaces model). In the present paper, we

R. Wan et al. (Eds.): Bifurcations, Instabilities and Degradations in Geomaterials, SSGG, pp. 91–111.
springerlink.com

will present a new modeling framework for over-consolidated soil and structured soil, looking to Asaoka and his colleague's work for cues. One-dimensional description of over-consolidated soil and structured soil are first presented to easily understand the fundamental concept of the present models. By introducing a state variable to represent the influence of density, a simple method to describe the elastoplastic consolidation behavior of over-consolidated soil is shown. To describe the consolidation behavior of structured soil, not only the state variable of density but also a state variable of bonding effect are introduced. This is because the real density and the bonding can be considered to be the main factors that affect a structured soil. Finally, these one-dimensional models are extended to ones which can describe three-dimensional stress-strain behavior of over-consolidated soil and structured soil in general stress conditions.

2 One-Dimensional Model for Over-Consolidated Soil

Figure 1 shows the $e - \ln \sigma$ relation in over-consolidated soil schematically. Even in the over-consolidation region, there occurs elastoplastic deformation, and the void ratio of soil gradually approaches to the normally consolidation line (NCL) with increasing stress. Figure 2 shows the change of void ratio when the stress condition moves from the initial state I (σ=σ_0) to the current state P (σ=σ). Here, e_0 and e are the initial and current void ratios of the over-consolidated soil, and e_{N0} and e_N are the corresponding void ratios on the normally consolidation line (NCL). The difference of void ratios between normally and over-consolidated soils is expressed as the change from ρ_0 (=e_{N0}-e_0) to ρ (=e_N-e). Here, it can be assumed that the recoverable change of void ratio Δe^e (elastic component) for over-consolidated soils is the same as that for normally consolidated soils and given by the following expression using the swelling index κ:

$$(-\Delta e)^e = \kappa \ln \frac{\sigma}{\sigma_0} \tag{1}$$

So, the plastic change of void ratio Δe^p for over-consolidated soil is obtained on referring to Fig. 2 and thus,

$$\begin{aligned}(-\Delta e)^p &= (-\Delta e) - (-\Delta e)^e \\ &= \left\{(e_{N0} - e_N) - (\rho_0 - \rho) - (-\Delta e)^e\right\} \\ &= \lambda \ln \frac{\sigma}{\sigma_0} - (\rho_0 - \rho) - \kappa \ln \frac{\sigma}{\sigma_0}\end{aligned} \tag{2}$$

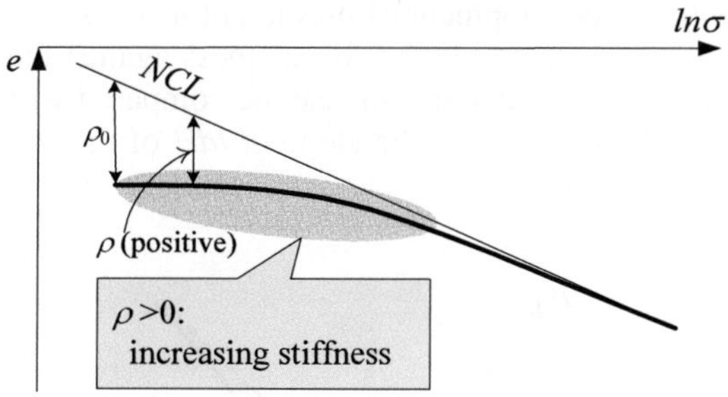

Fig. 1. Void ratio (e) - lnσ relation in OC clay

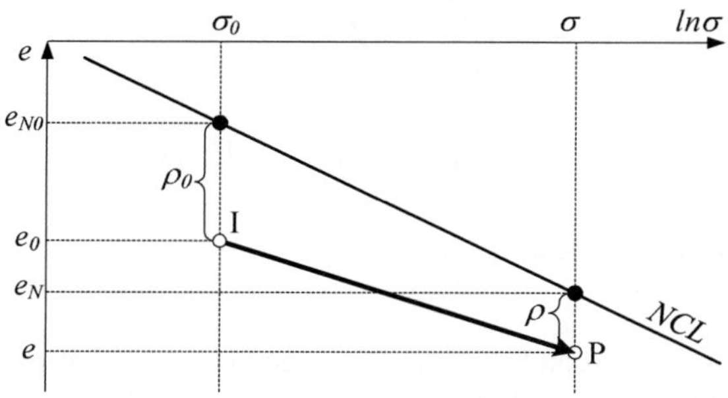

Fig. 2. Change of void ratio in OC clay

Here, λ is the compression index. It can be considered that the difference of void ratios ρ is the state variable which represents the influence of density. By defining the terms of the stress and void ratio as

$$F = (\lambda - \kappa)\ln\frac{\sigma}{\sigma_0} \tag{3}$$

$$H = (-\Delta e)^p \tag{4}$$

Eq. (2) can be written as follows:

$$F + \rho = H + \rho_0$$
or
$$f = F - \{H + (\rho_0 - \rho)\} = 0 \tag{5}$$

The solid line in Fig. 3 shows the relation between F and $(H+\rho_0)$ as expressed in Eq. (5). This line is approaching the broken line ($F=H$) of normally consolidated soil with the development of plastic deformation. The value of state variable ρ, which represents the difference of void ratios between over-consolidated soil and normally consolidated soil at the same stress condition, decreases monotonously from ρ_0 to zero with the development of plastic deformation. The tangential slope dF/dH of the solid line gives an idea of the stiffness against the plastic change of void ratio for over-consolidated soil. This can be compared with the stiffness of a normally consolidated soil given by the slope dF/dH of the broken line, which is always unity in this diagram.

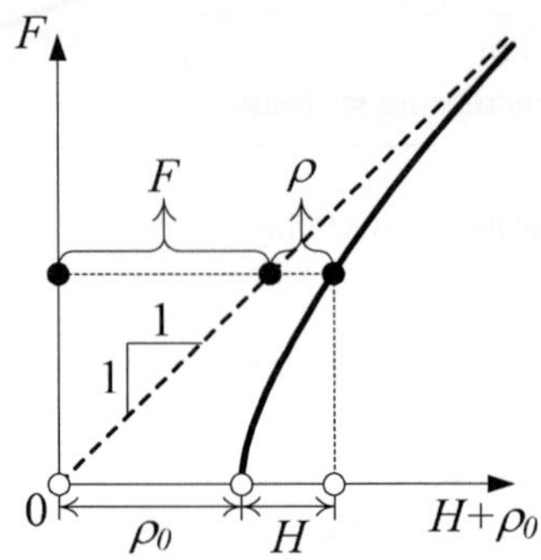

Fig. 3. Explanation of F and H in OC clay

From the consistency condition (df=0) at the occurrence of plastic deformation with satisfying Eq. (5), the following equation is obtained:

$$\begin{aligned} df &= dF - \{dH - d\rho\} \\ &= (\lambda - \kappa)\frac{d\sigma}{\sigma} - \{d(-\Delta e)^p - d\rho\} = 0 \end{aligned} \tag{6}$$

From this equation, the increment of the plastic change of void ratio is expressed as

$$d(-\Delta e)^p = (\lambda - \kappa)\frac{d\sigma}{\sigma} + d\rho \tag{7}$$

As shown in Figs. 1 and 3, it can be assumed that the state variable ρ representing the density decreases ($d\rho$<0) with the development of plastic deformation (volume contraction) and finally becomes zero (normally consolidated state). Further, it can be considered that the larger the value of ρ is, the faster the rate of degradation of ρ is. Therefore, the evolution rule of ρ can be given in the following form:

$$d\rho = -G(\rho) \cdot d(-\Delta e)^p \tag{8}$$

Here, $G(\rho)$ should be an increasing function of ρ with satisfying $G(0)$=0, from the condition above mentioned.

From Eqs. (1), (7) and (8), the increment of total change of void ratio is given by the summation of the plastic component and the elastic component, i.e.

$$d(-\Delta e) = d(-\Delta e)^p + d(-\Delta e)^e = \left\{ \frac{\lambda - \kappa}{1 + G(\rho)} + \kappa \right\} \frac{d\sigma}{\sigma} \quad (9)$$

Equation (9) means that $G(\rho)$ increases the stiffness of soil, and its effect becomes large with the increase of the value of ρ. The method to consider the influence of density presented here corresponds to a one-dimensional interpretation of the subloading surface concept by Hashiguchi (1980).

3 One-Dimensional Model for Structured Soil

The solid curve in Fig. 4 shows a typical $e - \ln \sigma$ relation of natural clay schematically. Natural clay behaves intricately compared with remolded clay which is used in laboratory tests, because natural clay develops a complex structure in its deposition process. Such structured soil can exist in a region where its void ratio is greater than that of non-structured normally consolidated soil under the same stress condition.

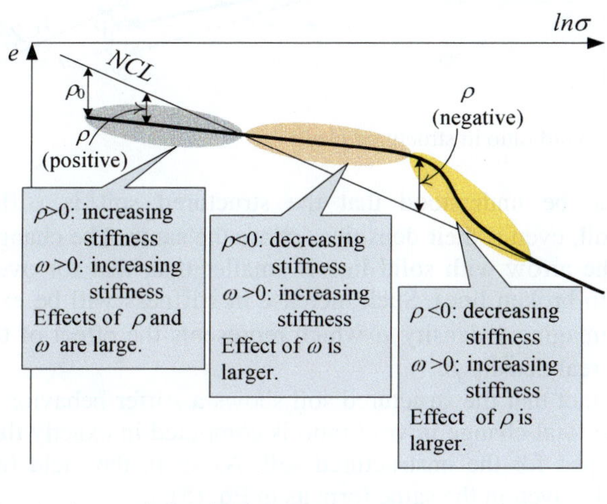

Fig. 4. Void ratio (e) - $\ln\sigma$ relation in structured clay

This type of structured soil shows more brittle and more compressive behavior than non-structured soil. Asaoka et al. (2002) and Asaoka (2005) developed a model to describe such structured soils, introducing subloading surface and superloading surface concepts to the Cam-clay model. In their modeling, a factor related to the over-consolidation ratio (corresponding to an imaginary density) has been introduced to increase the stiffness, and a factor related to the soil skeleton structure has been introduced to decrease the stiffness. By controlling the evolution rules of these factors, they described various features of consolidations and shear behaviors of structured soils.

In the present work, attention is focused on the real density and the bonding as the main factors that affect the behavior of structured soil, because it can be considered that the soil skeleton structure in a looser state than that of a normally consolidated soil is formed by bonding effects.

Figure 5 shows the change of void ratio when the stress condition moves from the initial state I ($\sigma=\sigma_0$) to the current state P ($\sigma=\sigma$) in the same way as that in Fig. 2. Here, e_0 and e are the initial and current void ratios of structured soil, and e_{N0} and e_N are the corresponding void ratios on the normally consolidation line. The arrow with broken line denotes the same change of void ratio as that in Fig. 2 for over-consolidated soil.

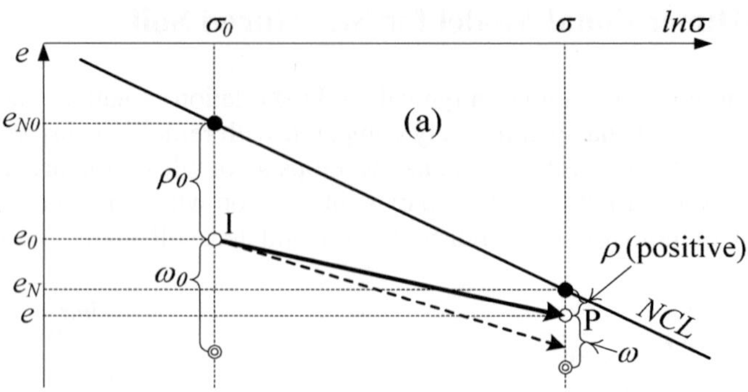

Fig. 5. Change of void ratio in structured clay

Now, it can be understood that the structured soil is stiffer than over-consolidated soil, even if their densities (ρ) are the same. The change of void ratio indicated by the arrow with solid line is smaller than that for over-consolidated soil (arrow with broken line). Such increase in stiffness will be expressed by introducing the imaginary density ω which represents the effect of the bonding, in addition to the real density ρ.

Despite the fact that the structured soil shows a stiffer behavior up to a certain stress level, the total change in void ratio is computed in exactly the same way as the one developed for the unstructured soil. As such, the yield function for the structured soil is given in the same form as in Eq. (5), i.e.

$$f = F - \{H + (\rho_0 - \rho)\} = 0 \tag{10}$$

The main difference resides in the evolution law for the real density ρ, which now can also assume negative values, as illustrated in Fig. 6, depending on the magnitude of the bonding effects represented by the imaginary density ω. The solid line in Fig. 6 shows the relation between F and ($H+\rho_0$) for a structured soil. When the degradation of the bonding effect ω becomes faster with the development of plastic deformations, the solid line monotonically approaches the broken line ($F=H$) of the normally consolidated (NC) soil as shown in Fig. 6a, which is similar to an over-consolidated soil as seen in Fig. 3. On the other hand, when the degradation of ω is not that fast, the solid line reaches the broken line before complete debonding ($\omega = 0$), so that the solid line enters in the region of $\rho < 0$ as shown in Fig. 6b. If it is assumed that a negative ρ has an effect of decreasing the stiffness contrary to a positive ρ, the solid line finally approaches the broken line from the region of $\rho < 0$.

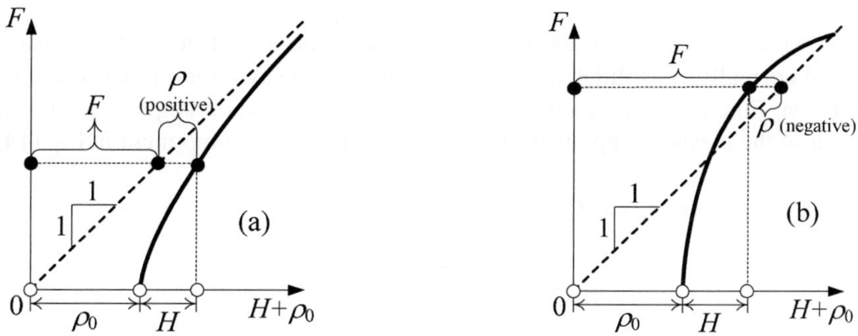

Fig. 6. Explanation of functions F and G for structured clay: (a) degradation of ω is faster, (b) degradation of ω is slower

From the consistency condition ($df = 0$) at the occurrence of plastic deformation and satisfying the above yield function, the following equation is obtained in the same way as that for non-structured over consolidated soil, i.e.

$$\begin{aligned} df &= dF - \{dH - d\rho\} \\ &= (\lambda - \kappa)\frac{d\sigma}{\sigma} - \{d(-\Delta e)^p - d\rho\} = 0 \end{aligned} \tag{11}$$

Next, it is necessary to account for the effect of bonding on the evolution rule for the density variable ρ. This should still be dependent on the development of plastic deformations in the structured soil, so that $d\rho \propto d(-e)^p$. Furthermore, it will be considered that the degree of degradation of ρ is determined not only by the state variable ρ related to the real density, but also by the state variable ω related to the imaginary increase of density due to bonding. This will be introduced through an extra function $Q(\omega)$ with an additive effect on the already defined function $G(\rho)$. Therefore, the evolution rule of ρ can be given in the following form:

$$d\rho = -\{G(\rho) + Q(\omega)\} \cdot d(-\Delta e)^p \tag{12}$$

An additional evolution rule must also be introduced for the imaginary density. The evolution rule for ω is described using the same function $Q(\omega)$ as follows:

$$d\omega = -Q(\omega) \cdot d(-\Delta e)^p \tag{13}$$

Here, $Q(\omega)$ is a monotonously increasing function of ω satisfying $Q(0)=0$, such as $Q(\omega)=b\omega$ (b: material parameter). Substituting Eq. (12)into Eq. (11) gives

$$d(-\Delta e)^p = \frac{\lambda-\kappa}{1+G(\rho)+Q(\omega)} \cdot \frac{d\sigma}{\sigma} \tag{14}$$

As shown in this equation, the increase in the stiffness of structured soil from remolded normally consolidated state is governed by the evolution functions of $G(\rho)$ and $Q(\omega)$. Finally the increment of total change of void ratio is given by the summation of the elastic component in Eq. (1) and the plastic component in Eq. (14):

$$\begin{aligned} d(-\Delta e) &= d(-\Delta e)^p + d(-\Delta e)^e \\ &= \left\{ \frac{\lambda-\kappa}{1+G(\rho)+Q(\omega)} + \kappa \right\} \frac{d\sigma}{\sigma} \end{aligned} \tag{15}$$

If $G(\rho)=0$ and $Q(\omega)=0$, Eq. (15) expresses the relation for remolded normally consolidated soil (abbreviated as NC soil). It is also apparent that positive values of $G(\rho)$ and $Q(\omega)$ have the effect of increasing the stiffness of the soil, while a negative value of $G(\rho)$ has the effect of decreasing its stiffness.

Now, the increment of the plastic change of void ratio for structured soil is given by $d\rho$ and $d\omega$ in addition to the same component as that of the normally consolidated soil. Therefore, the increment of real density ($\Delta\rho$) for structured soil is given not only by the effect of current density ρ but also by the effect of imaginary density ω due to bonding as follows:

$$\Delta\rho = d\rho = -\{G(\rho)+Q(\omega)\} \cdot d(-\Delta e)^p \tag{16}$$

The value of the state valuable ρ can also be updated as the difference between the void ratio corresponding to the normally consolidation line (NCL) at the current stress and the current void ratio.

Here, we will briefly discuss the consolidation behavior of structured soil in Fig. 4. Assume the initial state with positive ρ_0 and ω_0. During the first stage ($\rho>0$ and $\omega>0$), the stiffness of the soil is much larger than that of NC soil, because of the positive values of $G(\rho)$ and $Q(\omega)$. When the current void ratio becomes the same as that on NCL ($\rho=0$), the stiffness of the soil is still larger than that of NC soil ($\omega>0$), so it is possible to have the state of structured soil to be looser than that on NCL. In this stage ($\rho<0$ and $\omega>0$), the effect of increasing the stiffness by a positive value of ω is larger than the effect of decreasing the stiffness by a negative value of ρ. After this stage, the effect of ω becomes small with the development of plastic deformations. On the other hand, the effect of ρ toward decreasing the stiffness becomes prominent because of the negative value of ρ. Finally the void ratio approaches to that on NCL, because both ρ and ω converge to zero.

The one-dimensional loading conditions of over-consolidated and structured soils are presented assuming no occurrence of the plastic volume expansion as:

$$\begin{cases} d(-\Delta e)^p \neq 0 & \text{if} \quad d(-\Delta e)^p > 0 \\ d(-\Delta e)^p = 0 & \text{if} \quad d(-\Delta e)^p \leq 0 \end{cases} \tag{17}$$

In order to check the validity of the present model, numerical simulations of one-dimensional compression tests are carried out. The functions of $G(\rho)$ and $Q(\omega)$ in Eqs. (8) and (13) are given by the simple linear functions of ρ and ω as shown in Fig. 7. Here, though $Q(\omega)$ is depicted in the positive side alone, $G(\rho)$ is depicted both in positive and negative sides. This is because the values of ρ may become negative as mentioned above. Assuming Fujinomori clay which is used in the previous experimental verification of constitutive models (e.g., Nakai and Hinokio, 2004; Nakai, 2007), following material parameters are employed in the numerical simulations – compression index λ=0.104, swelling index κ=0.010 and void ratio on NCL at σ= 98kPa (atmospheric pressure) N=0.83.

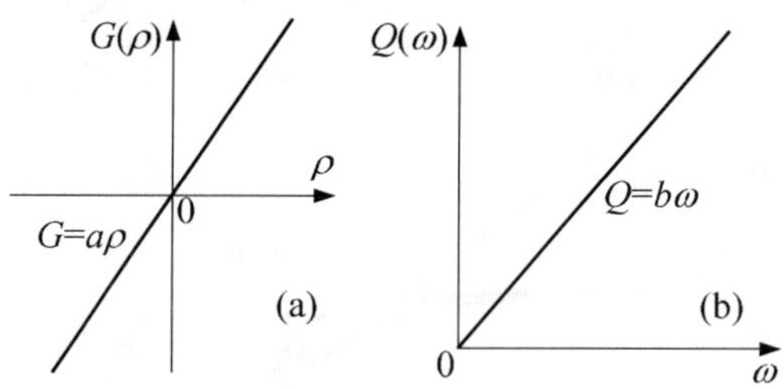

Fig. 7. Evolution of $G(\rho)$ and $Q(\omega)$ - linear functions of ρ

Figure 8(a) shows the calculated e – ln σ relation of the one-dimensional compression using initial bonding parameter (ω_0=0.2) and different initial void ratios. Figure 8(b) shows the calculated results using the same initial void ratio (ρ_0=0.1) and different bonding parameters. Though the density parameter ρ (=e_N - e), which is represented by the vertical distance between current void ratio and void ratio at NCL, of the clay without bonding (ω_0=0) decreases monotonically and converges to NCL, the density parameter of the clays with bonding (ω_0>0) decreases to some negative values and converges to NCL from negative side of ρ with a sharp reduction of the bulk stiffness. In these figures, the parameters (a and b) which represent the degradation rate of ρ and ω are fixed. It can be seen from these figures that it is possible to describe the deformation of structured clay only considering the effect of density and bonding and their evolution rules. Further, Figure 8(c) shows the results in which the initial void ratio and the initial bonding are the same but the parameter b is different. We can see that the result with large value of b (=100) describes void ratio-stress relation with strain softening.

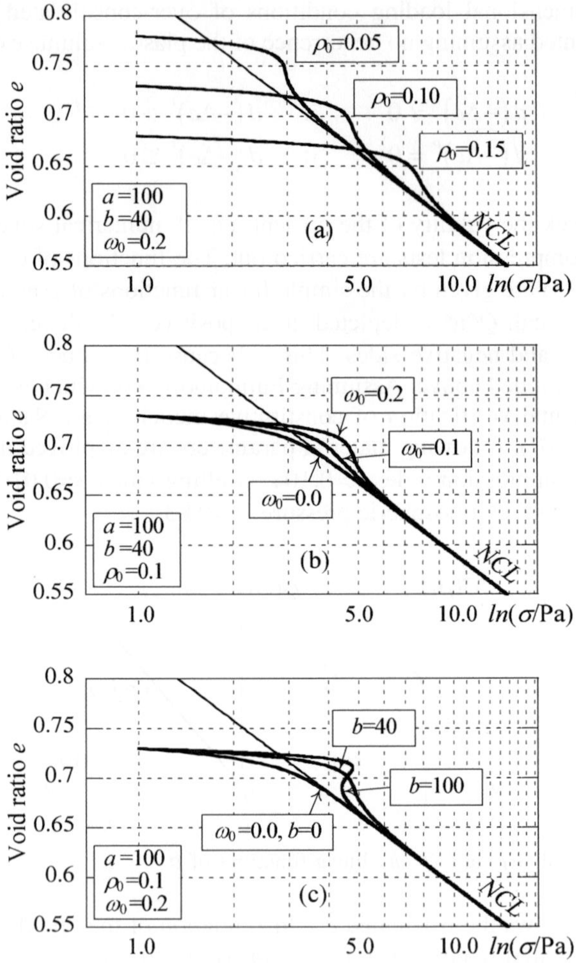

Fig. 8. Calculated results of clay with different ρ_0, ω_0 and b

4 Three-Dimensional Model for Over-Consolidated and Structured Soils

4.1 Ordinary Method to Formulate Three-Dimensional Model

In most of the three-dimensional models such as Cam clay model, their yield functions are formulated using the stress invariants (mean stress p and deviatoric stress q) or (p and $\eta=q/p$). Here, p and q correspond to the normal and parallel components of σ_{ij} to the octahedral plane as shown in Fig. 9 and are given by Eqs. (18) and (19) using three principal stresses (σ_1, σ_2 and σ_3) or stress tensor σ_{ij}, i.e.

$$p = \sqrt{\frac{1}{3}}\overline{\mathrm{ON}} = \frac{1}{3}(\sigma_1 + \sigma_2 + \sigma_3) = \frac{1}{3}\sigma_{ij}\delta_{ij} \tag{18}$$

$$\begin{aligned} q = \sqrt{\frac{3}{2}}\overline{\mathrm{NP}} &= \frac{1}{\sqrt{2}}\sqrt{(\sigma_1 - \sigma_2)^2 + (\sigma_2 - \sigma_3)^2 + (\sigma_3 - \sigma_1)^2} \\ &= \sqrt{\frac{3}{2}(\sigma_{ij} - p\delta_{ij})(\sigma_{ij} - p\delta_{ij})} \end{aligned} \tag{19}$$

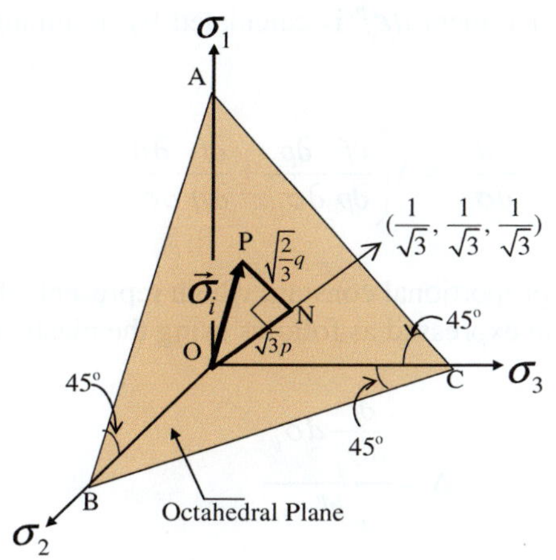

Fig. 9. Octahedral plane in σ_{ij} space and definition of p and q

The yield functions of ordinary elastoplastic models are formulated as follows using these stress invariants:

$$f = f(p, \eta = q/p,\ p_1) = \ln\frac{p}{p_0} + \varsigma(\eta) - \ln\frac{p_1}{p_0} = 0 \tag{20}$$

where, $\zeta(\eta)$ is an increasing function of η which satisfies the condition $\zeta(0)=0$.

Figure 10 shows the shape of the initial yield surface (broken curve) and the current yield surface (solid curve) of an ordinary model in p-q plane when stress condition moves from point I to point P. Mean stresses p_0 and p_1 in Eq. (20) are the values on p-axes of the initial and current yield surfaces, as shown in Fig. 10.

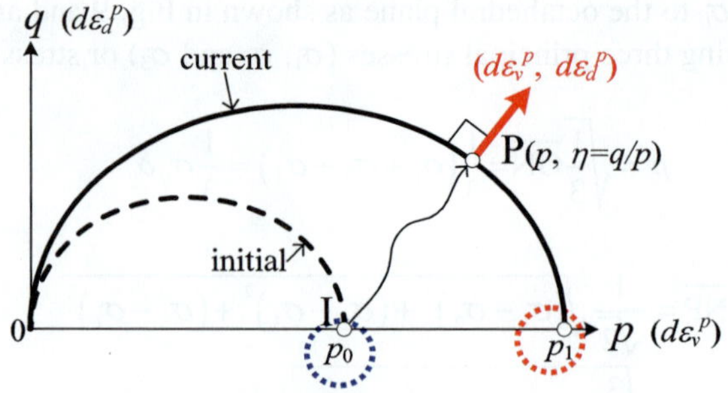

Fig. 10. Yield surface on (p, q) plane and direction of plastic strain increment

The plastic stain increment $d\varepsilon_{ij}^p$ is calculated by assuming the flow rule in σ_{ij} space as:

$$d\varepsilon_{ij}^p = \Lambda \frac{\partial f}{\partial \sigma_{ij}} = \Lambda \left(\frac{\partial f}{\partial p} \frac{\partial p}{\partial \sigma_{ij}} + \frac{\partial f}{\partial \eta} \frac{\partial \eta}{\partial \sigma_{ij}} \right) \tag{21}$$

Here, Λ is a positive proportional constant which represents the magnitude of plastic stain increment and expressed as follows using the plastic modulus h^p, i.e.

$$\Lambda = \frac{\frac{\partial f}{\partial \sigma_{ij}} d\sigma_{ij}}{h^p} \tag{22}$$

However, it is known that constitutive models formulated in such a way cannot consider the influence of intermediate principal stress on the deformation and strength of soils properly (e.g., Nakai and Mihara, 1984; Nakai and Hinokio, 2004; Nakai, 2008).

4.2 Method to Formulate Three-Dimensional Model Based on t_{ij} Concept

For considering the influence of intermediate principal stress automatically, the concept of modified stress t_{ij} is proposed (Nakai and Mihara, 1984). The modified stress t_{ij} is defined by the product of a_{ik} and σ_{kj}, using a symmetric non-dimensional tensor a_{ik}, i.e.

$$t_{ij} = a_{ik} \sigma_{kj} \tag{23}$$

and its principal values are given as follows using the principal values of t_{ij} and a_{ij}:

$$t_1 = a_1\sigma_1,\ t_2 = a_2\sigma_2,\ t_3 = a_3\sigma_3 \tag{24}$$

Here, a_{ij} is the symmetric tensor whose principal values (a_1, a_2 and a_3) are given by the direction cosines of the normal to the spatially mobilized plane (abbrev., SMP; Matsuoka and Nakai, 1974) such that:

$$a_1 = \sqrt{\frac{I_3}{I_2\sigma_1}},\ a_2 = \sqrt{\frac{I_3}{I_2\sigma_2}},\ a_3 = \sqrt{\frac{I_3}{I_2\sigma_3}} \tag{25}$$

$(I_2, I_3$: 1st and 2nd stress invariants of $\sigma_{ij})$

The invariants of modified stress (t_N and t_S) used in t_{ij} concept are defined as the normal and parallel components of t_{ij} to the SMP as shown in Fig. 11.

$$t_N = \overline{\mathrm{ON}} = t_1a_1 + t_2a_2 + t_3a_3 = t_{ij}a_{ij} \tag{26}$$

$$t_S = \overline{\mathrm{NT}} = \sqrt{t_1^2 + t_2^2 + t_3^2 - (t_1a_1 + t_2a_2 + t_3a_3)^2} = \sqrt{t_{ij}t_{ij} - (t_{ij}a_{ij})^2} \tag{27}$$

The yield function based on t_{ij} is formulated using the stress invariants defined by the stress invariants (t_N and t_S) or (t_N and $X=t_S/t_N$) in Eqs. (26) and (27) instead of (p and q) or (p and $\eta=q/p$) in the same form as Eq. (20), i.e.

$$f = f\left(t_N,\ X = t_S/t_N,\ t_{N1}\right) = \ln\frac{t_N}{t_{N0}} + \varsigma(X) - \ln\frac{t_{N1}}{t_{N0}} = 0 \tag{28}$$

Figure 12 shows the shape of the initial yield surface (broken curve) and the current yield surface (solid curve) of the model based on the t_{ij} concept in t_N-t_S plane, in the same way as that of the ordinary model in Fig. 10. Here, t_{N0} and t_{N1} in Eq. (28) are the values on t_N-axes of the initial and current yield surfaces in Fig. 12. The plastic stain increment $d\varepsilon_{ij}^p$ is obtained by assuming the flow rule not in σ_{ij} space but in t_{ij} space. Thus,

$$d\varepsilon_{ij}^p = \Lambda\frac{\partial f}{\partial t_{ij}} = \Lambda\left(\frac{\partial f}{\partial t_N}\frac{\partial t_N}{\partial t_{ij}} + \frac{\partial f}{\partial X}\frac{\partial X}{\partial t_{ij}}\right) \tag{29}$$

As mentioned above, employing the t_{ij} concept in which the yield function is formulated using the stress invariants (t_N and t_S) and assuming the flow rule in t_{ij} space, the influence of intermediate principal stress can be automatically taken into consideration.

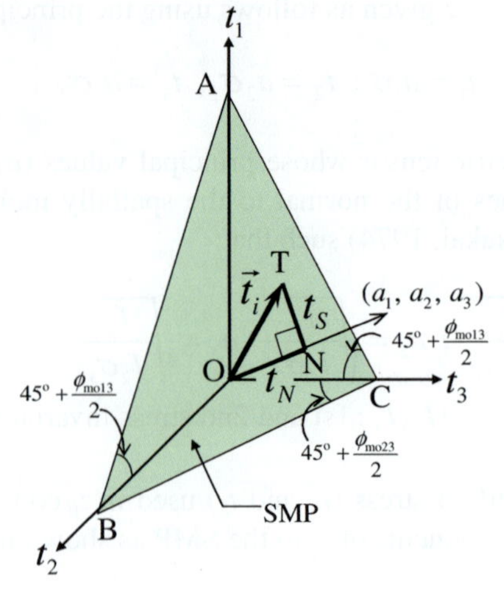

Fig. 11. Spatially mobilized plane (SMP) in t_{ij} space and definition of t_N and t_S

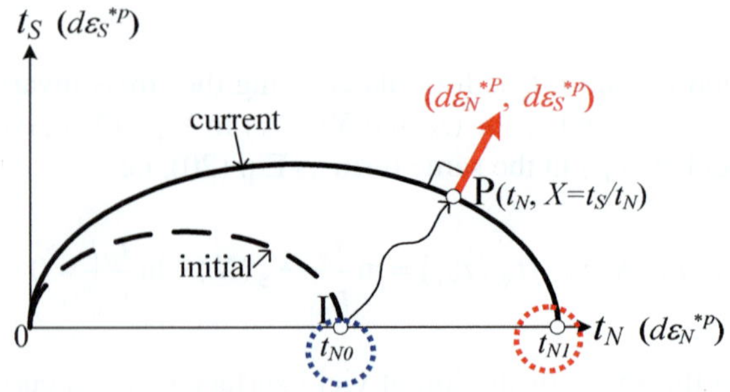

Fig. 12. Yield surface on (t_N, t_S) plane and direction of plastic strain increment

4.3 Extension of One-Dimensional Model to Three-Dimensional Model

We will here describe a simple method to extend directly the one-dimensional constitutive model for over-consolidated soil and structured soil shown in the previous section to three-dimensional models based on t_{ij} concept.

As described in the previous section, the yield function for the one-dimensional models for over-consolidated soil and structured soils are expressed as in Eqs. (5) and (10), respectively, using the terms F and H. Here, F is the function of stress σ, and H represents the plastic change of void ratio Δe^p (see Eqs. (3) and (4)). Now, it can be assumed that initial stress σ_0 and current stress σ in the one-dimensional models correspond to the terms t_{N1} and t_{N0} in three-dimensional models. Since there holds Eq. (28), the term F is given by replacing σ and σ_0 in Eq. (3) with t_{N1} and t_{N10}, respectively:

$$F = (\lambda - \kappa)\ln\frac{t_{N1}}{t_{N0}} = (\lambda - \kappa)\left\{\ln\frac{t_N}{t_{N0}} + \varsigma(X)\right\} \tag{30}$$

Also, H is described in the same way as that in the one-dimensional model, i.e.

$$H = (-\Delta e)^p = (1 + e_0)\cdot\varepsilon_v^p \tag{31}$$

Hereafter we will explain the formulation of model for structured soil alone, because the model without the bonding results into a model for over-consolidated soil. The yield function in three-dimensional models for structured soil is given by the same equation as in Eq. (10). We can then obtain the following equation from the consistency condition (df=0) at the occurrence of plastic deformation:

$$\begin{aligned} df &= dF - \{dH - d\rho - d\omega\} \\ &= dF - \{d(-\Delta e)^p - d\rho - d\omega\} \\ &= dF - \left\{(1+e_0)\Lambda\frac{\partial F}{\partial t_{ii}} - d\rho - d\omega\right\} = 0 \end{aligned} \tag{32}$$

Since it can be assumed that the positive values of ρ and ω decrease with the development of plastic strain in the same way as the one-dimensional model, the evolution rules of ρ and ω in the three-dimensional model are given as follows using monotonously increasing functions $G(\rho)$ and $Q(\omega)$:

$$d\rho = -(1+e_0)\frac{G(\rho)}{t_N}\Lambda \tag{33}$$

$$d\omega = -(1+e_0)\frac{Q(\omega)}{t_N}\Lambda \tag{34}$$

Here, though the evolution rule of ρ and ω are related to $d(-\Delta e)^p$ as shown in Eqs. (8) and (13) in the one-dimensional model, they should be related to Λ in the three-dimensional model. Considering the dimension of ρ and ω, i.e., these values have no unit, these evolution equations are divided by mean stress t_N. Substituting Eqs. (33) and (34) into Eq. (32), the proportional constant Λ is expressed as

$$\Lambda = \frac{dF}{(1+e_0)\left\{\dfrac{\partial F}{\partial t_{ii}} + \dfrac{G(\rho)}{t_N} + \dfrac{Q(\omega)}{t_N}\right\}} = \frac{dF}{h^p} \tag{35}$$

The loading conditions in the three-dimensional model are given as

$$\begin{cases} d\varepsilon_{ij}^p \neq 0 & \text{if} \quad \Lambda > 0 \\ d\varepsilon_{ij}^p = 0 & \text{if} \quad \Lambda \leq 0 \end{cases} \tag{36}$$

Then, using the symbol < > which denotes the Macaulay bracket, i.e., $<A>=A$ if $A>0$; otherwise $<A>=0$, the plastic strain increment is finally expressed as

$$d\varepsilon_{ij}^p = \langle \Lambda \rangle \frac{\partial F}{\partial t_{ij}} \tag{37}$$

The present model for structured soil is the same as those presented before (Nakai, 2007; Nakai et al., 2008), and the model in which the effect of bonding disappears corresponds to the subloading t_{ij} model by Nakai and Hinokio (2004).

4.4 Numerical Simulation of Structured Soil

In order to check the validity of the proposed three-dimensional model, numerical simulations of oedometer tests and undrained triaxial compression and extension tests for structured soil (for Fujinomori clay) are carried out. The values of material parameters for structured Fujinomori clay are shown in Table 1. In the three-dimensional model, the evolution rules for both ρ and ω are given by the relations as shown in Fig. 13. The material parameters except for b are the same as those of the subloading t_{ij} model for non-structured soil (Nakai and Hinokio, 2004).

Figure 14 shows the results of simulation using the three-dimensional model for the oedometer tests on structured clays which have the same initial void ratio but have different initial bonding effects, arranged in terms of the relations between void ratio and vertical stress in log scale. It is seen that three-dimensional model can also describe the typical one-dimensional consolidation behavior of structured soils in the same way as the one-dimensional model in Fig.8.

Table 1. Values of material parameters in three-dimensional model for structured Fujinomori clay and their meanings

λ	0.104	Same parameters as Cam clay model
κ	0.010	
$N = e_{NC}$ at $p = 98$ kPa & $q = 0$ kPa	0.83	
$R_{cs} = (\sigma_1/\sigma_3)_{cs(comp.)}$	3.5	
ν_e	0.2	
β	1.5	Shape of yield surface (same as original Cam clay at $\beta = 1$)
$a/(\lambda\text{-}\kappa)$	500	Influence of density and confining pressure
$b/(\lambda\text{-}\kappa)$	40	Degradation parameter of bonding

Figure 15 shows the results of numerical simulations of undrained triaxial compression and extension tests on structured clays. Diagram (a) refers to the results of effective stress paths, and diagram (b) to the results of stress-stain curves. In these figures, the upper part shows the results under triaxial compression condition, and lower part the results under triaxial extension condition. The straight lines from the origin in diagram (a) represent the critical state lines (CSL) in p-q plane. Under undrained shear conditions, clays with bonding are stiffer and have higher strength than clays without bonding. It is also seen that over-consolidated clay without bonding shows strain hardening with the decrease and the subsequent increase of mean stress, whereas clays with bonding show not only stress hardening with the decrease and the increase of mean stress, but also strain softening with the decrease of mean stress and deviatoric stress under undrained conditions. These are known to be typical behaviors of structured soil.

Figure 16 shows the results of numerical simulations of isotropic compression and subsequent undrained shear tests on structured Fujinomori clay. Diagram (a) refers to the results of isotropic compression test, while diagrams (b) and (c) show the results of effective stress paths and stress- strain curves in undrained triaxial compression tests on clays which are sheared from stress conditions (A), (B) and (C) in diagram (a). The model simulates well typical undrained shear behavior of structured soil. i.e., the differences of stress paths and stress-strain curves depending on the magnitude of confining pressure, 'rewinding' of stress path after increasing of mean stress and deviatoric stress and others.

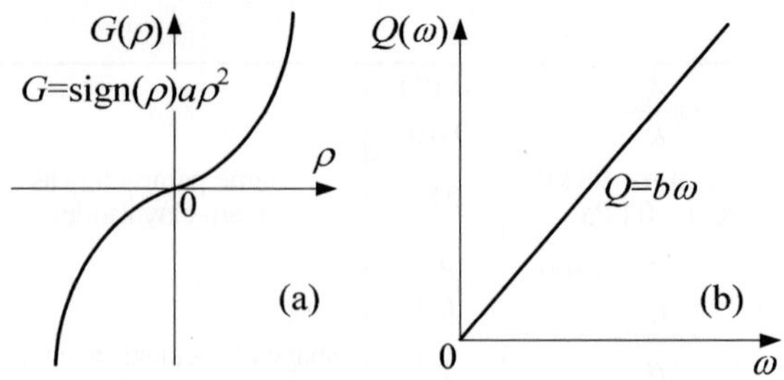

Fig. 13. Evolution of $G(\rho)$ and $Q(\omega)$ – square and linear functions of ρ and ω

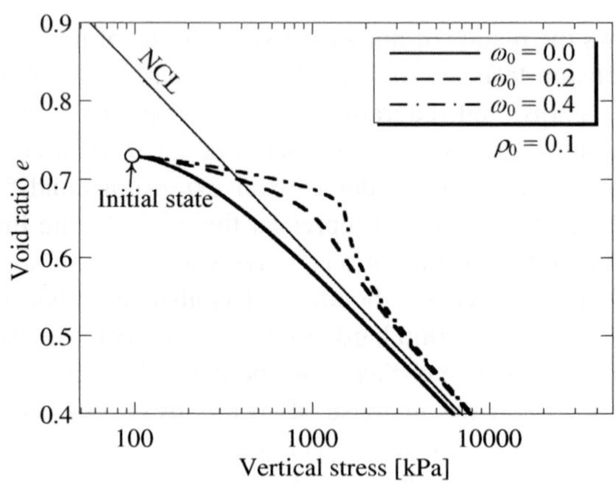

Fig. 14. Calculated results of oedometer tests on clays with the same initial void ratio but different initial bonding

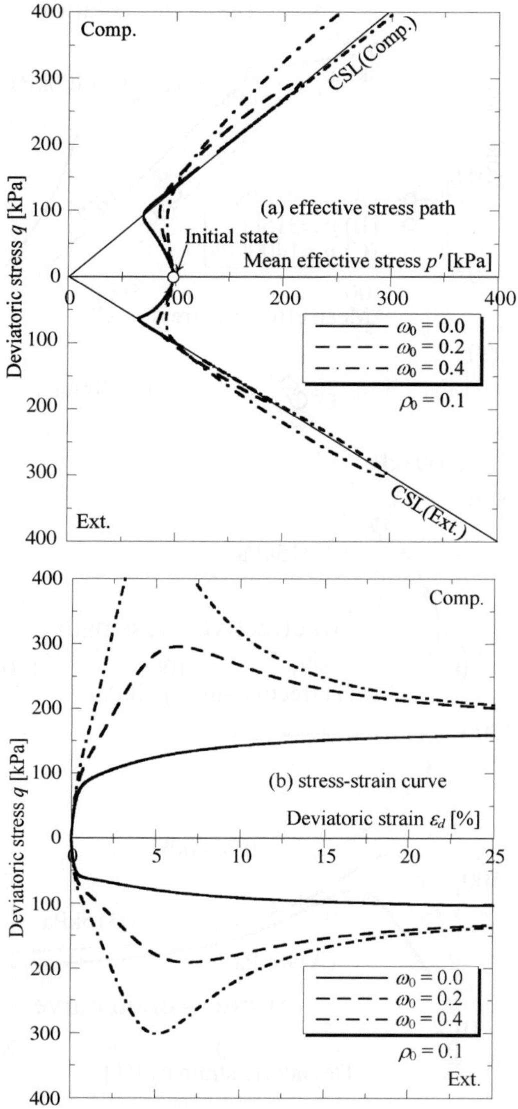

Fig. 15. Calculated results of undrained triaxial compression and extension tests on clays with the same initial void ratio but different initial bonding

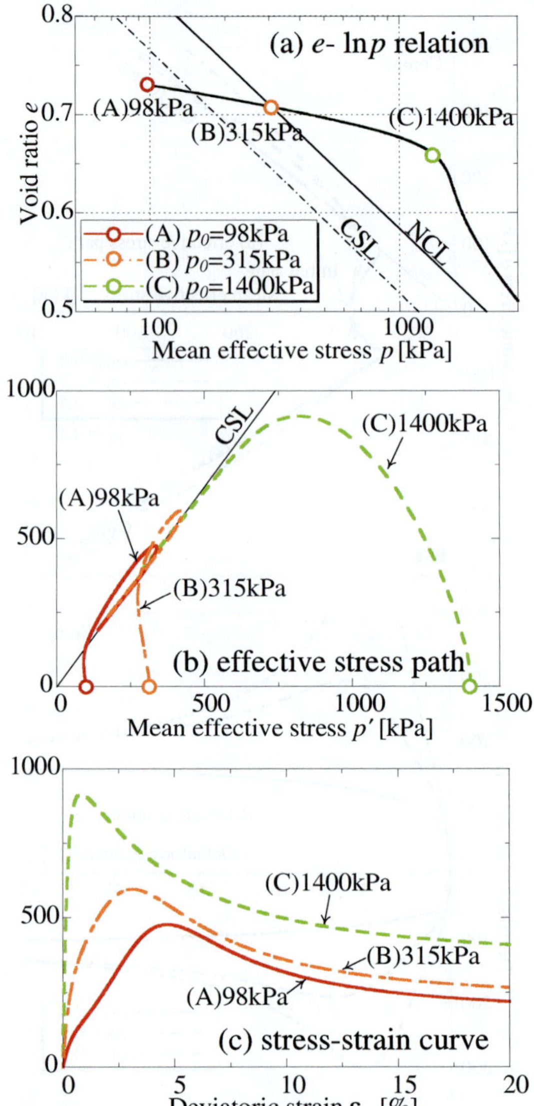

Fig. 16. Calculated results of isotropic compression and succeeding undrained shear tests on structured clay

5 Conclusions

As a one-dimensional model, a method to describe the behavior of over-consolidated soil is presented by using the state valuable of density and its monotonous evolution rule. Next, introducing the state valuable which represents the effect of bonding as well as the state variable of density, a model to describe the behavior of structured soil is developed. Furthermore, it is shown that by introducing the t_{ij} concept proposed before, these models can easily be extended to three-dimensional ones. The validities of these models have been checked by the simulations of one-dimensional compression and undrained shear tests on structured soils with different densities and bonding effects.

References

Asaoka, A.: Consolidation of clay and compaction of sand – an elastoplastic description. In: Proc of 12th Asian Regional Conf. on Soil Mech. and Geotechnical Eng., Keynote Paper, Singapore, vol. 2, pp. 1157–1195 (2005)

Asaoka, A., Noda, T., Yamada, E., Kaneda, K., Nakano, M.: An elasto-plastic description of two distinct volume change mechanisms of soils. Soils and Foundations 42(5), 47–57 (2002)

Hashiguchi, K.: Constitutive equation of elastoplastic materials with elasto-plastic transition. Jour. of Appli. Mech., ASME 102(2), 266–272 (1980)

Matsuoka, H., Nakai, T.: Stress-deformation and strength characteristics of soil under three different principal stresses. Proc. of JSCE 232, 59–70 (1974)

Nakai, T.: Modeling of soils behavior based on t_{ij} concept. In: Proc. of 13th Asian Regional Conf. on Soil Mech. and Geotechnical Eng., Keynote Paper, Kolkata, pp. 1–25 (2007) (preprint)

Nakai, T., Hinokio, T.: A simple elastoplastic model for normally and over-consolidated soils with unified material parameters. Soils and Foundations 44(2), 53–70 (2004)

Nakai, T., Mihara, Y.: A new mechanical quantity for soils and its application to elastoplastic constitutive models. Soils and Foundations 24(2), 82–94 (1984)

Nakai, T., Zhang, F., Kyokawa, H., Kikumoto, M., Shahin, H.M.: Modeling the influence of density and bonding on geomaterials. In: Proc. of 2nd International Symposium on Geotechnics of Soft Soils, Keynote Paper, Glasgow, pp. 65–76 (2008)

Schofield, A.N., Wroth, C.P.: Critical State Soil Mechanics. McGrow-Hill, London (1968)

Cyclic Mobility of Sand and Its Simulation in Boundary Value Problems

F. Zhang[1], Bin Ye[2], Y.J. Jin[1], and T. Nakai[1]

[1] Nagoya Institute of Technology, Nagoya, Japan
[2] Geo-Research Institute, Osaka, Japan

Abstract. In this paper, a new model is proposed to describe the mechanical behaviors of soils under different loading conditions, in which new evolution equations for stress-induced anisotropy and density of soils are proposed. By combining systematically the above two evolution equations with the evolution equation for the structure of soil, the newly proposed model is able to describe not only the mechanical behavior of soils under monotonic loading, but also under cyclic loading with different drained conditions. For given sand with different densities, very loose sand may liquefy without cyclic mobility, medium dense sand will liquefy with cyclic mobility while dense sand will not liquefy, which is just controlled by the density, the structure and the anisotropy of the sand. The proposed model can uniquely describe this behavior without changing its parameters. Shaking-table tests on saturated sandy ground with repeated liquefaction-consolidation process are then simulated with finite element-finite difference method (FE-FD) based on the newly proposed model.

1 Introduction

Much research has been done on the liquefaction of sand experimentally, empirically, and mathematically. Cyclic mobility of sand is one of typical mechanical behaviors of sand during liquefaction. Research related to testing methods and modeling of the cyclic mobility of sand can be found in many publications, e.g., the work by Oka et al (1992, 1999); LIQCA program by Yashima et al (1991 and Oka et al (1994). In recent years, research on constitutive model for soils has been developing very quickly. For instance, the concept "initial anisotropy" proposed by Sekiguchi (1977), the concept of "subloading" proposed by Hashiguchi and Ueno (1977), the concept of "superloading" proposed by Asaoka et al. (1998), make it possible not only to describe remolded soils (Roscoe et al., 1963 and Schofield and Wroth 1968), but also naturally deposited soils in which overconsolidation, structure and anisotropy of soils play a very important rule in determining the mechanical behaviors of the soil.

In this contribution, the authors propose a new approach for describing the stress-induced anisotropy together with a new evolution rule for changes of overconsolidation, by which the mechanical behavior of soils subjected to cyclic loading under drained and undrained conditions, including the cyclic mobility of medium dense sand, can be uniquely described. Meanwhile, experimental results

R. Wan et al. (Eds.): Bifurcations, Instabilities and Degradations in Geomaterials, SSGG, pp. 113–132.
springerlink.com

of shaking-table tests on saturated sandy ground with repeated liquefaction-consolidation process are presented. Furthermore, a finite element-finite difference method (FE-FD) based on the newly proposed model and two-phase field theory proposed by Oka et al. (1994) is conducted. Comparisons between the experiment and the numerical simulation show that the numerical simulation is capable of reproducing almost all main characteristics of the repeated liquefaction-consolidation of sandy grounds with different densities, such as the mechanical behavior pre- and during liquefaction, the settlement in post-liquefaction consolidation and the influence of density on the accumulation of excessive pore water pressure (EPWP) in repeated strong motions.

2 Modeling of Cyclic Mobility

The model proposed here, is based on the concepts of subloading and superloading as described in the work by Asaoka et al (1998). Here, we give just a brief description of the yielding surfaces shown in Fig. 1.

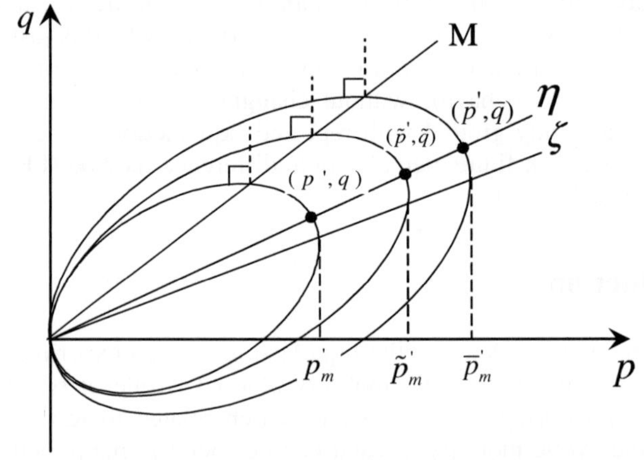

Fig. 1. Subloading, normal and superloading yield surfaces in *p-q* plane

The similarity ratio of the superloading surface to normal yield surface R^* and the similarity ratio of the superloading surface to subloading surface R are the same as those in the work by Asaoka et al. (1998), namely,

$$R^* = \frac{\tilde{p}'}{\bar{p}'} = \frac{\tilde{q}}{\bar{q}}, \; 0 < R^* \le 1 \; and \; \frac{\tilde{q}}{\tilde{p}'} = \frac{\bar{q}}{\bar{p}'} \tag{1}$$

$$R = \frac{p'}{\bar{p}'} = \frac{q}{\bar{q}}, 0 < R \le 1, \; and \; \frac{\bar{q}}{\bar{p}'} = \frac{\tilde{q}}{\tilde{p}'} = \frac{q}{p'} \tag{2}$$

where, (p', q), $(\tilde{p}', \tilde{q})$ and $(\overline{p}', \overline{q})$ represent the present stress state, the corresponding normally consolidated stress state and the structured stress state at *p*-*q* plane, respectively. The normally yield surface is given in the following form as:

$$f(\tilde{p}',\tilde{\eta}^*,\zeta)+\int_0^t J\,\mathrm{tr}\mathbf{D}^p d\tau = MD\ln\frac{\tilde{p}'}{\tilde{p}'_0}+MD\ln\frac{M^2-\zeta^2+\tilde{\eta}^{*2}}{M^2-\zeta^2}+\int_0^t J\,\mathrm{tr}\mathbf{D}^p d\tau=0 \quad (3)$$

where, $\tilde{\eta}^*=\eta^*$, and the other variables involved in Eqs. (1), (2) and (3) are defined as:

$$\eta^*=\sqrt{3\hat{\boldsymbol{\eta}}\cdot\hat{\boldsymbol{\eta}}}/\sqrt{2},\quad \hat{\boldsymbol{\eta}}=\boldsymbol{\eta}-\boldsymbol{\beta},\quad \boldsymbol{\eta}=\mathbf{S}/p',\quad \mathbf{S}=\mathbf{T}'+p'\mathbf{I},\quad p'=-\mathrm{tr}\mathbf{T}'/3 \qquad (4)$$

$$\zeta=\sqrt{2/3}\sqrt{\boldsymbol{\beta}.\boldsymbol{\beta}}\,,\qquad \eta=\sqrt{2/3}\sqrt{\boldsymbol{\eta}\cdot\boldsymbol{\eta}} \qquad (5)$$

where, $\mathbf{S}$ is the deviatoric stress tensor; $\boldsymbol{\beta}$ is the anisotropic stress tensor, and $\mathbf{T}'$ is the Cauchy effective stress tensor and is assumed to be positive in tension. It is clear from both the definition and Fig. 2 that the aspect ratio of the elliptical yield surface adopted in the newly proposed model changes with the value of anisotropy.

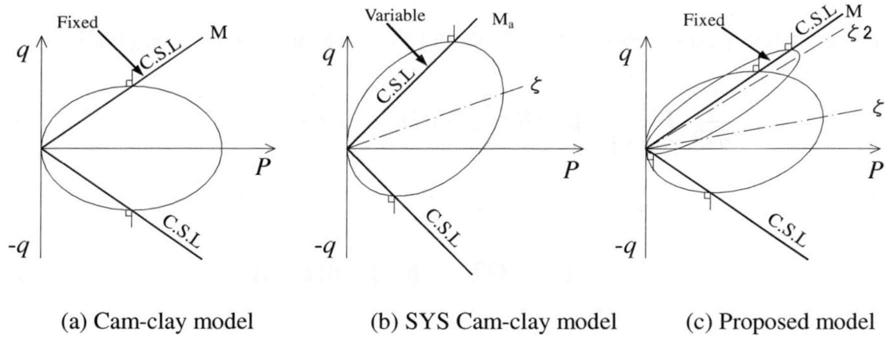

Fig. 2. Changes in the subloading yielding surfaces at different anisotropy ζ

The larger the stress-induced anisotropy ζ is, the larger the eccentric ratio of the ellipse will be. In Eq. (3), J is the Jacobian determination of deformation gradient tensor $\mathbf{F}$ and can be expressed as:

$$J=\det\mathbf{F}=(1+e)/(1+e_0)=v/v_0 \qquad (6)$$

where v and v_0 are the specific volume at the current time (t) and the specific value at the reference time (t=0). D is the dilatancy parameter which can be expressed by $\tilde{\lambda}, \tilde{\kappa}$, the compression and the swelling index, respectively, as follows:

$$D = \frac{\tilde{\lambda} - \tilde{\kappa}}{M(1+e_0)} = \frac{\tilde{\lambda} - \tilde{\kappa}}{Mv_0} \tag{7}$$

$\mathbf{D}^p$ denotes the plastic component of stretching $\mathbf{D}$ which is assumed to be positive in tension, and is related to the plastic volumetric strain rate in the following form under the condition that the compressive of the volumetric strain is supposed to be positive:

$$\varepsilon_v^p = -\int_0^t J \,\mathrm{tr}\mathbf{D}^p d\tau \tag{8}$$

By substituting Eqs. (1) and (2) into Eq. (3), the subloading yield surface can be obtained in the following equation as:

$$\begin{aligned} & f(p', \eta^*, \zeta) + MD\ln R^* - MD\ln R + \int_0^t J \,\mathrm{tr}\mathbf{D}^p d\tau \\ & = MD\ln\frac{p'}{p'_0} + MD\ln\frac{M^2 - \zeta^2 + \eta^{*2}}{M^2 - \zeta^2} + MD\ln R^* - MD\ln R + \int_0^t J \,\mathrm{tr}\mathbf{D}^p d\tau = 0 \end{aligned} \tag{9}$$

An associated flow rule is employed in the present model, namely,

$$\mathbf{D}^p = \lambda \partial f / \partial \mathbf{T}' \tag{10}$$

The consistency equation for the subloading yield surface can then be given as:

$$\frac{\partial f}{\partial \mathbf{T}'} \cdot \overset{\circ}{\mathbf{T}}' + \frac{\partial f}{\partial \boldsymbol{\beta}'} \cdot \overset{\circ}{\boldsymbol{\beta}} + MD\frac{\dot{R}^*}{R^*} - MD\frac{\dot{R}}{R} + J\,\mathrm{tr}\mathbf{D}^p = 0 \tag{11}$$

where

$$\overset{\circ}{\mathbf{T}}' = \dot{\mathbf{T}}' + \mathbf{T}'\boldsymbol{\Omega} - \boldsymbol{\Omega}\mathbf{T}', \quad \overset{\circ}{\boldsymbol{\beta}} = \dot{\boldsymbol{\beta}} + \boldsymbol{\beta}\boldsymbol{\Omega} - \boldsymbol{\Omega}\boldsymbol{\beta} \tag{12}$$

in which, $\overset{\circ}{\mathbf{T}}'$ and $\overset{\circ}{\boldsymbol{\beta}}$ are the Green-Naghdi rates of stress tensor $\mathbf{T}'$ and anisotropic stress tensor $\boldsymbol{\beta}$, respectively. $\boldsymbol{\Omega}$ is material spin tensor. The following differentials are useful in deriving the constitutive equation for the stress tensor and the stretching tensor:

$$\frac{\partial(\eta^{*2})}{\partial p'} = -\frac{3\hat{\boldsymbol{\eta}} \cdot \boldsymbol{\eta}}{p'}, \quad \frac{\partial(\eta^{*2})}{\partial \mathbf{S}} = \frac{3\hat{\boldsymbol{\eta}}}{p'} \tag{13}$$

$$\frac{\partial(\eta^{*2})}{\partial \boldsymbol{\beta}} = -3\hat{\boldsymbol{\eta}}, \quad \frac{\partial(\zeta^2)}{\partial \boldsymbol{\beta}} = 3\boldsymbol{\beta} \tag{14}$$

Based on Eqs. (13) and (14), it is easy to obtain the following relations:

$$\frac{\partial f}{\partial p'}=MD(\frac{1}{p'}+\frac{\partial(\eta^{*2})}{\partial p'}/(M^2-\zeta^2+\eta^{*2}))=MD\frac{M^2-\eta^2}{(M^2-\zeta^2+\eta^{*2})p'} \tag{15}$$

$$\frac{\partial f}{\partial \mathbf{S}}=MD\frac{3\hat{\boldsymbol{\eta}}}{(M^2-\zeta^2+\eta^{*2})p'} \tag{16}$$

$$\frac{\partial f}{\partial \mathbf{T}'}=\frac{\partial f}{\partial \mathbf{S}}+\frac{1}{3}\frac{\partial f}{\partial p'}\mathbf{I}=MD(\frac{3\hat{\boldsymbol{\eta}}}{(M^2-\zeta^2+\eta^{*2})p'}+\frac{1}{3}\frac{M^2-\eta^2}{(M^2-\zeta^2+\eta^{*2})p'}\mathbf{I}) \tag{17}$$

$$\frac{\partial f}{\partial \boldsymbol{\beta}}=MD(\frac{-\frac{\partial(\zeta^2)}{\partial\boldsymbol{\beta}}+\frac{\partial(\eta^{*2})}{\partial\boldsymbol{\beta}}}{M^2-\zeta^2+\eta^{*2}}-\frac{-\frac{\partial(\zeta^2)}{\partial\boldsymbol{\beta}}}{M^2-\zeta^2})=MD\frac{3(-\mathrm{M}^2\hat{\boldsymbol{\eta}}+\eta^{*2}\boldsymbol{\beta}+\zeta^2\hat{\boldsymbol{\eta}})}{(M^2-\zeta^2+\eta^{*2})(M^2-\zeta^2)} \tag{18}$$

From Eq. (15), it is clear that the *C.S.L*, defined by the condition in which $\partial f/\partial p'=0$, always satisfies the relation $\eta=M$, implying that the *C.S.L*, as the threshold between plastic compression and plastic expansion, does not move with the changes in the anisotropy.

2.1 Evolution Rule for Stress-Induced Anisotropic Stress Tensor $\boldsymbol{\beta}$

Different from the work by Hashiguchi and Chen (1998) and Asaoka et al (2002), the following evolution rule for the anisotropic stress tensor is defined as:

$$\overset{\circ}{\boldsymbol{\beta}}=\frac{J}{D}b_r(b_l M-\zeta)\sqrt{2/3}\left\|\mathbf{D}_s^p\right\|\frac{\hat{\boldsymbol{\eta}}}{\left\|\hat{\boldsymbol{\eta}}\right\|} \tag{19}$$

in which, an artificial limitation on the development of anisotropy originally proposed by Hashiguchi and Chen (1998) is no longer necessary. This is because firstly it more or less lacks physical evidence and secondly the stress-induced anisotropic stress tensor $\boldsymbol{\beta}$ also represents the stress history that the soil experienced and it will not exceed the *C.S.L*, which provides us with a natural physical limitation $\zeta<M$, that can be easily accepted. From the evolution Eq. (19), it is also known that development of anisotropy will stop at the state when $\boldsymbol{\eta}=\boldsymbol{\beta}$.

The plastic component of stretching tensor $\mathbf{D}_s^p$ can be calculated as follows:

$$\mathbf{D}_s^p=\mathbf{D}^p-(\mathrm{tr}\mathbf{D}^p)\mathbf{I}/3=\lambda\partial f/\partial\mathbf{S} \tag{20}$$

$$\left\|\mathbf{D}_s^p\right\|=\lambda\sqrt{\frac{\partial f}{\partial\mathbf{S}}\cdot\frac{\partial f}{\partial\mathbf{S}}}=\lambda MD\frac{3\left\|\hat{\boldsymbol{\eta}}\right\|}{(M^2-\zeta^2+\eta^{*2})p'} \tag{21}$$

Substituting Eq. (21) into Eq. (19), the evolution rule for the anisotropic stress tensor can be rewritten as:

$$\overset{\circ}{\boldsymbol{\beta}} = \lambda \frac{J\, M\sqrt{6}b_r(b_l M - \zeta)\hat{\boldsymbol{\eta}}}{(M^2 - \zeta^2 + \eta^{*2})p'} \tag{22}$$

Noting Eqs. (18) and (22), it is easy to calculate the increment in anisotropy as follows:

$$\frac{\partial f}{\partial \boldsymbol{\beta}} \cdot \overset{\circ}{\boldsymbol{\beta}} = \lambda MD \frac{JM\sqrt{6}b_r(b_l M - \zeta)\eta^{*2}(-2M^2 + 3\boldsymbol{\eta}\cdot\boldsymbol{\beta})}{(M^2 - \zeta^2 + \eta^{*2})^2(M^2 - \zeta^2)p'} \tag{23}$$

which plays a very important role in the evolution rule for the overconsolidation that will be discussed later. From Eq. (23), it is clear that if $\boldsymbol{\eta}\cdot\boldsymbol{\beta} \le 0$, which means the angle between the deviatoric stress tensor and the anisotropic tensor is larger than 90°, then $(\partial f / \partial \boldsymbol{\beta}) \cdot \overset{\circ}{\boldsymbol{\beta}}$ will always be less than zero.

*2.2 Evolution Rule for Degree of Structure R**

The following evolution rule for degree of structure R^*, which is the same as in the work by Asaoka et al. (2002), is adopted:

$$\dot{R}^* = JU^*\sqrt{2/3}\left\|\mathbf{D}_s^p\right\| \tag{24}$$

where,

$$U^* = aR^*(1 - R^*)/\mathrm{D} \qquad (0 < R^* \le 1) \tag{25}$$

in which a is the parameter that controls the rate of the collapse of the structure during shearing. From the definition, it is clear that the structure of a soil will never be regained once it has been lost. This seems natural because, based on the physical process, the structure of a soil is accumulated during the sedimentary process over a long period time and it would not be easy to regain it within a short period of time without any chemical processes. Substituting Eqs. (21) and (25) into Eq. (24), the rate of R^*can be evaluated as:

$$\dot{R}^* = \lambda \frac{JM\, 2aR^*(1 - R^*)\eta^*}{(M^2 - \zeta^2 + \eta^{*2})p'} \tag{26}$$

2.3 Evolution Rule for Degree of Overconsolidation R

In the present model, the changing rate of overconsolidation is assumed to be controlled by two factors, namely, the plastic component of stretching that was employed as the only factor in the work by Asaoka et al (2002), and the increment in anisotropy, in other words,

$$\dot{R} = JU\left\|\mathbf{D}^p\right\| + \frac{R}{MD}\frac{\eta}{M}\frac{\partial f}{\partial \boldsymbol{\beta}} \cdot \overset{\circ}{\boldsymbol{\beta}} \tag{27}$$

in which, by the definition of $\overset{\circ}{\boldsymbol{\beta}}$ in Eq. (19), $\overset{\circ}{\boldsymbol{\beta}}$ is proportional to the norm of the plastic component of stretching $\left\|\mathbf{D}_s^p\right\|$. Therefore, Eq. (27) is a strict evolution rule for the degree of overconsolidation. U is given by the following relation as:

$$U = -\frac{m}{D}\left(\frac{p'}{p'_0}\right)^2 \ln R \quad (p'_0 = 98.0\text{k Pa, reference stress}) \tag{28}$$

Using Eqs. (10) and (17), it is easy to obtain the following relation:

$$\left\|\mathbf{D}^p\right\| = \sqrt{\mathbf{D}^p \cdot \mathbf{D}^p} = \lambda MD\left(\sqrt{6\eta^{*2} + (M^2 - \eta^2)^2/3} / (M^2 - \zeta^2 + \eta^{*2})/p'\right. \tag{29}$$

Substituting Eqs. (28) and (29) into Eq. (27), we obtain:

$$\dot{R} = \lambda \frac{-mJ\, M \ln R \sqrt{6\eta^{*2} + (M^2 - \eta^2)^2/3}}{(M^2 - \zeta^2 + \eta^{*2})p'}\left(\frac{p'}{p'_0}\right)^2 + \frac{R}{MD}\frac{\eta}{M}\frac{\partial f}{\partial \boldsymbol{\beta}} \cdot \overset{\circ}{\boldsymbol{\beta}} \tag{30}$$

The plastic volumetric strain rate can be evaluated as:

$$-\dot{\varepsilon}_v^p = J\,\mathrm{tr}\mathbf{D}^p = J\lambda \mathrm{tr}\frac{\partial f}{\partial \mathbf{T}'} = J\,MD\,\lambda \frac{M^2 - \eta^2}{(M^2 - \zeta^2 + \eta^{*2})p'} \tag{31}$$

Substituting Eqs. (23), (26), (30) and (31) into Eq. (**11**), the positive multiplier λ can then be determined as:

$$\lambda = \frac{\partial f}{\partial \mathbf{T}'} \cdot \overset{\circ}{\mathbf{T}}' / \left(J \frac{MD}{(M^2 - \zeta^2 + \eta^{*2})p'}(M_s^2 - \eta^2)\right) \tag{32}$$

where

$$\begin{aligned} M_s^2 = M^2 - \frac{mM \ln R}{R}\left(\frac{p'}{p'_0}\right)^2 \sqrt{6\eta^{*2} + \frac{1}{3}(M^2 - \eta^2)^2} \\ -2aM(1 - R^*)\eta^* + \left(1 - \frac{\eta}{M}\right)\frac{\sqrt{6}Mb_r(b_l M - \zeta)\eta^{*2}(2M^2 - 3\boldsymbol{\eta} \cdot \boldsymbol{\beta})}{(M^2 - \zeta^2 + \eta^{*2})(M^2 - \zeta^2)} \end{aligned} \tag{33}$$

If the stretching is divided into elastic and plastic components, and the elastic components follow as

$$\overset{\circ}{\mathbf{T}}' = \mathbf{E}\mathbf{D}^e, \quad \mathbf{D} = \mathbf{D}^e + \mathbf{D}^p, \quad \overset{\circ}{\mathbf{T}}' = \mathbf{E}\mathbf{D} - \Lambda \mathbf{E}\frac{\partial f}{\partial \mathbf{T}'} \tag{34}$$

then by substituting Eqs (23), (26), (30), (31), and (34) into Eq. (11), we can obtain another expression for the following positive multiplier, namely, $\Lambda(\Lambda = \lambda)$:

$$\Lambda = \frac{\partial f}{\partial \mathbf{T}'}\mathbf{E}\mathbf{D} \Big/ \left(\frac{\partial f}{\partial \mathbf{T}'}\mathbf{E}\frac{\partial f}{\partial \mathbf{T}'} + J \frac{MD}{(M^2 - \zeta^2 + \eta^{*2})p'}(M_s^2 - \eta^2)\right) \tag{35}$$

3 Performance of the Proposed Model

Eight parameters are involved in the proposed model, among which five parameters, M, N, $\tilde{\lambda}$, $\tilde{\kappa}$, and ν are the same as in the Cam-clay model. The other three parameters are: a = parameter which controls the collapse rate of structure when the soil is subjected to shearing or compression; m = parameter which controls the losing rate of overconsolidation when the soil subjected to shearing or compression; b_r = parameter which controls the developing rate of stress-induced anisotropy when the soil subjected to shearing or compression. These three parameters can be easily determined from conventional triaxial compression tests.

We now consider a set of sands with different densities similar to the work by Nakai (2005), which is prepared from a sand whose material parameters and initial condition are listed in Tables 1 and 2.

Table 1. Material parameters of sand

Compression index $\tilde{\lambda}$	0.050
Swelling index $\tilde{\kappa}$	0.012
Critical state parameter M	1.0
Void ratio N (p'=98 kPa on *N.C.L.*)	0.98
Poisson's ratio ν	0.30
Degradation parameter of overconsolidation state m	0.10
Degradation parameter of structure a	2.2
Evolution parameter of anisotropy b_r	1.5

Table 2. Reference conditions of sand before vibrating compaction

Reference void ratio e_r	1.29
Reference mean effective stress p_r' (kPa)	10.0
Reference degree of structure R_r^*	0.00625
Reference degree of overconsolidation $1/R_r$	1.00
Reference anisotropy ζ_r	0.00

From the reference values for R_r, R_r^*, andζ_r, it is understood that the sand is originally a normally consolidated highly structured loose sand without stress-induced anisotropy and with a very large void ratio. The set of sands with different densities are just prepared by vibration compaction with different numbers of compactions, as shown in Fig. 3. The amplitude of the vibration is 2.3kPa. After the compaction, these sands with different densities are then isotropically consolidated to a prescribed confining pressure of 294 kPa. Table 3 lists the initial values for these sands before they are subjected to cyclic loading under undrained condition. All these sands have the same five parameters which are listed in the upper part of Table 1.

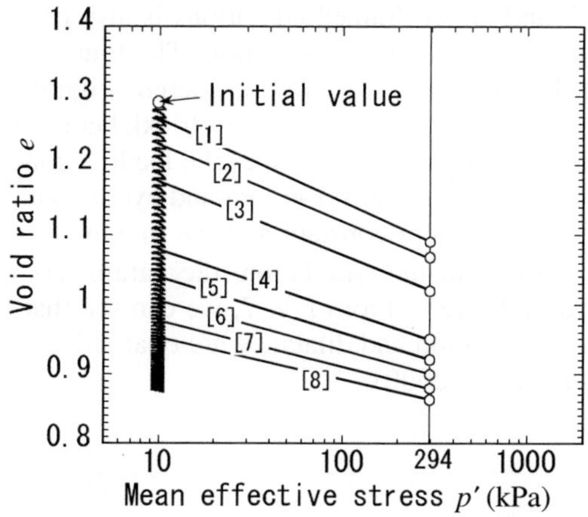

Fig. 3. Preparation of a set of sands with different densities from the same loose sand

Figure 4 shows the mechanical behavior of the set of sands with different densities in undrained cyclic loading tests. It is found from the figure that a very loose sand will fail along the way towards the zero effective stress state before cyclic mobility has a chance to occur. For medium dense sand, however, cyclic mobility does occur. Dense sand will never show cyclic mobility. Figure 5 shows the mechanical behavior of the set of sands with different densities in undrained triaxial compression tests. We believe that these results are very familiar to our readers who are interested in soil mechanics. The above results mean that the mechanical behavior of sand, subjected to monotonic/cyclic loading under undrained conditions, can be uniquely and properly described by the proposed model no matter what density it may have.

Table 3. Initial conditions of sand before cyclic loading

	Vibration numbers n	Initial void ratio e_0	Initial degree of structure R_0^*	Initial degree of over-consolidation $1/R_0$	Initial anisotropy ζ_0
[1]	2	1.09	0.00950	1.19	1.5e-5
[2]	5	1.07	0.0160	1.51	1.6e-5
[3]	10	1.02	0.037	2.12	1.8e-5
[4]	20	0.947	0.150	3.73	2.6e-5
[5]	25	0.920	0.241	4.79	4.6e-5
[6]	30	0.897	0.341	6.07	1.1e-4
[7]	35	0.879	0.434	7.60	3.1e-4
[8]	40	0.865	0.516	9.40	8.1e-4

The behavior of sand under drained conditions is also discussed. We prepared the same set of sands as in the previous section. The four sands listed in Table 3, namely, sands [1], [5], [6], and [8], are considered. Figure 6 shows the stress-strain-dilatancy relations of the sands in drained triaxial compression tests.

Finally, behaviors of sand under drained cyclic loading condition are simulated. The confining pressure of the sand is 196 kPa and cyclic loading condition is that the mean effective stress is kept constant and the maximum principal stress ratio (σ_1/σ_3) is restricted to 4. In the simulation, the parameters of the sand are the same as those listed in Table 1. From Fig. 7, we can see that the tendency of the changes in stress-strain relation and dilatancy are qualitatively the same as the test results reported by Hinokio (2000).

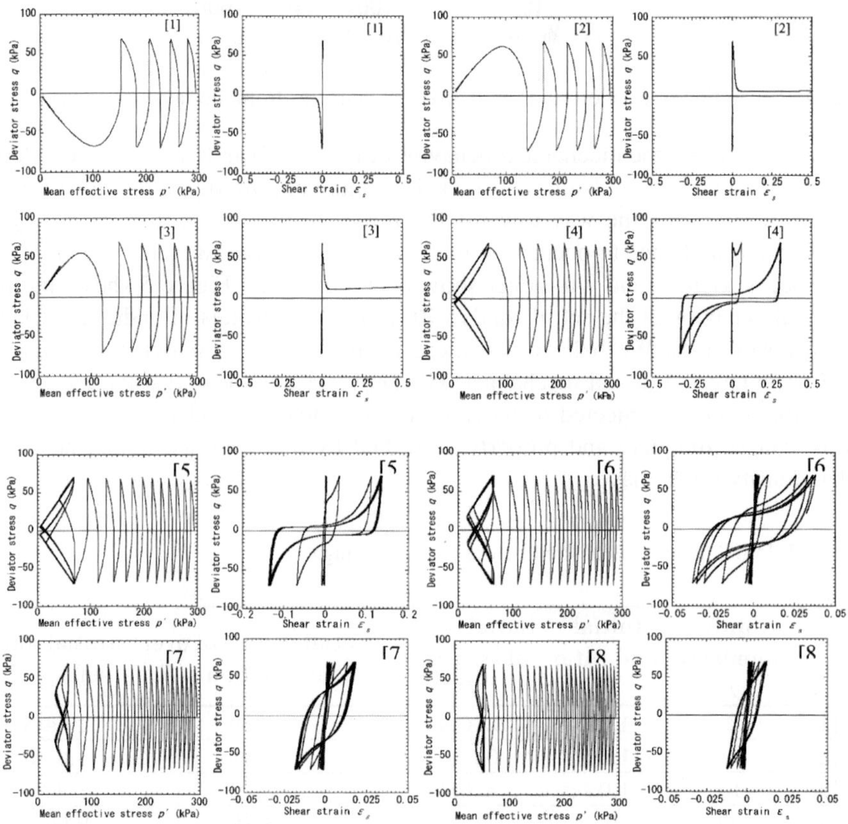

Fig. 4. Mechanical behaviors of the sands with different densities in undrained triaxial cyclic loading tests

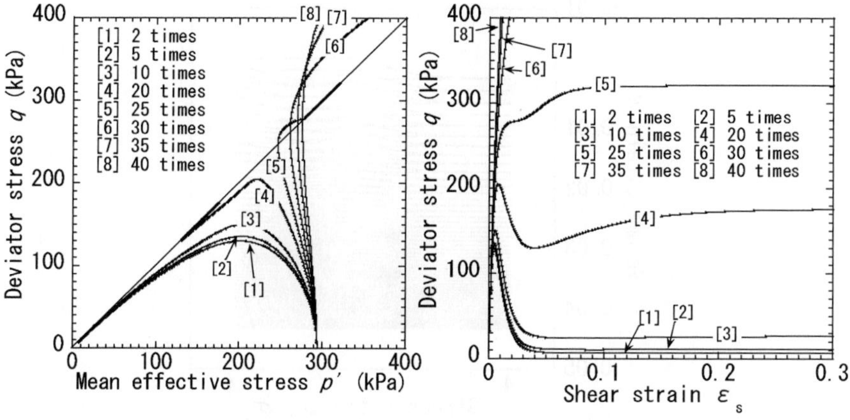

Fig. 5. Effective stress paths and stress-strain relation of the sands with different densities in undrained triaxial compression test

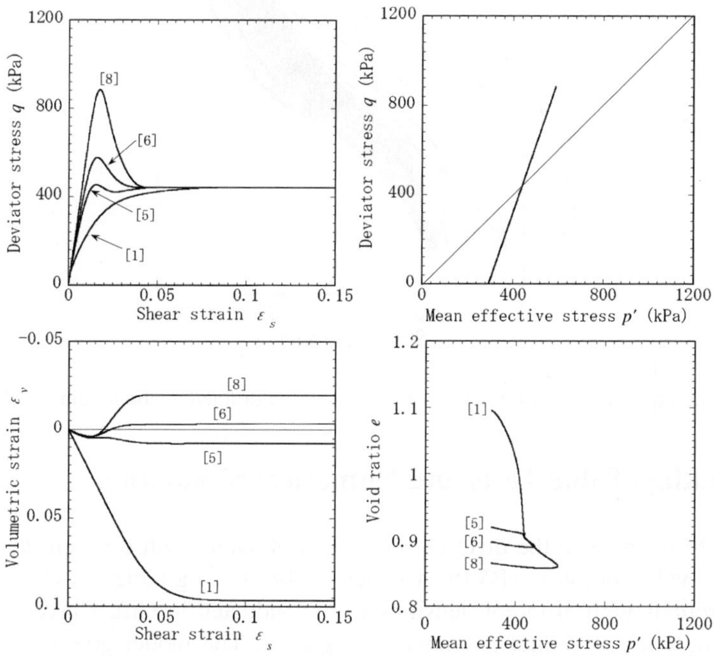

Fig. 6. Stress-strain relations of the sands with different densities in drained triaxial compression tests

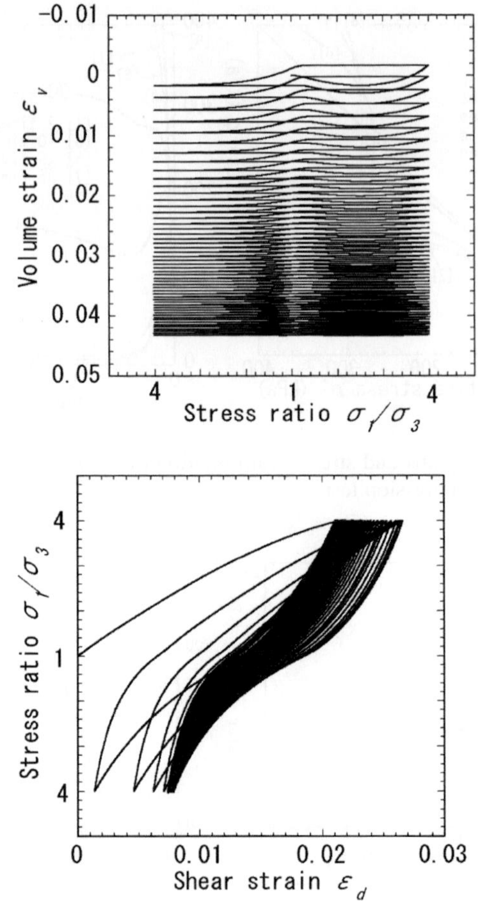

Fig. 7. Simulated stress-strain relation of dense sand in drained cyclic loading

4 Shaking-Table Tests and Numerical Simulation

In order to investigate the different behaviors of sands with different densities in boundary value problem (BVP), a series of 1g shaking-table tests on saturated sandy grounds with different densities were conducted in Gifu University (Ye et al., 2006). The test model is shown in Figure 8. The model ground is made of Toyoura Sand, which is widely used in geotechnical experiments in Japan. Accelerations and excessive pore water pressure (EPWP) were measured at different depths of the model ground.

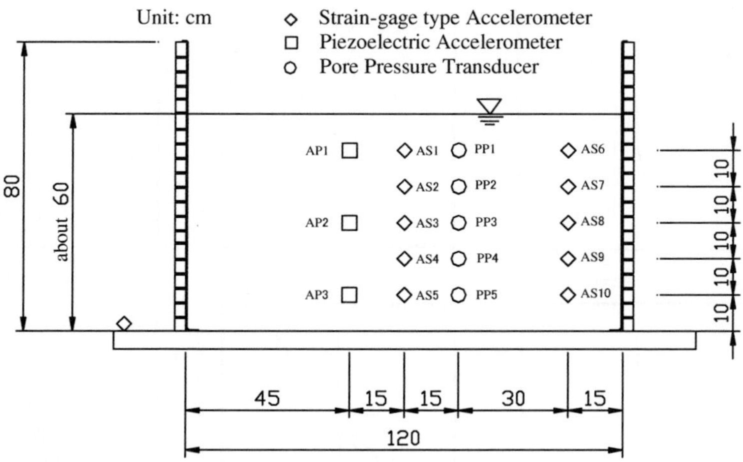

Fig. 8. Shaking-table tests

An initial loose sandy ground, with void ratio of about 0.78 and relative density of 53%, was prepared carefully by 'water sedimentation method', in which, firstly, water with a depth of 10cm was poured into the shear box, and then saturated sand was poured into the shear box slowly and carefully by scoop beneath water level. The model ground was built up until the ground reached the height of about 60cm.

The prepared ground was shaken three times in succession, marked by Case 1, Case 2 and Case 3, and the corresponding material parameters of Toyoura Sand are listed in Table 4. Figure 9 shows the simulated element behaviors of Toyoura sand with different densities at confining pressure of p_0'=298kPa. The shakings in Case 2 and Case 3 were applied to the model ground after the EPWP built up in the previous case had dissipated completely. The model ground became denser and denser after the repeated vibrations due to the dissipation of pore water in the model ground during and after each shaking. The input waves in all cases are the same, and is a sweep wave increased from zero to 300 Gal, as shown in Figure 10.

The simulation is conducted with a program called DBLEAVES (Ye et al, 2007). This 2D and 3D FEM program is capable of solving repeated static and dynamic soil-water coupled BVP by using Oka's two-phase field theory (Oka et al., 1994) and FE-FD soil-water coupled method. The numerical simulation aims to reproduce the three cases of shaking-table tests. In the numerical calculation, the process during shaking is simulated by a dynamic soil-water coupled analysis, while the process of dissipation of EPWP is simulated by a static consolidation analysis. The whole process of the model tests, which consists of three shakings and three consolidations, is a continuous one and therefore it should be simulated in sequential continuously. The simulation is therefore divided into six stages, i.e. three stages of dynamic analysis and three stages of consolidation analysis. The dynamic analysis lasts for 20 seconds until shaking stops, and the consolidation analysis lasts for 1000 seconds until EPWP dissipates completely.

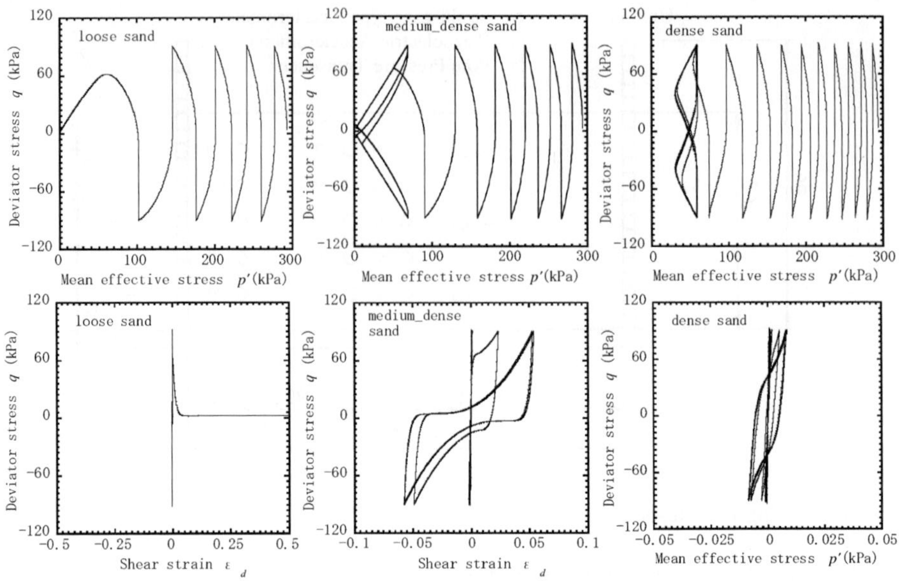

Fig. 9. Element simulation of Toyoura sand with different densities (p_0'=298kPa)

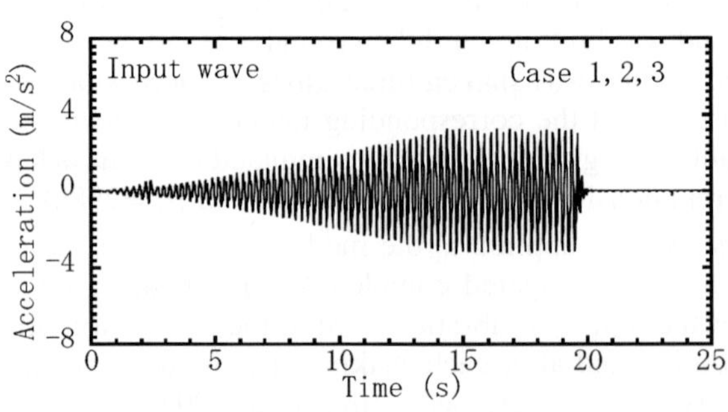

Fig. 10. Input wave

The values of material parameters, listed in Table 4, are the same in all 6 stages. Five initial state parameters before shaking are listed in Table 5. The state of soil elements represented by these five state parameters is only prescribed at the very beginning and the values of these parameters will be delivered automatically to the next stage of the analysis in the whole process, which is totally the same as what been done in the model test. The initial stress field is calculated based on a self-weight static analysis carried out in advance using the same finite element mesh as employed in dynamic analyses, shown in Fig. 11.

Table 4. Material parameters of Toyoura Sand

Compression index λ	0.05
Swelling index κ	0.0064
Critical state parameter M	1.30
Void ratio N (p'=98 kPa on *N.C.L.*)	0.74
Poisson's ratio ν	0.30
Degradation parameter of overconsolidation state m	0.10
Degradation parameter of structure a	2.2
Evolution parameter of anisotropy b_r	1.5

Table 5. Initial conditions of sands in numerical simulation

Initial void ratio e_0	0.78
Initial mean effective stress p' (kPa)	2.50
Initial degree of structure R_0^*	0.80
Initial degree of overconsolidation $1/R_0$	25.0
Initial anisotropy ζ_0	0.00

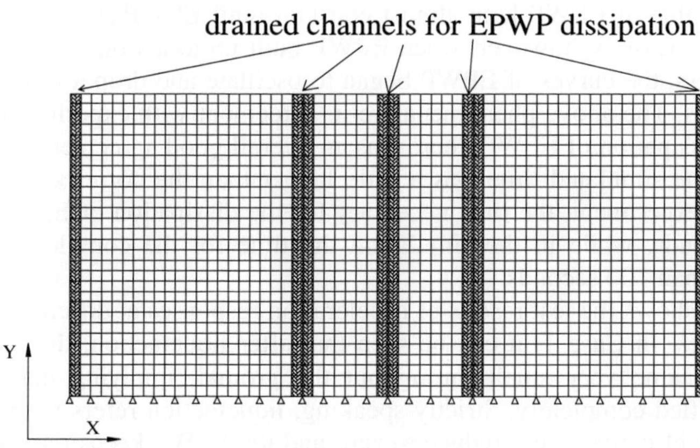

Fig. 11. FEM mesh for shaking-table tests

Figure 12 shows the calculated settlement of the ground surface. The white circles represent the measured settlements at the end of every case. The calculated settlement was larger than the measured settlement in Case 1, but agreed with the measured settlements in Cases 2 and 3. This disagreement is probably due to the inaccurate evaluation of the stress condition for loose sand at very low confining pressure that cannot be easily confirmed with conventional triaxial compression tests.

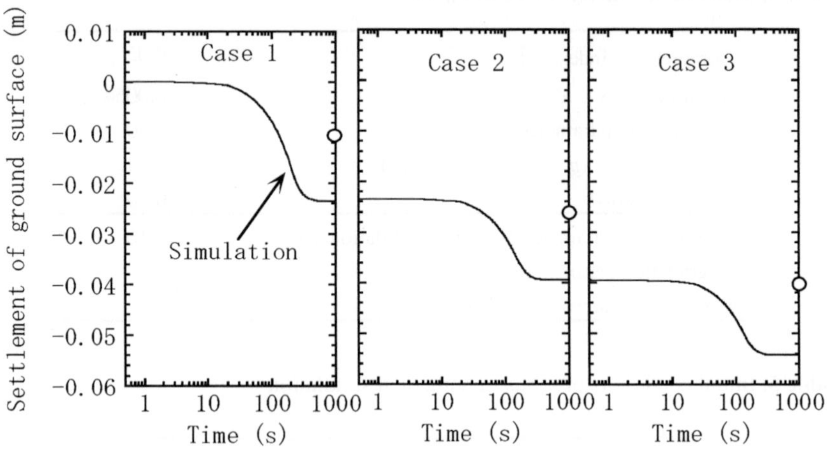

Fig. 12. Comparison of the settlement of ground surface (dots represent test results)

Figure 13(a) shows the EPWP measured at different depths in different cases. It is noticeable that the dissipation times of EPWP were different for all cases. The denser the ground was, the faster the dissipation of EPWP. Figure 13 (b) shows the simulated result of EPWP at different depths in different cases. In Case 1, EPWP built up quickly to the highest value that is equal to the vertical effective stress. The value of EPWP kept almost constant until EPWP started to dissipate. In Case 2 and Case 3, however, when EPWP built up to a value close to vertical effective stress, the curves of EPWP began to oscillate and drop a little especially at deep places. The above phenomena are in agreement with experimental results. As for the dissipation of EPWP, there was an overall good agreement with the test results, but the simulated time was a little longer than the corresponding test results. In the experiment, the time necessary for the dissipation of EPWP was less than 30 seconds, while in the simulation the time for the complete dissipation lasted more than 100 seconds.

Figure 14 shows the test results of acceleration responses at different depths for different cases. In Case 1, it can be seen that after liquefaction, the acceleration responses became very small throughout the ground, implying that the loose ground liquefied completely. Strictly speaking, liquefaction refers to the situation in which, the effective stress reduce to zero and the EPWP keeps the value of the initial effect stress during shaking. This can be verified with the time change of EPWP. It can also be predicted indirectly with the time change of the responding acceleration of surface ground because after liquefaction the responding acceleration of the ground above the liquefied soil layer became very small. In Case 2 and Case 3, the acceleration responses remained small in the ground near surface but stayed large in deeper places, showing that after experiencing liquefaction, the densities of sands increased and the sands in deeper places may not liquefy in the next vibration.

Figure 15 shows the simulated results of acceleration responses at different depths for different cases. It can be seen that acceleration responses of ground increased with the ground becoming denser and denser case by case, which shows the same tendency as the experimental results. In every case, acceleration was small near ground surface, and increased along with the increase of depth. Experimental results also show this trend in Case 2 and Case 3. However, in Case 1, acceleration of experiment became very small throughout the ground after liquefaction, while simulated result still gave a rather large value of acceleration in deep place. This disagreement might be caused by stiffness proportional damping used in dynamic analysis which is very difficult to be determined for liquefaction analysis.

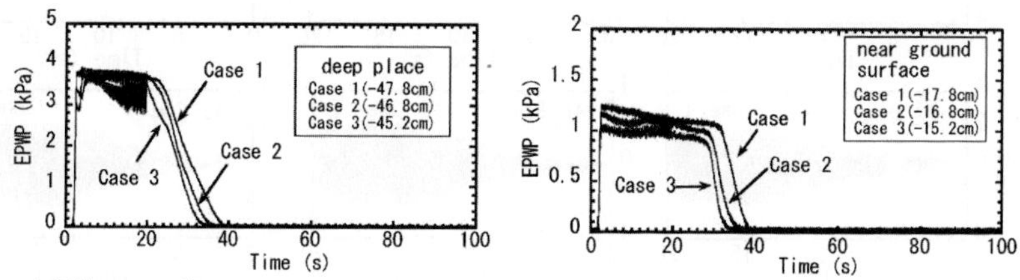

(a) Test results

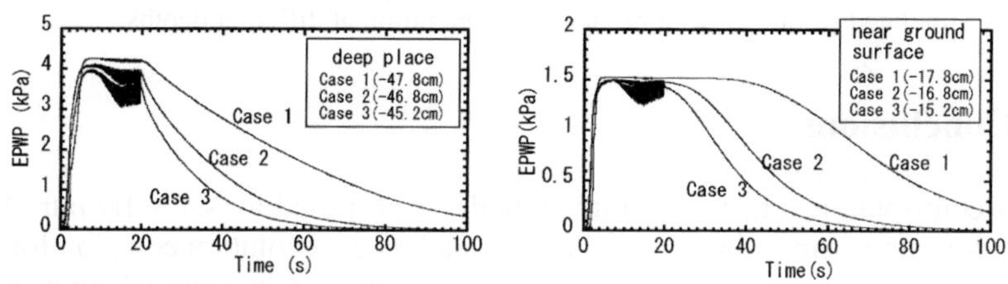

(b) computed results

Fig. 13. Comparison of EPWP in different cases

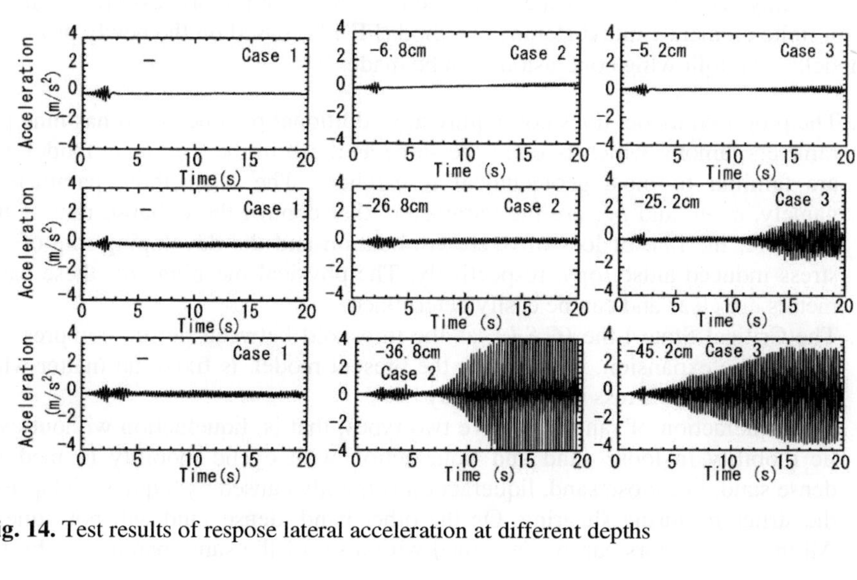

Fig. 14. Test results of respose lateral acceleration at different depths

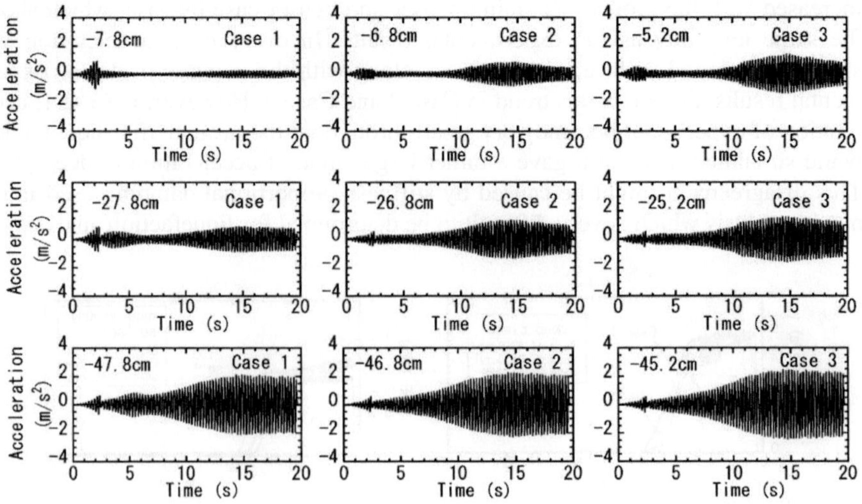

Fig. 15. Simulated results of response lateral acceleration at different depths

5 Conclusions

In this contribution, a new constitutive model is proposed for sand. By introducing a new concept of stress-induced anisotropy and a new evolution equation for overconsolidation, it is possible to describe not only the mechanical behavior of soils under monotonic loading, but also the behavior of soils under cyclic loading under drained and undrained condition uniquely. Meanwhile, shaking-table tests on saturated sandy ground with repeated liquefaction-consolidation process are simulated with finite element-finite difference method (FE-FD) based on the newly proposed model. The following conclusions can be made:

1. The proposed model does not require any additional parameters. It has nine parameters, among which five are the same as those in the Cam-clay model and are familiar to most geotechnical researchers. The other three parameters, namely, a, m, and b_r , are the parameters that control the collapse rate of the structure, the rate of loss of overconsolidation and the developing rate of the stress-induced anisotropy, respectively. The physical meanings for these parameters are clear and can be easily determined.
2. The Critical State Line (*C.S.L*), as the threshold between plastic compression and plastic expansion, proposed in the present model, is fixed, no matter what kind of effective stress path there may be.
3. The liquefaction of sand may have two types, that is, liquefaction without cyclic mobility in loose sand and liquefaction with cyclic mobility in medium dense sand. For loose sand, liquefaction is mainly caused by a quick collapse of the structure during shearing. On the other hand. dense sand will not liquefy. All these behaviors can be described with a set of the same parameters by the

model. It is not necessary to assign, in advance, which sand will be liquefied or not. It is simply dependent on the state, namely, the overconsolidation ratio, the anisotropy, and the structure, once the material parameters have been fixed, in other words, once the sand has been selected.

4. The model can describe the phenomenon whereby liquefied soils may liquefy again if they are subjected to strong motions after the excessive pore water pressure has dissipated.
5. Once liquefied loose sandy ground will get denser in its consolidation process, it is still possible to liquefy again in the next strong motion. A sandy ground gets denser and denser in the repeated liquefaction-consolidation process, and correspondingly acceleration response of the ground will increase while the dissipation time of EPWP after liquefaction gets shorter and shorter.
6. The analysis conducted in this contribution is capable of reproducing uniquely the different responses of liquefied grounds with different densities during repeated shaking and consolidation.
7. The different mechanical behaviors of sands with different densities subjected to repeated shaking and consolidation are only dependent on the state of the sand, not on the values of material parameters.
8. In simulating the whole process of repeated liquefaction and consolidation of sandy ground, although the state of the sand changes constantly, all initial input material parameters are kept the same for all analyses. This feature makes the numerical simulation very meaningful.

References

Asaoka, A., Nakano, M., Noda, T.: Super loading yield surface concept for the saturated structured soils. In: Proc. of the Fourth European Conference on Numerical Methods in Geotechnical Engineering-NUMGE 1998, pp. 232–242 (1998)

Ye, B., Ye, G.L., Zhang, F., Yashima, A.: Experiment and numerical simulation of repeated liquefaction-consolidation of sand. Soils and Foundations 47(3), 547–558 (2007)

Hashiguchi, K., Ueno, M.: Elastoplastic constitutive laws of granular material, Constitutive Equations of Soils. In: Murayama, S., Schofield, A.N. (eds.) Pro. 9th Int. Conf. Soil Mech. Found. Engrg., Spec. Ses., vol. 9, pp. 73–82. JSSMFE, Tokyo (1977)

Hinokio, M.: Deformation characteristic of sand subjected to monotonic and cyclic loadings and its application to bearing capacity problem, Doctoral Thesis, Nagoya Institute of Technology, 141-144 (2000) (in Japanese)

Nakai, K.: An elasto-plastic constitutive modeling of soils based on the evolution laws describing collapse of soil skeleton structure, loss of overconsolidation and development of anisotropy, Doctoral Thesis, Nagoya University (2005) (in Japanese)

Oka, F., Yashima, A., Kato, M., Sekiguchi, K.: A constitutive model for sand based on the non-linear kinematic hardening rule and its application. In: Proc. 10th World Conf. Earthquake Engineering, Madrid, Balkema, vol. 5, pp. 2529–2534 (1992)

Oka, F., Yashima, A., Shibata, T., Kato, M., Uzuoka, R.: FEM-FDM coupled liquefaction analysis of a porous soil using an elasto-plastic model. Applied Scientific Research 52, 209–245 (1994)

Oka, F., Yashima, A., Tateishi, A., Taguchi, Y., Yamashita, S.: A cyclic elastoplastic constitutive model for sand considering a plastic-strain dependence of the shear modulus. Geotechnique 49(5), 661–680 (1999)

Roscoe, K.H., Schofield, A.N., Thurairajah, A.: Yielding of clays in states wetter than critical. Geotechnique 13(3), 250–255 (1963)

Schofield, A.N., Wroth, C.P.: Critical State Soil Mechanics. McGraw-Hill, New York (1968)

Sekiguchi, H.: Rheological characteristics of clays. In: Proc. 9th Int. Conf. Soil Mech., Found. Eng., Tokyo, vol. 1, pp. 289–292 (1977)

Yashima, A., Oka, F., Shibata, T., Uzuoka, R.: Liquefaction analysis by LIQCA. In: Proceedings of JGS Conference on Liquefaction of Ground and its Counter measure, pp. 165–174 (1991) (in Japanese)

Ye, B., Yokawa, H., Kondo, T., Yashima, A., Zhang, F., Yamada, H.: Investigation on Stiffness Recovery of Liquefied Sandy Ground after Liquefaction using Shaking-Table Tests. Soil and Rock Behavior and Modeling, ASCE Geotechnical Special Publication 150, 482–489 (2006)

An Updated Hypoplastic Constitutive Model, Its Implementation and Application

Xuetao Wang and Wei Wu

Institut für Geotechnik, Universität für Bodenkultur, Austria
e-mail: xuetao.wang@boku.ac.at, wei.wu@boku.ac.at

Abstract. The paper presents an updated hypoplastic constitutive model by introducing a new term. Moreover, an extension to include cohesion into the model is made. The model is implemented in a finite difference code. The procedure for parameter identification is presented. The numerical implementation is verified by some element tests. Finally, the constitutive model is used to simulate excavation and support of a shallow tunnel.

1 Introduction

The idea of hypoplasticity was developed by Kolymbas (1985)[4] by using nonlinear tensorial function of the rate-type. A general hypoplastic constitutive equation was presented by Wu and Kolymbas (1990)[8]. Based on the general hypoplastic constitutive equation, a simple hypoplastic constitutive model was proposed by Wu and Bauer (1994)[5], which is composed of two linear and two nonlinear terms in strain rate. However, this model shows excessive contraction (volume reduction) in triaxial extension. This was remedied by merging the two nonlinear in one term (Bauer 1995 [2]). An ensuing problem is that the constitutive equation has only three parameters, which severely restrict its applicability. In this paper an update is presented by including a new term into the constitutive equation.

2 Framwork of Hypoplasticity

2.1 *Hypoplastic Constitutive Equation*

The simple hypoplastic constitutive model proposed by Wu and Bauer is shown as follows

$$\overset{\circ}{\sigma} = \mathrm{c}_1(\mathrm{tr}\sigma)\dot{\varepsilon} + \mathrm{c}_2\frac{\mathrm{tr}(\sigma\dot{\varepsilon})}{\mathrm{tr}\sigma}\sigma + \left(\mathrm{c}_3\frac{\sigma^2}{\mathrm{tr}\sigma} + \mathrm{c}_4\frac{\sigma^{*2}}{\mathrm{tr}\sigma}\right)\|\dot{\varepsilon}\| \tag{1}$$

where c_i, $(i = 1,2,3,4)$ are dimensionless parameters. The deviatoric stress tensor σ^* in the above equation is given by $\sigma^* = \sigma - 1/3(\mathrm{tr}\sigma)\mathbf{1}$ with $\mathbf{1}$ being the unit tensor. The four parameters can be identified with a single triaxial compression test. A calibration procedure using the initial tangential stiffness E_i, the initial Poisson ratio μ, the friction angle ϕ and the dilatancy angle ψ was given by Wu and Bauer (1994)[5]. It turned out that the parameters calibrated for triaxial compression test

R. Wan et al. (Eds.): Bifurcations, Instabilities and Degradations in Geomaterials, SSGG, pp. 133–143.
springerlink.com

do not necessarily lead to critical states for other stress paths, e.g. triaxial extension. It was found out that critical state is reached for all paths if the two nonlinear terms are merged into one term by letting (Bauer, 1996)[2]

$$\mathrm{c}_3 = -\mathrm{c}_4 \tag{2}$$

As a consequence, the number of parameters in equation (1) reduces from four to three.

$$\mathring{\boldsymbol{\sigma}} = \mathrm{c}_1(\mathrm{tr}\boldsymbol{\sigma})\dot{\boldsymbol{\varepsilon}} + \mathrm{c}_2\frac{\mathrm{tr}(\boldsymbol{\sigma}\dot{\boldsymbol{\varepsilon}})}{\mathrm{tr}\boldsymbol{\sigma}}\boldsymbol{\sigma} + \mathrm{c}_3(\boldsymbol{\sigma}+\boldsymbol{\sigma}^*)\|\dot{\boldsymbol{\varepsilon}}\| \tag{3}$$

This severely restricts the adaptability of the model. For instance, the initial Poisson ratio cannot be varied. To reslove this problem, a new term $(\mathrm{tr}\dot{\boldsymbol{\varepsilon}})\boldsymbol{\sigma}$ is added to the above equation so that the number of parameters regains four:

$$\mathring{\boldsymbol{\sigma}} = \mathrm{c}_1(\mathrm{tr}\boldsymbol{\sigma})\dot{\boldsymbol{\varepsilon}} + \mathrm{c}_2(\mathrm{tr}\dot{\boldsymbol{\varepsilon}})\boldsymbol{\sigma} + \mathrm{c}_3\frac{\mathrm{tr}(\boldsymbol{\sigma}\dot{\boldsymbol{\varepsilon}})}{\mathrm{tr}\boldsymbol{\sigma}}\boldsymbol{\sigma} + \mathrm{c}_4(\boldsymbol{\sigma}+\boldsymbol{\sigma}^*)\|\dot{\boldsymbol{\varepsilon}}\| \tag{4}$$

Note that the same notations for the four parameters are retained in the above equation. Obviously this new term vanishes in critical state with $\mathrm{tr}\dot{\boldsymbol{\varepsilon}} = 0$.

2.2 *Failure Surface*

By the definition of failure, the stress rate at failure vanishes, that is $\mathring{\boldsymbol{\sigma}} = \mathbf{0}$. Based on this, the failure surface can be derived from constitutive equation (4). For simplicity, we may set $\mathrm{c}_1 = 1$. Equation (4) can be separated into a spherical part and a deviatoric part. Let us first consider the spherical part, which can be obtained by taking the trace of both sides of equation (4)

$$\mathrm{tr}\mathring{\boldsymbol{\sigma}} = (\mathrm{tr}\boldsymbol{\sigma})(\mathrm{tr}\dot{\boldsymbol{\varepsilon}}) + \mathrm{c}_2(\mathrm{tr}\dot{\boldsymbol{\varepsilon}})(\mathrm{tr}\boldsymbol{\sigma}) + \mathrm{c}_3\frac{\mathrm{tr}(\boldsymbol{\sigma}\dot{\boldsymbol{\varepsilon}})}{\mathrm{tr}\boldsymbol{\sigma}}(\mathrm{tr}\boldsymbol{\sigma}) + \mathrm{c}_4(\mathrm{tr}\boldsymbol{\sigma}+\mathrm{tr}\boldsymbol{\sigma}^*)\|\dot{\boldsymbol{\varepsilon}}\| \tag{5}$$

Note that $\mathrm{tr}\mathring{\boldsymbol{\sigma}} = 0$ and $\mathrm{tr}\dot{\boldsymbol{\varepsilon}} = 0$ in a critical state and make use of the relation $\boldsymbol{\sigma}^* = \boldsymbol{\sigma} - \frac{1}{3}(\mathrm{tr}\boldsymbol{\sigma})\mathbf{1}$ and $\mathrm{tr}(\boldsymbol{\sigma}^*\dot{\boldsymbol{\varepsilon}}^*) = \|\boldsymbol{\sigma}^*\|\|\dot{\boldsymbol{\varepsilon}}^*\|\cos\theta$, with θ being the angle between $\boldsymbol{\sigma}^*$ and $\dot{\boldsymbol{\varepsilon}}^*$ and $\dot{\boldsymbol{\varepsilon}}^*$ is the deviatoric strain rate. The following equation can be obtained by letting $\mathrm{tr}\mathring{\boldsymbol{\sigma}} = 0$

$$\mathrm{c}_3\|\boldsymbol{\sigma}^*\|\|\dot{\boldsymbol{\varepsilon}}^*\|\cos\theta + \mathrm{c}_4\mathrm{tr}\boldsymbol{\sigma}\|\dot{\boldsymbol{\varepsilon}}\| = 0 \tag{6}$$

Let r_c denotes the stress ratio $\|\boldsymbol{\sigma}^*\|/\mathrm{tr}\boldsymbol{\sigma}$ in a critical state. From above equation we can get

$$\mathrm{c}_3 r_c\cos\theta + \mathrm{c}_4 = 0 \tag{7}$$

The term $\cos\theta$ represents the flow direction with reference to the stress in a critical state. As will be shown thereafter, we have $\cos\theta = 1$. In this case, a relationship between c_3 and c_4 is obtained from the equation (7)

$$-\frac{\mathrm{c}_4}{\mathrm{c}_3} = r_c \tag{8}$$

It is clear that equation (8) represents the failure criterion of Drucker-Prager. Now, let us turn our attention to the deviatoric part of the consitutive equation (4)

$$\mathring{\sigma}^* = (\mathrm{tr}\sigma)\dot{\varepsilon}^* + c_2(\mathrm{tr}\dot{\varepsilon})\sigma^* + c_3\frac{\|\sigma^*\|\|\dot{\varepsilon}^*\|}{\mathrm{tr}\sigma}\cos\theta\sigma^* + 2c_4\sigma^*\|\dot{\varepsilon}\| = 0 \tag{9}$$

We make use of $\mathrm{tr}\dot{\varepsilon} = 0$ and $\mathrm{tr}(\sigma\dot{\varepsilon}) = \mathrm{tr}(\sigma^*\dot{\varepsilon}^*) = \|\sigma^*\|\|\dot{\varepsilon}^*\|\cos\theta$ and $\|\dot{\varepsilon}\| = \|\dot{\varepsilon}^*\|$ to get

$$\frac{\dot{\varepsilon}^*}{\|\dot{\varepsilon}^*\|} = -(c_3 r_c\cos\theta + 2c_4)\frac{\sigma^*}{\mathrm{tr}\sigma} \tag{10}$$

The above equation indicates that the stress tensor and the strain rate tensor are coaxial in a critical state. This can be also expected for the failure surface of Durcker-Prager. Note that $(\dot{\varepsilon}^*/\|\dot{\varepsilon}^*\|) : (\dot{\varepsilon}^*/\|\dot{\varepsilon}^*\|) = 1$ we have

$$c_3 {r_c}^2\cos\theta + 2c_4 r_c - 1 = 0 \tag{11}$$

Again, equaion (11) is quardratic in r_c. Combining equation (8) and equaion (11), we can solve for c_3 to get

$$c_3 = -\frac{1}{{r_c}^2} \tag{12}$$

The paramters c_3 can be set into equation (8) to give c_4

$$c_4 = \frac{1}{r_c} \tag{13}$$

Then equation (4) can be rewritten as follows after the parameters are specified by equations (12) and (13)

$$\|\sigma^*\| = r_c(\mathrm{tr}\sigma) \tag{14}$$

It is noteworthy that equation (14) is similar to the failure formula proposed by Bardet(1990)[1]. This formula has the advantage that it encompassed two widely used failure criteria Matsuoka-Nakai (1974) and Lade-Duncan (1975)[7].

2.3 Calibartion Procedure of Parameters

Four parameters are required for the hypoplastic constitutive equation, viz.c_i, $(i = 1,2,3,4)$. A calibration procedure for these four parameters is described by Wu and Bauer (1994)[5]. This procedure based on a triaxial test is briefly given below.

Figure 1 shows a typical triaxial compression test on sand. The test started from point A and reached failure at point B. The four parameters c_i, $(i = 1,2,3,4)$ are related to the four knowns E_i, β_A, β_B, $(\sigma_1 - \sigma_2)_{max}$ from triaxial test.

The two angles β_A and β_B can be expressed by volumetric strain rate $\dot{\varepsilon}_v$ and axial strain rate $\dot{\varepsilon}_1$ at point A and B to give

$$\beta_A = \arctan\left(\frac{\dot{\varepsilon}_v}{\dot{\varepsilon}_1}\right)_A = \arctan\left(\frac{\dot{\varepsilon}_1 + 2\dot{\varepsilon}_2}{\dot{\varepsilon}_1}\right)_A = \arctan\left(1 + 2\frac{\dot{\varepsilon}_2}{\dot{\varepsilon}_1}\right)_A \tag{15}$$

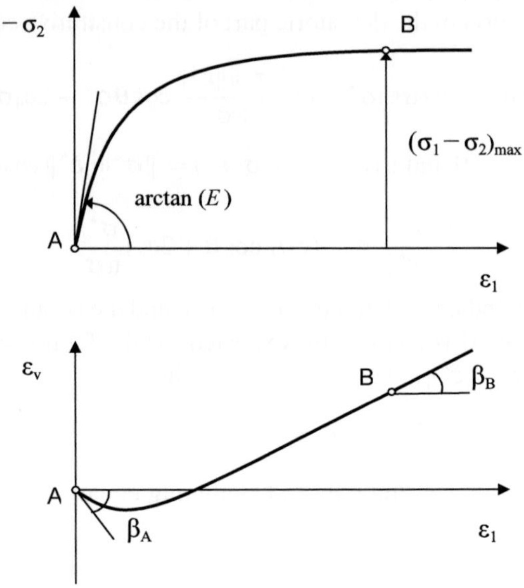

Fig. 1. Typical triaxial compression test

$$\beta_B = \arctan\left(\frac{\dot{\varepsilon}_v}{\dot{\varepsilon}_1}\right)_B = \arctan\left(\frac{\dot{\varepsilon}_1 + 2\dot{\varepsilon}_2}{\dot{\varepsilon}_1}\right)_B = \arctan\left(1 + 2\frac{\dot{\varepsilon}_2}{\dot{\varepsilon}_1}\right)_B \tag{16}$$

Because the hypoplastic constitutive model is rate-independent, we can set $\dot{\varepsilon}_1 = 1$ to obtain

$$\dot{\varepsilon}_2 = \dot{\varepsilon}_3 = \frac{1}{2}(1 - \tan\beta) \tag{17}$$

So the strain rates at point A and B can be shown to be

$$\dot{\varepsilon}_{A/B} = \begin{pmatrix} -1 & 0 & 0 \\ 0 & \frac{1}{2}(1-\tan\beta_{A/B}) & 0 \\ 0 & 0 & \frac{1}{2}(1-\tan\beta_{A/B}) \end{pmatrix} \tag{18}$$

We consider a triaxial test under constant confining pressure σ_c. The stress tensors at point A and B can be readily written out

$$\sigma_A = \begin{pmatrix} \sigma_{11A} & 0 & 0 \\ 0 & \sigma_{22A} & 0 \\ 0 & 0 & \sigma_{33A} \end{pmatrix} = \begin{pmatrix} \sigma_c & 0 & 0 \\ 0 & \sigma_c & 0 \\ 0 & 0 & \sigma_c \end{pmatrix} \tag{19}$$

$$\sigma_B = \begin{pmatrix} \sigma_{11B} & 0 & 0 \\ 0 & \sigma_{22B} & 0 \\ 0 & 0 & \sigma_{33B} \end{pmatrix} = \begin{pmatrix} \sigma_{11B} & 0 & 0 \\ 0 & \sigma_c & 0 \\ 0 & 0 & \sigma_c \end{pmatrix} \tag{20}$$

The stress σ_{11B} can be expressed by friction angle and confining pressure as follows

$$\sigma_{11B} = \sigma_c \left(\frac{1+\sin\phi}{1-\sin\phi} \right) \tag{21}$$

And the stress rate $\dot{\sigma}$ at point A and B are

$$\dot{\sigma}_A = \begin{pmatrix} -E_i & 0 & 0 \\ 0 & 0 & 0 \\ 0 & 0 & 0 \end{pmatrix} \quad \dot{\sigma}_B = \begin{pmatrix} 0 & 0 & 0 \\ 0 & 0 & 0 \\ 0 & 0 & 0 \end{pmatrix} \tag{22}$$

Finally, the parameters c_1, c_2, c_3, c_4 can be determined by solving the following equation system

$$\begin{pmatrix} \mathrm{tr}(\sigma_A)\dot{\varepsilon}_{11A} & \mathrm{tr}(\dot{\varepsilon}_A)\sigma_{11A} & \frac{\mathrm{tr}(\sigma_A\dot{\varepsilon}_A)}{\mathrm{tr}(\sigma_A)}\sigma_{11A} & s_{11A} \\ \mathrm{tr}(\sigma_A)\dot{\varepsilon}_{33A} & \mathrm{tr}(\dot{\varepsilon}_A)\sigma_{33A} & \frac{\mathrm{tr}(\sigma_A\dot{\varepsilon}_A)}{\mathrm{tr}(\sigma_A)}\sigma_{33A} & s_{33A} \\ \mathrm{tr}(\sigma_B)\dot{\varepsilon}_{11B} & \mathrm{tr}(\dot{\varepsilon}_B)\sigma_{11B} & \frac{\mathrm{tr}(\sigma_B\dot{\varepsilon}_B)}{\mathrm{tr}(\sigma_B)}\sigma_{11B} & s_{11B} \\ \mathrm{tr}(\sigma_B)\dot{\varepsilon}_{33B} & \mathrm{tr}(\dot{\varepsilon}_B)\sigma_{33B} & \frac{\mathrm{tr}(\sigma_B\dot{\varepsilon}_B)}{\mathrm{tr}(\sigma_B)}\sigma_{33B} & s_{33B} \end{pmatrix} \begin{pmatrix} c_1 \\ c_2 \\ c_3 \\ c_4 \end{pmatrix} = \begin{pmatrix} -E_i \\ 0 \\ 0 \\ 0 \end{pmatrix} \tag{23}$$

where $s_{11A} = \sigma_{11A} + \sigma^*_{11A}$, $s_{33A} = \sigma_{33A} + \sigma^*_{33A}$, $s_{11B} = \sigma_{11B} + \sigma^*_{11B}$, $s_{33B} = \sigma_{33B} + \sigma^*_{33B}$. The following relationship is useful to relate the angle β_A with the initial Poisson ratio μ_i

$$\tan\beta_A = (1+2\mu_i) \tag{24}$$

and β_B is the dilatancy angle ψ. Hence we relate the four parameters c_1, c_2, c_3, and c_4 to the parameters initial tangential stiffness E_i, the inital Poisson's ratio μ_i, the friction angle ϕ and the dilatancy angle ψ. Equation (23) can be easily solved by using the symbolic computational program *Mathematica*.

3 Implementation of the Updated Hypoplastic Model

The hypoplastic constitutive equation (1) is originally developed for sand. In practice, however, most soils show cohesion to some extent. Therefore, it is important to take cohesion into consideration. For cohesive materials, the constitutive equation (4) is extended by replacing the stress tensor σ with the following translated stress tensor

$$\eta = \sigma - c\mathbf{I} \tag{25}$$

where c is cohesion, $\mathbf{I}$ is unit matrix. Then the hypoplastic constitutive equation (4) can be rewritten as follows

$$\mathring{\sigma} = c_1(\mathrm{tr}\eta)\dot{\varepsilon} + c_2(\mathrm{tr}\dot{\varepsilon})\eta + c_3\frac{\mathrm{tr}(\eta\dot{\varepsilon})}{\mathrm{tr}\eta}\eta + c_4(\eta+\eta^*)\|\dot{\varepsilon}\| \tag{26}$$

Equation (26) can be implemented in the numerical code FLAC3D, which is a widely used in geotechnical engineering. FLAC3D is a finite difference program, which provides friendly interface for users to implement user-defined models[3]. For numerical implementation in FLAC3D, the constitutive equation must be written in incremental form. Equation (26) can be rewritten in the following incremental form by considering time increment Δt

$$\Delta\sigma = c_1(\mathrm{tr}\eta)\Delta\varepsilon + c_2(\mathrm{tr}\Delta\varepsilon)\eta + c_3\frac{\mathrm{tr}(\eta\Delta\varepsilon)}{\mathrm{tr}\eta}\eta + c_4(\eta+\eta^*)\|\Delta\varepsilon\| \tag{27}$$

We proceed to write out equation (27) explicitly

$$\begin{aligned}
\Delta\sigma_{11} &= c_1(\mathrm{tr}\eta)\Delta\varepsilon_{11} + c_3\frac{\mathrm{tr}(\eta\Delta\varepsilon)}{\mathrm{tr}\eta}\eta_{11} + c_4(\eta_{11}+\eta_{11}^*)\|\Delta\eta\| \\
\Delta\sigma_{22} &= c_1(\mathrm{tr}\eta)\Delta\varepsilon_{22} + c_3\frac{\mathrm{tr}(\eta\Delta\varepsilon)}{\mathrm{tr}\eta}\eta_{22} + c_4(\eta_{22}+\eta_{22}^*)\|\Delta\eta\| \\
\Delta\sigma_{33} &= c_1(\mathrm{tr}\eta)\Delta\varepsilon_{33} + c_3\frac{\mathrm{tr}(\eta\Delta\varepsilon)}{\mathrm{tr}\eta}\eta_{33} + c_4(\eta_{33}+\eta_{33}^*)\|\Delta\eta\| \\
\Delta\sigma_{12} &= c_1(\mathrm{tr}\eta)\Delta\varepsilon_{12} + c_3\frac{\mathrm{tr}(\eta\Delta\varepsilon)}{\mathrm{tr}\eta}\eta_{12} + c_4(\eta_{12}+\eta_{12}^*)\|\Delta\eta\| \\
\Delta\sigma_{13} &= c_1(\mathrm{tr}\eta)\Delta\varepsilon_{13} + c_3\frac{\mathrm{tr}(\eta\Delta\varepsilon)}{\mathrm{tr}\eta}\eta_{13} + c_4(\eta_{13}+\eta_{13}^*)\|\Delta\eta\| \\
\Delta\sigma_{23} &= c_1(\mathrm{tr}\eta)\Delta\varepsilon_{23} + c_3\frac{\mathrm{tr}(\eta\Delta\varepsilon)}{\mathrm{tr}\eta}\eta_{23} + c_4(\eta_{23}+\eta_{23}^*)\|\Delta\eta\|
\end{aligned} \tag{28}$$

In FLAC3D, the user-defined constitutive model is written in C++ and compiled as DLL (dynamic link library) [3]. The model can be then loaded into FLAC3D for calculations. The user-defined models written in this way are as efficient as the built-in models. Afterawards, the code need be verified against some benchmark problems.

4 Numerical Simulation

The implementated hypoplastic constitutive model is used to simulate some laboratory tests and engineering problems. In this paper, two simulation examples are considered. The first example deals with triaxial tests, while the second is the simulation of excavation and support in shallow tunnel.

4.1 Simulation of Triaxial Tests

Triaxial test is often used for studying the stress-strain properties of soil. We consider drained triaxial tests under constant confining pressure. The specimen is modelled as a cube in FLAC3D. The properties of hypoplastic material are shown in Table 1.

Table 1. Parameters of hypoplastic material for the triaxial test

E [MPa]	μ [-]	ψ [°]	ϕ [°]	c [kPa]
17	0.2	15	30	0/100

Starting from an isotropic stress state, triaxial tests are simulated by applying the vertical velocity while keeping the lateral stress constant. Negative velocity (downward) defines triaxial compression and positive velocity (upward) for triaxial extension. For the hypoplastic model with cohesion, an unconfined compression test with a cohesion of 100 kPa (see Table 1) is compared with a triaxial compression test without cohesion under a confining pressure of 100 kPa.

Figure 2 and Figure 3 show the numerical results of triaxial compression and extension test. The FLAC3D simulations in Figure 2 and Figure 3 are well corraborated by numerical simulations of one element as described by Wu and Bauer (1994). Figure 4 can be appreciated by considering the Mohr-Coulomb failure criterion. The following relationship between cohesion and unconfined compressive strength can be easily shown

$$c = \frac{1-\sin\phi}{2\sin\phi}\sigma_p \tag{29}$$

where σ_p is the unconfined compressive strength. According the above relationship, an unconfined compression test with a given cohesion is equivalent to a triaxial compression test under certain confining pressure without cohesion. After equation (29), the unconfined compressive strength for a friction angle of 30° and a cohesion of 100 kPa is 200 kPa. As can be seen from Figure 4 similar stress-strain curves are obtained for the unconfined test and the triaxial test under a confining pressure of 100 kPa. This gives us some confidence to consider cohesive soil in the numerical calculations.

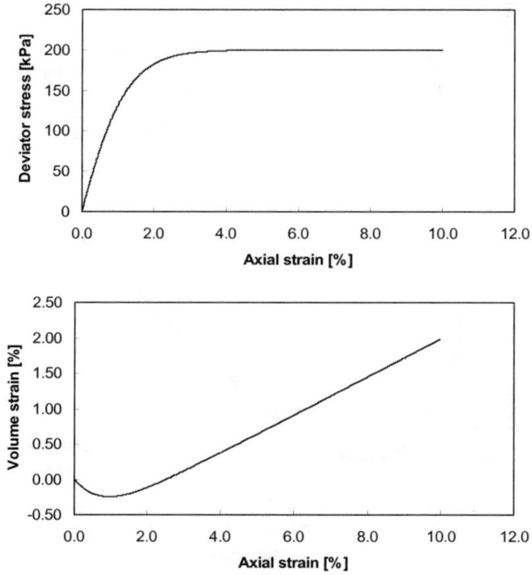

Fig. 2. FLAC3D simulation of triaxial compression test

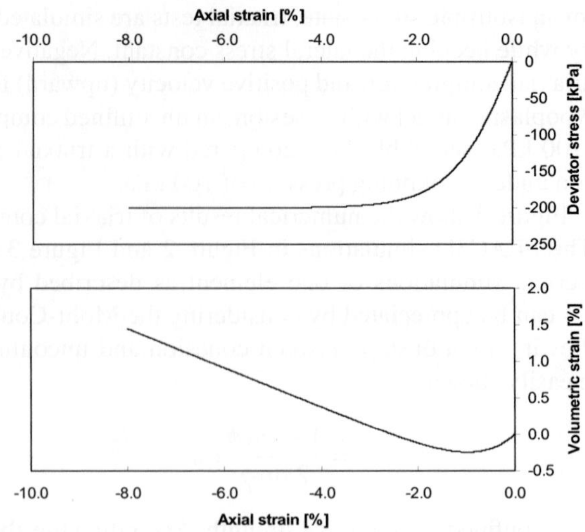

Fig. 3. FLAC3D simulation of triaxial extension test

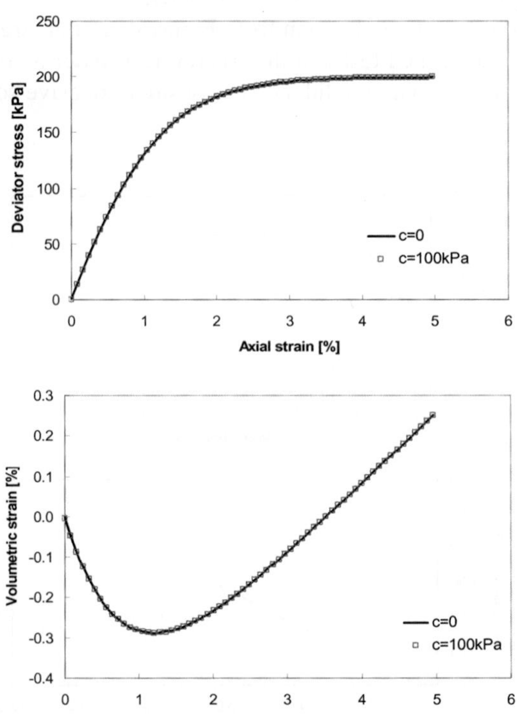

Fig. 4. Comparison between unconfined compression test and triaxial compression test with a confining pressure of 100 kPa

4.2 Simulation of Excavation and Support of Shallow Tunnel

We consider a shallow tunnel in soft ground in urban area. The tunnel is constructed according to the New Austrian Tunneling Method with a primary shortcrete lining and a final cast in-place concrete liner. It is important to minimize the impact of tunnelling on surface structures. Surface settlements depend on both excavation method and tunnel support. The surface settlement resulting from an advancing tunnel is three dimensional and ought to be treated as such.

The tunnel model with a length of 51 m is shown the following figure. By making use of the symmetry, only one-half of the tunnel is modelled. The base of the tunnel is located about 39.5 m below the ground surface. The tunnel crosssection consists of a half circle with a radius of 5 m. The properties of the material are shown in Table 2.

The primary shotcrete lining has a thickness of 30 cm and is modelled with shell structural elements because of its small thickness. The final concrete lining has a thickness of about 1.5 m and is modelled with zones because of its large thickness. The liners are assumed to be elastic with an elastic modulus of 31.4 GPa and a

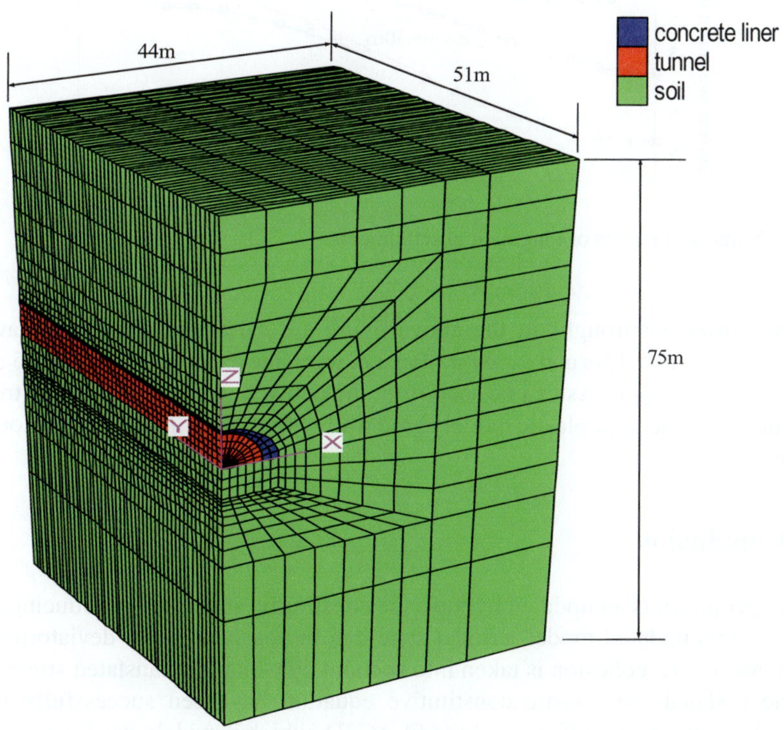

Fig. 5. Model grid of the tunnel in FLAC3D

Table 2. Parameters of hypoplastic material

E [MPa]	μ [-]	ψ [°]	ϕ [°]	c [kPa]
30	0.3	5	25	5

Poisson ratio of 0.25. The excavation and support installation are conducted incrementally with an excavation depth of 3 m. The shotcrete lining is installed immediately after the excavation. The final liner is installed 3 m behind the excavation face. A total of fifteen excavations are simulated. Figure 6 shows the numerical results of longitudinal settlements with the hypoplastic model. For comparison, the numerical results using Mohr-Coulomb model are presented too.

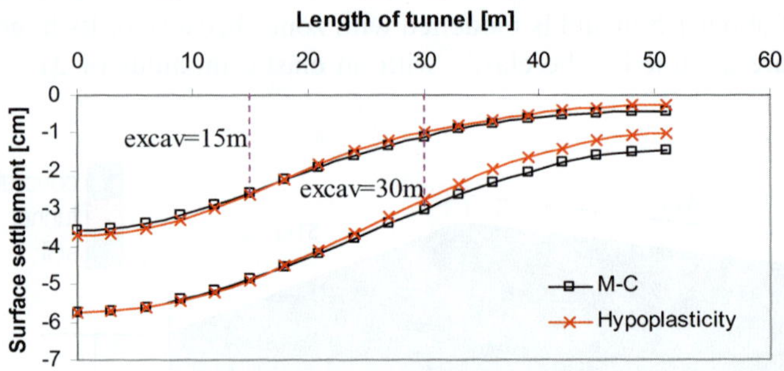

Fig. 6. Numerical results of longitudinal settlements

The settlement troughs in the above figure are calculated for two excavation stages, i.e. $excav = 15m$ and $excav = 30m$, i.e. when the excavation reaches 15 m and 30 m from the portal. As can be seen from the above figure, the settlement troughs obtained with the hypoplastic model agree well with those with the Mohr-Coulomb model.

5 Conclusion

The paper presents an updated hypoplastic constitutive model by introducing a new term. In this updated model, critical state can be reached for all deviatoric stress paths. Moreover, cohesion is taken into account by using the translated stress.

The updated hypoplastic constitutive equation has been successfully implemented into the finite difference code FLAC3D, which is widely used in geotechnical engineering. The model is verified by simulating triaixal tests and the excavation and support of a shallow tunnel.

Acknowledgement. The first author wishes to thank the Otto Pregl Foundation for Geotechnical Fundamental Research for the financial support.

References

1. Bardet, J.P.: Lode dependences for isotropic pressure-sensitive elastoplastic materials. J. Appl. Mech., ASME 57, 498–506 (1990)
2. Bauer, E.: Calibration of a comprehensive constutitive model for granual materials. Soils and Foundations 36, 13–26 (1995)
3. Itasca Consulting Group, Inc.: FLAC3D (Fast Lagrangian Analysis of Constinua in 3 Dimensions),Version 3.0. Minneapolis, MN 55401
4. Kolymbas, D.: An outline of hypoplasticity. Arch. Appl. Mech. 61, 143–151 (1996)
5. Wu, W., Bauer, E.: A simple hypoplastic constitutive model for sand. Int. J. Numer. Anal. Methods Geomech. 18, 833–862 (1994)
6. Wu, W.: On a simple critical state model for sand. In: Proc. of the seventh Int. Symp. on Numerical Models in Geomechanics-NUMOG, Graz, Austria, Balkema, pp. 47–52 (1999)
7. Wu, W., Niemunis, A.: Failure criterion, flow rule and dissipation function derived from hypoplasticity. Mech. Coheisive-Frictional Mater. 1, 145–163 (1996)
8. Wu, W., Kolymbas, D.: Numerical testing of the stability critierion for hypoplastic constitutive equations. Mech. Mater. 9, 245–253 (1990)

A Chemo-Thermo-Mechanically Coupled Analysis of Ground Deformation Induced by Methane Hydrate Dissociation

Sayuri Kimoto, Fusao Oka, and Tomohiko Fushita

Department of Civil and Earth Resources Engineering,
Kyoto University, Kyoto, Japan, 615-8540
e-mail: kimoto.sayuri.6u@kyoto-u.ac.jp, oka.fusao.2s@kyoto-u.ac.jp

Abstract. In the present paper, we have numerically analyzed the dissociation process of seabed ground and predicted the deformation of hydrate-bearing sediments. The simulation is conducted using the chemo-thermo-mechanically coupled model that takes into account the phase changes between hydrates and fluids during dissociation, the deformation behavior of the solid skeleton, and heat transfer simultaneously (Kimoto et al. 2007a). In addition, the dependency of the permeability coefficients for water and gas on hydrate saturation is introduced in the present analysis. From the analytical results, it has been found that ground deformation is induced by the dissipation and generation of water and gas and by a reduction of soil strength during the dissociation process.

1 Introduction

Methane hydrates are presently viewed as a potential energy resource for the 21st century because a large amount of methane gas is trapped mainly within ocean sediments and regions of permafrost. However, we have little knowledge about the performance of sediments under dissociation of hydrates. Therefore, a prediction of the ground performance of the ocean bed during the various production processes is required.

The dissociation process follows the phase changes that occur from solids to fluids, i.e., from hydrates to water and gas, and the ground will be under unsaturated conditions. In addition, heat transfer becomes important during the dissociation process, since the phase equilibrium is controlled by temperature and pressure, and the dissociation reaction is an endothermic reaction. Recently, some researchers have conducted experimental studies to investigate the characteristics of hydrate-bearing sediments. Wu and Grozic (2008) indicated that the dissociation of even a small percentage of gas hydrates will lead to failure from experimental evidence. Sakamoto et al. (2008) conducted experimental studies on the dissociation behavior of hydrate sediments, and showed that the increase of effective stress at the initial stage of depressurization is a dominant factor for deformations. Miyazaki et al. (2008) experimentally showed that the strength of hydrate sediments depends on hydrate saturation, and they exhibit rate dependency on the strength.

R. Wan et al. (Eds.): Bifurcations, Instabilities and Degradations in Geomaterials, SSGG, pp. 145–165.
springerlink.com

As for numerical studies, several numerical simulators have been developed to evaluate gas production levels in the past few years. Masuda et al. (2002) and Ahmadi et al. (2004) created numerical models using the finite difference method to predict gas and water flows which accompany hydrate dissociation. While they considered both fluid flow and heat transfer, the solid phase was assumed to be immobile. Although other numerical simulators have been developed, the solid phase was assumed to be rigid in most of them (e.g., Bejan et al. 2002; Tsypkin 2000; Bondarev et al. 1999). Klar and Soga (2005) presented a flow-deformation analysis for methane hydrate extraction problems using a finite difference code without considering heat transfer. Oka et al. (2005a) developed a numerical method for solving the thermo-mechanically coupled problem, and Oka et al. (2005b) and Kimoto et al. (2007a and 2007b) developed a numerical simulator for the deformation of soil containing methane hydrates due to the dissociation of gas hydrates based on the chemo-themo-mechanically coupled mixture theory. Rutqvist and Moridis (2008) presented a numerical simulator for analyzing the geomechanical performance of hydrate-bearing sediments. They combined numerical simulators for solving hydraulic and mechanical behaviors using a staggered technique.

The present work extends the previously developed theory (Kimoto et al. 2007a) by considering the dependency of the permeability coefficients for water and gas on hydrate saturation. Using the proposed method, we have numerically analyzed the dissociation process of the seabed ground and predicted the deformation of hydrate-bearing sediments.

2 Simulation Method

2.1 Multiphase Mixture Theory

Geomaterials generally fall into the category of multiphase materials. They are basically composed of soil particles, water, and air. The behavior of multiphase materials can be described within the framework of a macroscopic continuum approach through the use of the theory of porous media (Boer, 1998). The theory is considered to be a generalization of Biot's two-phase mixture theory for saturated soil (Biot, 1955).

Proceeding from the general geometrically non-linear formulation, the governing balance relations for multiphase materials can be obtained (e.g., Boer 1998; Loret & Khalili 2000; Ehlers et al. 2004). Mass conservation laws for the gas phase as well as for the liquid phase are considered in those analyses. In the field of geotechnics, air pressure is assumed to be zero in many research works (e.g., Sheng et al. 2003), since geomaterials usually exist in an unsaturated state near the surface of the ground. Considering hydrate dissociation, however, we have to deal with the high level of gas pressure that exists deep in the ground; this means that the mass balance for both phases must be considered. Oka et al. (2006) proposed an air-water-soil coupled finite element model in which the skeleton stress is used as a stress variable, the suction effect is introduced in the constitutive equation for soil, and the simulated compression behavior for unsaturated soils is under impermeable conditions for both water and gas flow.

Furthermore, the conservation of energy is required when there is a considerable change in temperature during the deformation process. Oka et al. (2005a) numerically simulated the thermal consolidation process. Since hydrate dissociation is an endothermic reaction, heat transfer plays an important role in both gas production and ground deformation.

2.2 General Setting

The material to be modelled is composed of four phases, namely, solid (S), water (W), and gas (G), which are continuously distributed throughout space, and hydrates (H). For simplicity, we assume that hydrates (H) move with the solid phase before dissociation. The total volume V is obtained from the sum of the partial volumes of the constituents, namely,

$$\sum_{\alpha} V^{\alpha} = V \quad (\alpha = S, W, G, H) \tag{1}$$

The volume of void V^{v}, which is composed of water, gas, and hydrates, is given by

$$\sum_{\beta} V^{\beta} = V^{v} \quad (\alpha = W, G, H) \tag{2}$$

The volume fraction n^{α} is defined as the local ratio of the volume element with respect to the total volume as:

$$n^{\alpha} = \frac{V^{\alpha}}{V} \tag{3}$$

$$\sum_{\alpha} n^{\alpha} = 1 \quad (\alpha = S, W, G, H) \tag{4}$$

The volume fraction of the void, n, is written as:

$$n = \sum_{\beta} n^{\beta} = \frac{V^{v}}{V} = \frac{V - V^{S}}{V} = 1 - n^{S} \quad (\beta = W, G, H) \tag{5}$$

In addition, the water saturation is required in the model, namely,

$$s_r = \frac{V^W}{V^W + V^G} = \frac{n^W}{n^W + n^G} = \frac{n^W}{n^F} \tag{6}$$

2.3 Skeleton Stress

In the theory of porous media, the concept of the effective stress tensor is related to the deformation of the soil skeleton and plays an important role. The effective stress tensor has been defined by Terzaghi (1943) for water-saturated soil; however, the effective stress needs to be redefined if the fluid is made of compressible materials. In the present study, skeleton stress tensor σ'_{ij} is defined and then used for the stress variable in the constitutive relation for the soil skeleton. Total stress tensor σ_{ij} is obtained from the sum of the partial stress values, σ_{ij}^{α}, namely,

$$\sum_{\alpha} \sigma_{ij}^{\alpha} = \sigma_{ij} \quad (\alpha = S, W, G) \tag{7}$$

$$\sigma_{ij}^{S} = \sigma'_{ij} - n^S P^F \delta_{ij} \tag{8}$$

$$\sigma_{ij}^{\alpha} = -n^{\alpha} P^{\alpha} \delta_{ij} \quad (\alpha = W, G) \tag{9}$$

where P^F is the average pressure of the fluids surrounding the solid skeleton (Schrefler, 1996) given by

$$P^F = s_r P^W + (1 - s_r) P^G \tag{10}$$

σ'_{ij} is called the skeleton stress in the present study. It is used as the stress variable in the constitutive relation for the soil skeleton such that

$$\sigma_{ij} = -P^F \delta_{ij} + \sigma'_{ij} \tag{11}$$

2.4 Conservation of Mass

The conservation of mass for the water and gas phases, $\beta(=W, G)$, is given in the following equation:

$$\frac{\partial}{\partial t}\left(\rho^{\beta} n^{\beta}\right) = -q_{Mi,i}^{\beta} + \dot{m}^{\beta} \quad (\beta = W, G) \tag{12}$$

in which ρ^{β} is the material density, q_{Mi}^{β} is the flux vector of the fluid, and $\dot{m}^{\beta}$ is the mass rate of change per unit volume generated by dissociation. The flux vector is expressed in terms of the relative velocity of the flow, V_i^{β}, with respect to the solid phase as

$$V_i^{\beta} = n^{\beta}(v_i^{\beta} - v_i^{S}) \quad (\beta = W, G) \tag{13}$$

$$q_{Mi}^{\beta} = \rho^{\beta} v_i^{\beta} \quad (\beta = W, G) \tag{14}$$

where v_i^{β} is the velocity of phase β. The conservation laws in Eq. (12) for the water and the gas phases are expressed with water saturation s_r and the volume fraction of void n.

The water density is assumed to be constant. As for describing changes in gas density, the equation of ideal gases is used, i.e.

$$\rho^G = \frac{M^G P^G}{R\theta} \tag{15}$$

$$\dot{\rho}^G = \frac{M^G}{R}\left(\frac{\dot{P}^G}{\theta} - \frac{P^G \dot{\theta}}{\theta^2}\right) \tag{16}$$

in which M^G is the molecular weight of gas, R is the gas constant, θ is the temperature, and tension is positive in the equation.

Dividing Eq. (16) by Eq. (15) yields

$$\frac{\dot{\rho}^G}{\rho^G} = \frac{\dot{P}^G}{P^G} - \frac{\dot{\theta}}{\theta} \tag{17}$$

The rate of change of volume fraction hydrate is given by

$$\dot{n}^H = \frac{\dot{V}^H}{V} = \frac{\dot{m}^H}{\rho^H} \tag{18}$$

where $\dot{n}^H < 0$ when dissociation occurs, and $\dot{m}^H$ is the rate of mass decrease of the hydrate.

The mass conservation law for the solid phase, namely, soil skeleton is given by

$$\frac{\partial}{\partial t}(n^s \rho^s) + (n^s \rho^s v_i^s)_{,i} = 0 \tag{19}$$

For the methane hydrate phase, the hydrate decomposes to form methane gas and water due to an increase in temperature and a decrease in pressure. Therefore, the mass of the hydrate decreases due to its dissociation following the chemical reaction equation. Since rehydration of the methane gas is not considered in the present paper, the dissociation process is not reversible.

2.5 Conservation of Momentum

Momentum balance is required for each phase, namely,

$$n^{\alpha} \rho^{\alpha} \dot{v}_i^{\alpha} = \sigma_{ij,j}^{\alpha} + \rho^{\alpha} n^{\alpha} \bar{F}_i - \tilde{P}_i^{\alpha} \quad (\alpha = S, W, G) \tag{20}$$

in which $\overline{F}_i$ is the gravity force and $\tilde{P}_i^{\alpha}$ is related to the interaction term given in

$$\tilde{P}_i^{\alpha} = \sum_{\gamma} D^{\alpha\beta}(v_i^{\alpha} - v_i^{\gamma}), \quad D^{\alpha\gamma} = D^{\gamma\alpha} \quad (\alpha, \gamma = S, W, G) \tag{21}$$

where $D^{\alpha\beta}$ are parameters which describe the interaction with each phase. The momentum balance equation for each phase is obtained with the following equations when the acceleration is disregarded. The parameters $D^{\alpha\beta}$ are given as

$$D^{WS} = \frac{(n^W)^2 \rho^W g}{k^W}, \qquad D^{GS} = \frac{(n^G)^2 \rho^G g}{k^G} \tag{22}$$

in which k^W and k^G are the permeability coefficients for the water phase and the gas phase, respectively.

When we assume that the space derivative of volume fraction $n_{,i}^{\alpha}$ is negligible and the interaction between water and gas phases D^{GW} and D^{WG} is zero, Darcy's law for the water phase and the gas phase is obtained from momentum balance for water and gas phases, respectively as:

$$V_i^W = n^W \left(v_i^W - v_i^S\right) = -\frac{k^W}{\rho^W g}\left(P_{,i}^W - \rho^W \overline{F}_i\right) \tag{23}$$

$$V_i^G = n^G \left(v_i^G - v_i^S\right) = -\frac{k^G}{\rho^G g}\left(P_{,i}^G - \rho^G \overline{F}_i\right) \tag{24}$$

The sum of Eqs. (27)-(29) leads to

$$\sigma'_{ji,j} + \rho^E \overline{F}_i = 0, \qquad \rho^E = \sum_{\alpha} n^{\alpha} \rho^{\alpha} \quad (\alpha = S, W, G) \tag{25}$$

The rate type of conservation for the momentum is given by

$$\dot{S}^t_{ji,j} = 0 \tag{26}$$

in which $\dot{S}^t_{ij}$ is the nominal stress rate tensor, and changes in the material density are ignored.

2.6 Conservation of Energy

The following energy conservation equation is applied in order to consider the heat conductivity and the heat sink rate associated with hydrate dissociation:

$$(\rho c)^E \dot{\theta} = D_{ij}^{vp} \sigma'_{ij} - h_{i,i} + \dot{Q}^H, \quad (\rho c)^E = \sum n^{\alpha} \rho^{\alpha} c^{\alpha} \left(\alpha = S, W, G, H\right) \tag{27}$$

where c^{α} is the specific heat, θ (K) is the temperature for all the phases, D_{ij}^{vp} is the viscoplastic stretching tensor, and $\dot{Q}^{H}$ is the dissociation heat rate per unit volume due to hydrate dissociation is

$$\dot{Q}^{H} = \frac{\dot{N}^{H} Q}{V} \tag{28}$$

where Q (kJ/kmol) is the dissociation heat per unit kilomole which changes with temperature, namely,

$$Q = 56599 - 16.744\theta \tag{29}$$

The heat flux h_i is given by

$$h_i = -\lambda^{E} \theta_{,i} \tag{30}$$

$$\lambda^{E} = \sum_{\alpha} n^{\alpha} \lambda^{\alpha} \quad (\alpha = S, W, G, H) \tag{31}$$

in which λ^{α} is the thermal conductivity.

2.7 *Soil-Water Characteristic Curve*

The relation between suction and saturation is given in the following equation proposed by van Genuchten (1980):

$$s_{re} = \left\{1 + \left(\alpha P^{C}\right)^{n}\right\}^{-m} \tag{32}$$

in which α, m, and n are material parameters, and the relation $m = 1 - 1/n$ is assumed. The term $P^{C} = (P^{G} - P^{W})$ is the suction and S_{re} is the effective saturation, namely,

$$s_{re} = \frac{s_r - s_{rmin}}{s_{rmax} - s_{rmin}} \tag{33}$$

where s_{rmax} and s_{rmin} are the maximum and minimum values of suction, respectively.

3 Dissociation of Hydrates

3.1 *Phase Equilibria of Methane Hydrates*

If the conditions for pore pressure and temperature shift to the unstable region given in the following equation (Bejan et al. 2002), gas hydrates dissociate into water and gas with the reaction expressed in Eq. (46). The average pressure of the fluids, P^F, is regarded as pore pressure P in the analysis, i.e.

$$P \le c\exp\left(a - \frac{b}{T}\right) \quad \text{(Unstable region)} \tag{34}$$

$$CH_4 \cdot nH_2O\ (hydrate) \rightarrow nH_2O\ (water) + CH_4\ (gas) \tag{35}$$

where a, b, and c are material parameters, and n is a hydrate number and is assumed to be equal to 5.75. The dissociation ratio $\dot{N}^H$ (kmol/s) is given by the following Kim-Bishnoi equation (Kim et al. 1987):

$$\dot{N}^H = -0.585 \times 10^{10} \times exp^{-\frac{9400}{\theta}} \left(P^e - P\right) N^{H_0 \frac{1}{3}} N^{H \frac{2}{3}} \tag{36}$$

in which N^H (kmol) is the number of moles of hydrates in volume V (m^3), N^{H_0} (kmol) is the number of moles of hydrates at the initial state, P (kPa) is the average pore pressure, and P^e is an equilibrium pressure at temperature θ (K). When dissociation occurs, the dissociation ratio is negative, i.e. $\dot{N}^H < 0$.

Water and gas generating ratios, $\dot{N}^W$ and $\dot{N}^G$ (kmol/s) are given by

$$\dot{N}^W = -5.75 \dot{N}^H \tag{37}$$

$$\dot{N}^G = -\dot{N}^H \tag{38}$$

Mass increasing ratios (t/sec/m^3) for hydrates, water, and the gas phase, required in the mass conservation law in Eqs. (17) and (18), can be calculated from the above equations.

3.2 Dependency of Hydrate Saturation on Permeability

The permeability coefficients for water and gas, k^W (m/s) and k^G (m/s), are dependent on hydrate saturation S_r^H (Masuda et al. 2002) according to:

$$k^\alpha = k_0^\alpha \exp\left(\frac{e - e_0}{2}\right)\left(1 - S_r^H\right)^N \quad (\alpha = W, G) \tag{39}$$

$$S_r^H = \frac{V^H}{V^v} \tag{40}$$

where V^v is the volume of the void, V^H is the volume of the hydrates, e is the current void ratio, e_0 is the initial void ratio, and k_0^α is the permeability when hydrate saturation S_r^H is equal to zero. From the experimental observations made by Sakamoto et al. (2004), the permeability decreases from 1/30 to 1/50 when $S_r^H = 0.4$ compared to that of $S_r^H = 0$. Subsequently, material parameter N is determined to be 6.

4 Constitutive Model for Soil

4.1 *Elasto-Viscoplastic Model for Unsaturated Soil*

Natural hydrates exist between soil particles and are considered to have a bonding effect which increases the solid phase strength. From this point of view, we introduced the effect of hydrate saturation in an elasto-viscoplastic model (Kimoto et al. 2004) as the shrinkage or expansion of the OC boundary surface and the static yield surface.

An elasto-viscoplastic model of the overstress type viscoplasticity with soil structure degradation for saturated soils has been extended to unsaturated soils using the skeleton stress and the newly introduced suction effect within the constitutive model (Oka et al. 2006).

It is assumed that the total stretching tensor consists of elastic stretching tensor D_{ij}^e and viscoplastic stretching tensor D_{ij}^{vp} as

$$D_{ij} = D_{ij}^e + D_{ij}^{vp} \tag{41}$$

The elastic stretching is given by a generalized Hooke type of law, i.e.,

$$D_{ij}^e = \frac{1}{2G}\dot{S}_{ij} + \frac{\kappa}{3(1+e)}\frac{\dot{\sigma}'_m}{\sigma'_m}\delta_{ij} \tag{42}$$

where S_{ij} is the deviatoric stress tensor, σ'_m is the mean skeleton stress, G is the elastic shear coefficient, e is the initial void ratio, κ is the swelling index, and the superimposed dot denotes the time differentiation.

4.2 *Overconsolidation Boundary Surface*

In this model, it is assumed that there is an overconsolidation (OC) boundary surface that delineates the normally consolidated (NC) region, $f_b \geq 0$, from the overconsolidated region (OC), $f_b < 0$, as follows:

$$f_b = \overline{\eta}^*_{(0)} + M^*_m \ln(\sigma'_m / \sigma'_{mb}) = 0 \tag{43}$$

$$\overline{\eta}^*_{(0)} = \left\{\left(\eta^*_{ij} - \eta^*_{ij(0)}\right)\left(\eta^*_{ij} - \eta^*_{ij(0)}\right)\right\}^{\frac{1}{2}}, \quad \eta^*_{ij} = S_{ij} / \sigma'_m \tag{44}$$

where M^*_m is the value of $\eta^* = \sqrt{\eta^*_{ij}\eta^*_{ij}}$ when the volumetric strain increment changes from compression to dilation, which is equal to ratio M^*_f at the critical state, σ'_{mb} is the hardening parameter, and b is a material parameter which is determined by the σ'_m -axis intercept of the critical state line. The stress ratio tensor at the end of the anisotropic consolidation is given by $\eta^*_{ij(0)}$.

4.3 Effect of Suction and Hydrate Saturation

The suction effect is introduced by relating it to the value of $\sigma'_{ma(s)}$ which controls the size of the OC boundary surface (Oka et al. 2006). In addition, the effect of hydrate saturation on the strength is introduced. It has been experimentally revealed that the strength of soils containing methane hydrates depends on the hydrate saturation in the void, since hydrates have a bonding effect between soil particles. Once hydrates dissociate, degradation of the soil structure occurs with decreasing hydrate saturation. Considering the effect of the suction and the hydrate saturation, the hardening-softening rule of σ'_{mb} is given as

$$\sigma'_{mb} = N_m N_s \sigma'_{ma}(z) \exp\left(\frac{1+e}{\lambda-\kappa}\varepsilon^{vp}_{kk}\right) \tag{45}$$

in which N_m and N_s denote the effects of hydrate saturation and suction respectively, $\sigma'_{ma}(z)$ refers to structural degradation with increasing viscoplastic strain, λ is the compression index, and κ is the swelling index. The parameter N_s is defined as

$$N_s = 1 + S_I \exp\left\{-s_d\left(\frac{P^C_i}{P^C} - 1\right)\right\} \tag{46}$$

where S_I is the strength ratio of unsaturated soils when the value of suction P^C equals to P^C_i, and s_d controls the decreasing ratio of strength with decreasing suction. The term P^C_i is set to be the maximum value for suction. At initial state when $P^C = P^C_i$, the strength ratio of the unsaturated soil to the saturated soil is $1+S_I$ and decreases with a decline in suction.

The parameter N_m is given as

$$N_m = 1 + n_m \exp\left\{-n_d\left(\frac{S^H_{ri}}{S^H_r} - 1\right)\right\} \tag{47}$$

where S_r^H is the hydrate saturation in the void defined as

$$S_r^H = \frac{V^H}{V^v} \tag{48}$$

In Eq. (60), n_m describes the strength ratio when the saturation of hydrate S_r^H equals to S_{ri}^H, and n_d is the stress decreasing ratio with decreasing hydrate saturation.

Finally, the term σ'_{ma} is a strain-softening parameter used to describe the degradation of the material caused by structural changes, namely,

$$\sigma'_{ma} = \sigma'_{maf} + \left(\sigma'_{mai} - \sigma'_{maf}\right)\exp(-\beta z) \tag{49}$$

$$z = \int_0^t \dot{z}dt\ ,\quad \dot{z} = \sqrt{\dot{\varepsilon}_{ij}^{vp}\dot{\varepsilon}_{ij}^{vp}} \tag{50}$$

in which σ'_{mai} and σ'_{maf} are the initial and the final values for σ'_{ma}, β is a material parameter which controls the rate of structural changes, and z is the accumulation of the second invariant of viscoplastic strain rate $\dot{\varepsilon}_{ij}^{vp}$. Details of the constitutive model with soil structure degradation are given in Kimoto et al. (2004).

4.4 Static Yield Surface

The static yield function is given by

$$f_y = \bar{\eta}_{(0)}^* + \tilde{M}^* \ln\left(\sigma'_m / \sigma'^{(s)}_{my}\right) = 0 \tag{51}$$

In the same way as for the OC boundary surface, the effects of suction and hydrate saturation are introduced in the value of $\sigma'^{(s)}_{my}$, i.e.

$$\sigma'^{(s)}_{my} = \frac{N_m N_s \sigma'_{ma}(z)}{\sigma'_{mai}} \sigma'^{(s)}_{myi} \exp\left(\frac{1+e}{\lambda - \kappa}\varepsilon_{kk}^{vp}\right) \tag{52}$$

4.5 Viscoplastic Potential Surface

The viscoplastic potential surface is described as

$$f_p = \bar{\eta}_{(0)}^* + \tilde{M}^* \ln\left(\sigma'_m / \sigma'_{mp}\right) = 0 \tag{53}$$

where $\tilde{M}^*$ is assumed to be constant in the NC region and to vary with the current stress in the OC region as

$$\tilde{M}^* = \begin{cases} M_m^* & : NC \quad region \\ -\dfrac{\sqrt{\eta_{ij}^*\eta_{ij}^*}}{ln\left(\sigma'_m / \sigma'_{mc}\right)} & : OC \quad region \end{cases} \tag{54}$$

where M_m^* is the value of $\sqrt{\eta_{ij}^*\eta_{ij}^*} / \sigma'_m$ at critical state, and σ'_{mc} denotes the mean skeleton stress at the intersection of the OC boundary surface and the σ'_m axis.

4.6 *Viscoplastic Flow Rule*

The viscoplastic stretching tensor is expressed by the following equation which is based on Perzyna's viscoplastic theory as

$$D_{ij}^{vp} = C_{ijkl}\left\langle \Phi_1\left(f_y\right)\right\rangle \frac{\partial f_p}{\partial \sigma'_{kl}} \tag{55}$$

$$C_{ijkl} = a\delta_{ij}\delta_{kl} + b\left(\delta_{ik}\delta_{jl} + \delta_{il}\delta_{jk}\right), \qquad C_1 = 2b,\ C_2 = 3a + 2b \tag{56}$$

in which $\langle\ \rangle$ are Macaulay's brackets; $\langle f(x)\rangle = f(x)$, if $x > 0$, =0, if $x \le 0$, C_1 and C_2 are the viscoplastic parameters for the deviatoric and the volumetric components, respectively, and Φ_1 indicates material function for the strain rate sensitivity. The dependency of viscoplastic property of soils on temperature is also introduced in viscoplastic parameter C_1 and C_2.

5 Simulation of Dissociation Process

5.1 *Initial and Boundary Conditions*

Weak forms of conservation of the mass for water and gas, conservation of momentum, conservation of energy are discretized in space and solved by the finite element method. For the finite element method, an updated Lagrangian method with the objective Jaumann rate of Cauchy stress is used (Kimoto et al. 2004, Oka et al. 2006). The unknown variables are nodal velocity, pore water pressure, pore gas pressure, and temperature. The backward finite difference method is used for the time discretization. We have simulated the production process by heating and depressurizing. Initial and final values of pore pressure and temperature are shown in Fig. 1 with the methane hydrate equilibrium curve.

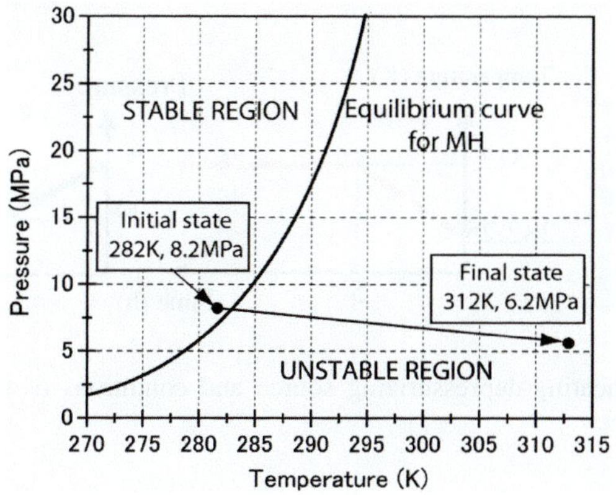

Fig. 1. Equilibrium curve for methane hydrate, and initial and final conditions of the simulation

The finite element mesh and the boundary conditions for the simulation are shown in Fig. 2, in which a plane strain condition is assumed. The seabed ground at 250 m depth from the bottom of the sea at a water depth of 650 m is modeled. The ground is assumed to consist of silty clay and hydrate-bearing sediment exists at a ground depth of 150-200 m. There is the drilling rig at the left side of the model, hence the left boundary is set to be impermeable to both water and gas with zero heat flow. The static pressures are given at the top, bottom and the right boundaries. A heating-depressurizing source is placed at the center of the MH (Methane Hydrate) layer.

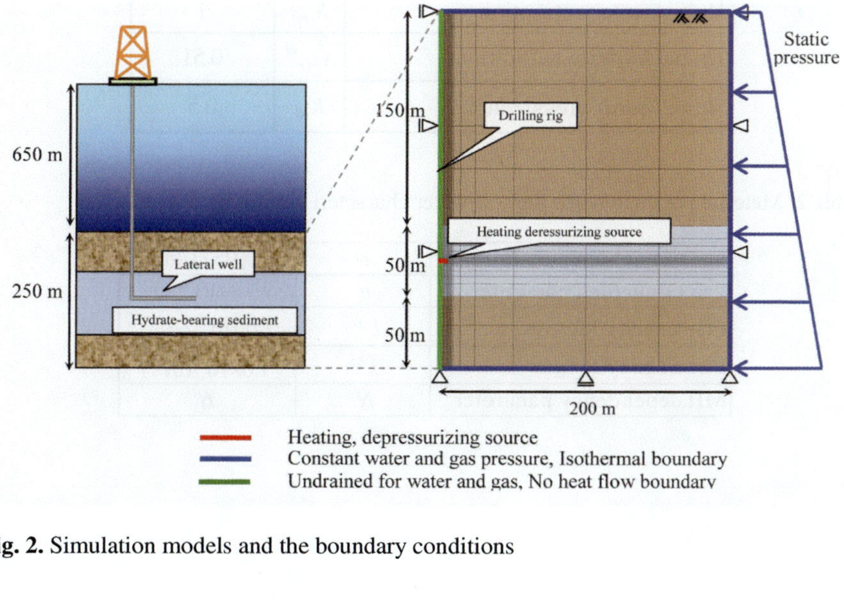

Fig. 2. Simulation models and the boundary conditions

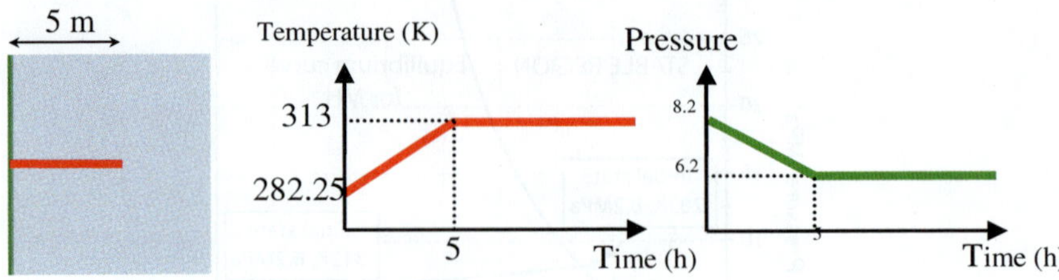

Fig. 3. Modelling heating-depressurizing source and conditions of temperature and pore pressure at the source

Figure 3 shows the model of the heating-depressurizing source and conditions of heating and depressurizing. The source is 5 meters long. The temperature increases from 282 to 313K in 5 hours, and at the same time the pressure decreases from 8.2 to 6.2 MPa. The initial conditions and material parameters are summarized in Tables 1, 2, and 3. The initial volume fraction of void, that is, the sum of water and hydrate phase, is 0.47, and initial hydrate saturation in the void is 0.51. Material parameters for the soil shown in Table 3 are mainly determined from the results of triaxial tests of samples obtained from the field research conducted along the Nankai Trough.

Table 1. Initial conditions

Initial volume fraction of void	n_0	0.47
Initial water saturation	S_{r0}	1
Initial hydrate saturation	S_{r0}^H	0.51
Initial earth pressre at rest	K_0	0.5

Table 2. Material parameters for the soil-water characteristic curve

van Genuchten parameter	α	0.0025 (1/kPa)
van Genuchten parameter	n	10
Permeability for water	k^W	1.0×10^{-5} (m/s)
Permeability for gas	k^G	1.0×10^{-4} (m/s)
MH dependency parameter	N	6

Table 3. Material parameters for the soil

Compression index	λ	0.169
Swelling index	κ	0.017
Initial shear elastic modulus	G_0	53800 (kPa)
Viscoplastic parameter	C_0	1.0×10^{-12}(1/s)
Viscopalstic parameter	m'	23
Stress ratio at failure	M'_m	1.08
Compression yield stress	σ'_{mbi}	1150 (kPa)
Structural parameter	σ'_{maf}	1150 (kPa)
Parameter for suction effect	P^C_i	100 (kPa)
Parameter for suction effect	S_I	0.2
Parameter for suction effect	s_d	0.25
Parameter for MH effect	S^H_{ri}	0.65
Parameter for MH effect	n_m	0.6
Parameter for MH effect	n_d	0.75

5.2 Simulation Results and Discussions

Figure 4 shows the distributions of hydrate saturation around the heating-depressurizing source. The figure represents an enlarged area around the heating-depressurizing source. When heating and depressurizing have been completed after 5 hours, dissociation first occurs in the area just around the source and then progresses around it. Figure 5 shows the distribution of temperature around the heating-depressurizing source. The temperature decreases to about 270 K around the dissociated elements after 5 hours. This is due to dissociation being an endothermic reaction, and also because of the self-preservation effect of the methane hydrate. Figure 6 shows the changes in pore water pressure in the whole ground at the initial state and after 5 hours of production. When depressurizing has completed after 5 hours, the pore water pressure decreases around the source. Figures 7 and 8 show the distributions of pore water pressure and excess pore gas pressure around the source respectively. It is found that produced gas exists in the dissociated elements, and the gas pressure is very small. Figures 9 and 10 show distributions of the mean skeleton stress in the whole ground and around the heating-depressurizing source respectively. The mean effective stress increases around the source mainly because of the depressurization. Figure 11 shows distributions of the volumetric strain and Fig. 12 illustrates the vertical settlement at ground surface. The compressive volumetric strain localizes around the dissociated area where the skeleton stress increases and the soil strength decreases due to the loss of hydrates. The settlement rapidly occurs in the initial state of the production and then gradually increases to a final total settlement of 5 cm as shown in Fig. 12. Figure 13 illustrates the volume of the produced gases over time.

In addition, we have simulated the production process by heating only without depressurizing. The results showed that the ground deformation for the heating method is rather smaller than that obtained by the heating-depressurizing method. This indicates that depressurizing causes considerable deformations.

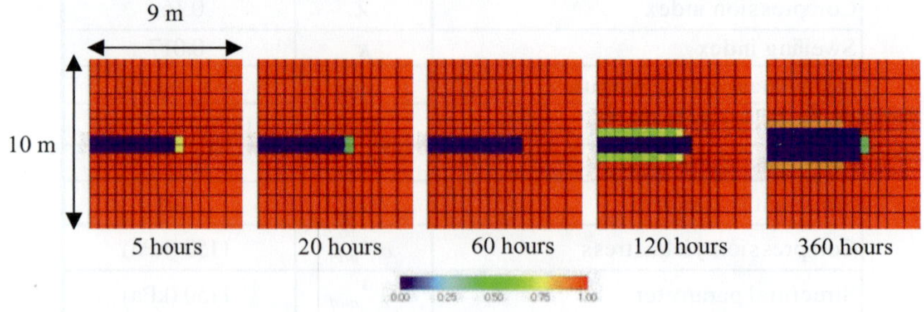

Fig. 4. Distributions of the hydrate saturation around the heating-depressurizing source

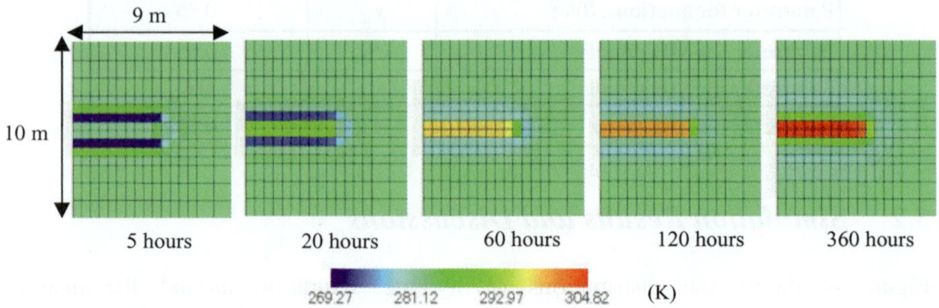

Fig. 5. Distributions of the temperature around the heating-depressurizing source

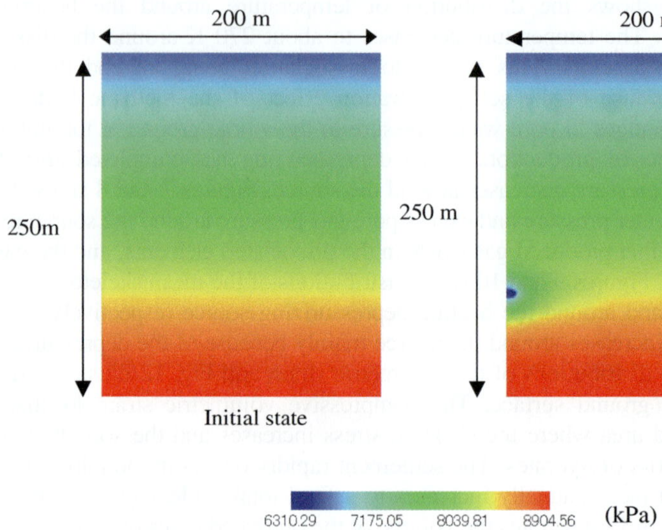

Fig. 6. Distributions of the pore water pressure in the whole ground

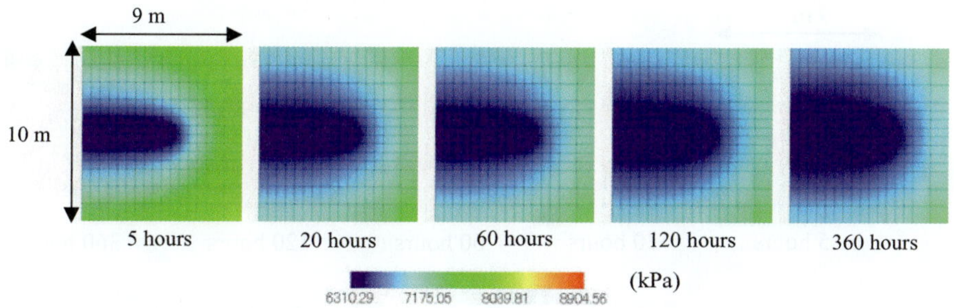

Fig. 7. Distributions of the pore water pressure around the heating-depressurizing source

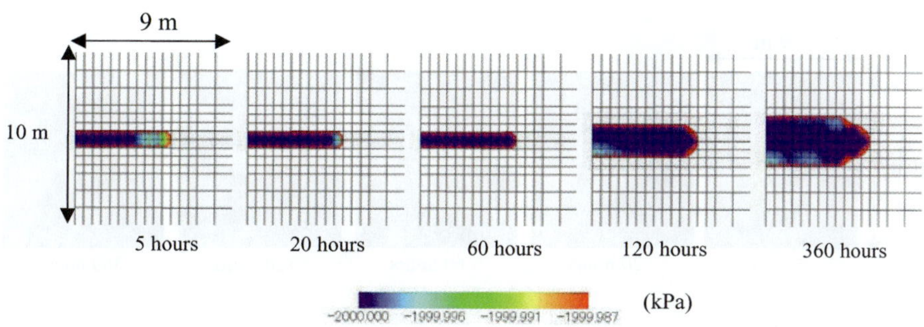

Fig. 8. Distributions of the excess gas pressure around the heating-depressurizing source

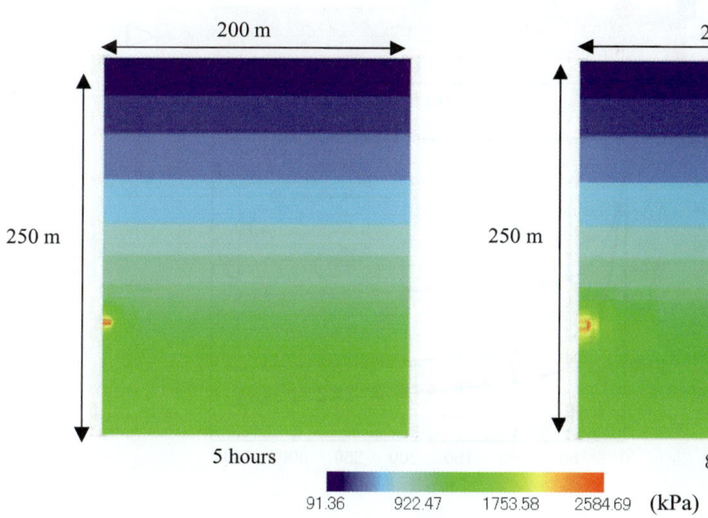

Fig. 9. Distributions of the mean effective stress in the whole ground

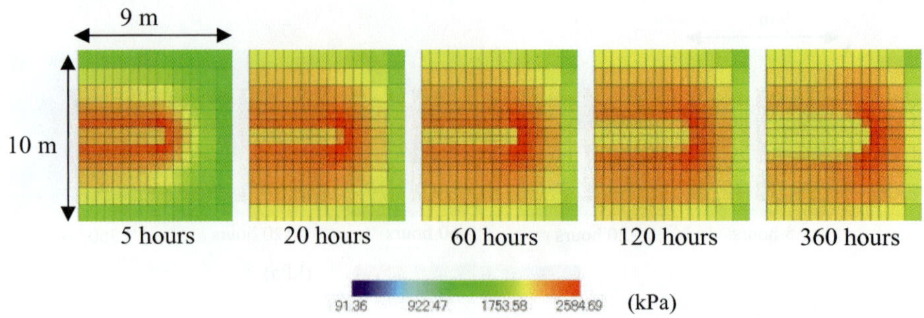

Fig. 10. Distributions of the mean effective stress around the heating-depressurizing source

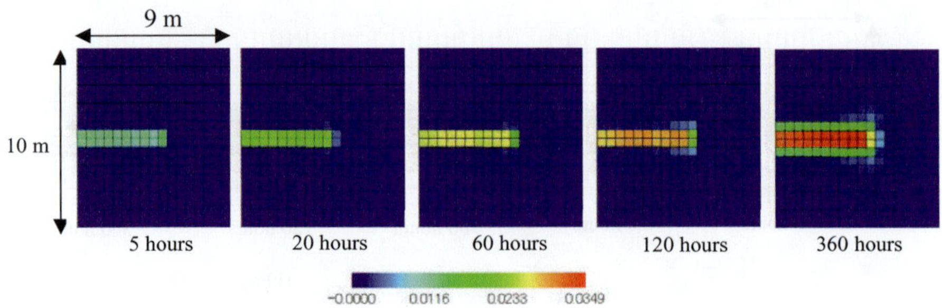

Fig. 11. Distributions of the volumetric strain around the heating-depressurizing source

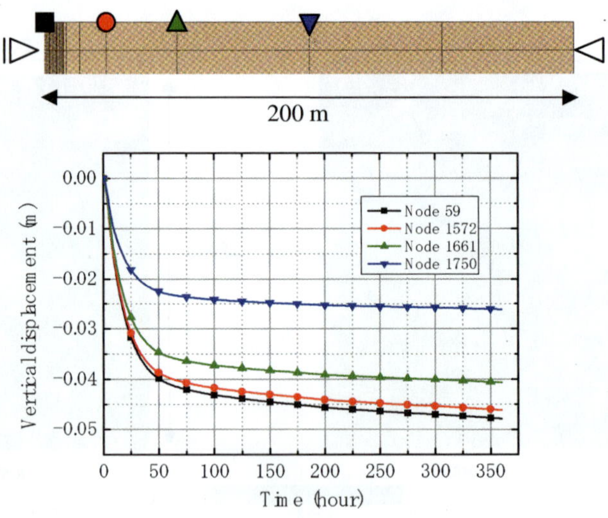

Fig. 12. Vertical settlement – time profile on the seabed surface

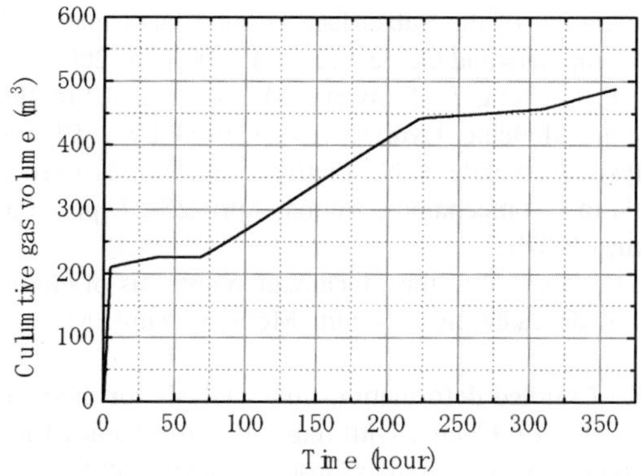

Fig. 13. Volume of the produced gases

6 Conclusion

We have successfully conducted the dissociation analysis of seabed ground and hence predicted the deformation of hydrate-bearing sediments. The simulations have been conducted by using a chemo-thermo-mechanically coupled analysis that accounts for coupling processes of dissociation, deformation and the heat transfer. From the numerical results, it is found that ground deformation is induced by both the generation and dissipation of water and gas, and by the reduction of soil strength due to the loss of hydrates.

References

Ahmadi, G., Ji, C., Smith, D.: Numerical solution for natural gas production from methane hydrate dissociation. J. Petroleum Science and Engineering 41, 269–285 (2004)

Bejan, A., Rocha, L.A.O., Cherry, R.S.: Methane hydrates in porous layers: Gas formation and convection. In: Ingham, B.D., Pop, I. (eds.) Transport Phenomena in Porous Media. Pergamon, Oxford (2002)

Biot, M.A.: Theory of elasticity and consolidation for a porous anisotropic media. J. Appl. Phys. 27, 459–467 (1955)

Boer, R.D.: Theory of porous media-past and present. Z. Angew. Math. Tech. 78(7), 441–466 (1998)

Bondarev, E.A., Kapitonova, T.A.: Simulation of multiphase flow in porous media accompanied by gas hydrate formation and dissociation. Russ. J. Eng. Thermophysics 9(1,2) (1999)

Ehlers, W., Graf, T., Ammann, M.: Deformation and localization analysis of partially saturated soil. Compt. Methods Appl. Mech. Engrg. 193, 2885–2910 (2004)

Englezos, P., Kalogerakis, N., Dohlabhai, P.D., Bishnoi, P.R.: Chem. Eng. Sci. 42, 2647–2658 (1987)

Kim, H.C., Bishnoi, P.R., Heidemann, R.A., Rizvi, S.S.H.: Kinetics of methane hydrate deconposition. Chemical Engineering Science 42(7), 645–1653 (1987)

Kimoto, S., Oka, F., Fushita, T., Fujiwaki, M.: A Chemo-Thermo-Mechanically Coupled Numerical Simulation of the Subsurface Ground Deformation due to Methane Hydrate Dissociation. Computers and Geotechnics 34, 216–228 (2007a)

Kimoto, S., Oka, F., Fushita, T., Fujiwaki, M.: Numerical Simulation of the Ground Deformation due to Methane Hydrate Dissociation by a Chemo-Thermo-Mechanically Coupled Analysis. In: Pande, G.N., Pietruszczak, S. (eds.) Proc. 9th Int. Symp. On Numerical Models in Geomechanics, Rodos, April 25-27, pp. 303–309. Taylor & Fransis Group, Balkema (2007b)

Kimoto, S., Oka, F., Higo, Y.: Strain localization analysis of elasto-viscoplastic soil considering structural degradation. Comput. Methods Appl. Mech. Engrg. 193, 2854–2866 (2004)

Klar, A., Soga, K.: Coupled deformation-flow analysis for methane hydrate production by depressurized wells. In: Cheng, Ulm (eds.) Poromechanics-Biot Centennial-, Abousleiman, pp. 653–659. Taylor & Francis Group, London (2005)

Loret, B., Khalili, N.: A three-phase model for unsaturated soils. Int. J. Numer. Anal. Meth. Geomech. 24, 893–927 (2000)

Masuda, Y., Kurihara, M., Ohuchi, H., Sato, T.: A field-scale simulation study on gas productivity of formations containing gas hydrates. In: Proc. 4th Int. Conf. on Gas Hydrate, Yokohama, Japan, pp. 40–46 (May 2002)

Miyazaki, K., Masui, A., Yamaguchi, T., Sakamoto, Y., Haneda, H., Ogata, Y., Aoki, K., Okubo, S.: Strain rate dependency of peak and residual strength of sediment containing synthetic methane hydrate. J. MMIJ 124, 619–625 (2008) (in Japanese)

Tsypkin, G.G.: Mathematical models of gas hydrates dissociation in porous media. Annals New York Ac. Sci. 912, 428–436 (2000)

Oka, F., Kimoto, S., Kim, Y.-S., Takada, N., Higo, Y.: A finite element analysis of the thermo-hydro-mechanically coupled problem of cohesive deposit using a thermo-elasto-viscoplastic model. In: Abousleiman, Y.N., Cheng, A.H.-D., Ulm, F.-J. (eds.) Poromechanics-Biot-centennial, Proc. 3rd Biot Conference on Poromechanics, Balkema, pp. 383–388 (2005a)

Oka, F., Kimoto, S., Kodaka, T., Takada, N., Fujita, Y., Higo, Y.: A finite element analysis of the deformation behavior of a multiphase seabed ground due to the dissociation of natural gas hydrates. In: Barla, G., Barla, M. (eds.) Proc. 11th Int. Conference of IACMAG, vol. 1, pp. 127–134. AGI, Patron Editore (2005b)

Oka, F., Kodaka, T., Kimoto, S., Kim, Y.-S., Yamasaki, N.: An Elasto-viscoplastic Model and Multiphase Coupled FE Analysis for Unsaturated Soil. In: Proc. 4th Int. Conf. on Unsaturated Soils, Carefree, Arizona, Geotechnical Special Publication, No. 147, April 2-6, vol. 2, pp. 2039–2050. ASCE (2006)

Rutqvist, J., Moridis, G.J.: Development of a numerical simulator for analyzing the geomechanical performance of hydrate-bearing sediments. In: The 42nd US Rock Mechanics Symposium and 2nd U.S-Canada Rock Mechanics Symposium, San Francisco, June 29-July 2 (2008)

Sakamoto, Y., Komai, T., Kawabe, T., Tenma, N., Yamaguchi, T.: Formation and dissociation behavior of methane hydrate in porous media–Estimation of permeability in methane hydrate reservoir, Part 1 120, 85–90 (2004) (in Japanese)

Sakamoto, Y., Shimokawara, M., Ohga, K., Miyazaki, K., Komai, T., Aoki, K., Yamaguchi, T.: Experimental study on consolidation behavior and permeability characteristics during dissociation of methane hydrate by depressurization process – Estimation of permeability in methane hydrate reservoir, Part 6-. J MMIJ 124, 498–507 (2008) (in Japanese)

Sheng, D., Sloan, W., Gens, A., Smith, D.W.: Finite element formulation and algorithms for unsaturated soils. Part I: Theory. Int. J. Numer. Anal. Meth. Geomech. 27, 745–765 (2003)

Schrefler, B.A., Gawin, D.: The effective stress principle: incremental or finite form. Int. J. Numer. Anal. Meth. Geomech. 20(11), 785–814 (1996)

Terzaghi, K.: Theoretical soil mechanics. John Wiley & Sons, Chichester (1943)

Van Genuchten, M.T.: A closed-form equation for predicting the hydraulic conductivity of unsaturated soils. Soil Science Society of America Journal 44, 892–898 (1980)

Wu, L., Grozic, J.L.H.: Laboratory analysis of carbon dioxide hydrate-bearing sand. J. Geotechnical and Geoenvironmental Eng. 134(4), 547–550 (2008)

Model for Pore-Fluid Induced Degradation of Soft Rocks

Marte Gutierrez[1] and Randall Hickman[2]

[1] Division of Engineering, Colorado School of Mines, Golden, CO 80401, USA
e-mail: mgutierr@mines.edu
[2] BP America, Inc., 501 Westlake Park Boulevard, Houston, TX 77079-2696, USA
e-mail: Randall.Hickman@bp.com

Abstract. Porous rocks, particularly chalk, are known to behave differently when saturated with different pore fluids. The mechanical behavior of these rocks varies with different pore fluid composition and additional deformation occurs when the pore fluid composition changes. In this article, we review the evidence that behavior of porous rocks is pore fluid dependent, and present a constitutive model for pore fluid dependent porous rocks. Our review indicates that theories of Unsaturated Soil Mechanics (USM) are not fully applicable to the modeling of the effects of pore fluid composition on soft rocks such as chalk. Instead of using USM, the paper proposes a model that is based on chemo-plasticity whereby the material response is dependent on the pore-fluid composition, and the material can degrade with changes in pore-fluid composition. Three degradation matrices are introduced, which are namely the elastic, elastoplastic and viscoplastic degradation matrices, to model, respectively, the reduced elastic stiffness, reduced shear strength, and the lower pore collapse strength and accelerated time-dependent deformation of soft rocks due to changes in pore-fluid composition. Comparisons of model predictions with published experimental data indicate that the model is capable of reproducing observed behavior of chalk under a variety of loading and pore fluid conditions.

1 Introduction

The sensitivity of rock mechanical behavior to the composition of the pore fluid is widely recognized for many rock types. The earliest experimental evidence of the effects of fluid composition on rock mechanical behavior was shown by Horne and Deere [1] who indicated that the frictional resistance between rock mineral surfaces is affected by the presence and type of fluid between the surfaces. From the 1960s, numerous experimental studies have shown that the shear strength and deformability of soft rocks are affected by the pore fluid composition [e.g., 2-5]. Several mechanisms have been proposed to explain the influence of pore fluid content on rock mechanical behavior. Gutierrez et al. [6] classify the different mechanisms into three groups, namely: (1) capillarity effects, (2) chemical effects and (3) physio-chemical effects. Capillary effects arise from the capillary forces

R. Wan et al. (Eds.): Bifurcations, Instabilities and Degradations in Geomaterials, SSGG, pp. 167–185.
springerlink.com

which develop at the contact between rock grain surface and that of the pore fluid. Chemical effects include mineral precipitation or dissolution. Mineral precipitation can lead to cementation of rock grains, and the filling up of the pore space, which in turn can make rocks stiffer and stronger. Dissolution on the other hand can lead to reduction in the grain sizes and weakening of the contacts between grain particles. Physio-chemical effects include stress-corrosion whereby stronger mineral bonds at crack tips are replaced by weaker bonds upon contact with fluids.

Of the different rock types, carbonate rocks (dolomites, limestones and chalks) appear to be the most affected by water. Recently, there has been growing particular interest in chalk and how it is affected by water in connection with production in several hydrocarbon fields. Chalk is a commonly found rock in hydrocarbon reservoirs most notably in North Sea hydrocarbon fields. One widely used technique to increase hydrocarbon production is to inject massive amounts of water into the reservoir. However, a negative consequence of the water injection in chalk reservoirs that has been observed is the weakening of the chalk. This weakening has resulted in increased reservoir compaction and seabed subsidence as exemplified by the case of the Ekofisk Field in the Norwegian sector of the North Sea. After nearly four decades of production, the Ekofisk field has undergone more than 10 m in seabed subsidence. It has been detected that the injection of water was subsequently followed by an increase in the subsidence rate, which was attributed to the water-induced weakening of the chalk.

Early attempts at explaining the effects of water on chalk centered on capillarity effects and the use of Unsaturated Soil Mechanics (USM) to model the interaction between chalk and water. For instance, the Barcelona Basic Model [7] was suggested by Delage et al. [8] and adopted by several researchers [9-11] to explain the behavior of chalk with a mixture of oil and water in its pores. The use of USM to explain the response of chalk to water is based on the idea that water is a wetting fluid for carbonate rocks. In oil-water system, oil is analogous to air as the non-wetting phase in unsaturated soil mechanics. As such, the invasion of the rock pore space by water destroys the capillarity and matric suction due to partial water saturation. The lost of matric suction then leads to the weakening and softening of the rock.

However, with the availability of more experimental data, it has now become more evident that the USM cannot fully explain the effects of water on chalk. There are three important factors which indicate why it is difficult to use unsaturated soil mechanics to completely explain the influence of water in chalk: (1) Water reduces the frictional resistance of chalk while the predominant view in unsaturated soil mechanics is that matric suction does not effect the frictional resistance, (2) The maximum suction that can be induced in chalk is in order of 0.2 MPa which is small compared to the strength of the carbonate rocks, with the unconfined compressive strengths in order of 100 MPa, and (3) Injection of a polar fluid in chalk, which does not create capillary pressures, also causes changes in the mechanical response of the chalk.

The sensitivity of chalk to water is now believed to be most possibly due to geochemical interactions between the calcium and carbonates in the chalk and water [6]. Specifically, this interaction involves the migration of ions (Ca^{2+} and CO_3^{2-}) which accompanies pressure solution resulting in grain volume reduction and changes on the surface characteristics of the rock minerals [12,13]. The reduction in grain volume accounts for the local compaction observed at the water front as water invades a chalk sample. The changes in the rock mineral surface characteristics can account for the reduction in the strength and stiffness of chalk when the pores are invaded by water.

Due to the importance of their geomechanical response in hydrocarbon production and other applications, several constitutive models have been developed for chalk. However, understanding of the mechanism responsible for chalk water interaction is crucial as it dictates the foundation for establishing a constitutive model that can account for the influence of pore-fluid rock interaction on the mechanical behavior of chalk. Since models allied with unsaturated soil mechanics cannot fully model the effects of water in chalk, there is a need to introduce other frameworks for establishing more suitable models. The objective of this paper is to introduce a framework for modeling the effects of pore fluid induced degradation on the mechanical response of chalk. The model is formulated directly from experimental observations on how the different constitutive parameters involved in the model are affected by water.

This paper is organized in the following sections: (1) The formulation of a general constitutive model for soft rocks is presented, (2) Experimental data are shown on the effects of fluid composition on the mechanical behavior of chalk, (3) Mathematical procedures to modify the constitutive model to account for pore fluid effects based on chemo-plasticity are formulated, and (4) The predictions from the pore-fluid dependent constitutive mode are compared with experimental results.

Although the focus of the paper is on chalk, results from the literature show that chalk is not alone among porous rocks in exhibiting pore fluid dependent behavior. For instance, Carles and Lapointe [14] show that the same water-weakening effects exist in a North Sea limestone reservoir rock as in chalk, while Papamichos et al. [15] show that Red Wildmoor sandstone, an analog North Sea reservoir rock, is stronger while oil-saturated than while water-saturated.

2 Constitutive Model for Soft Rocks

The main elements of the model for the dependent constitutive behavior of soft rocks are shown in Figs. 1 and 2. A three-surface model is used which divides the stress space into a volumetric elliptical cap region, a shear failure region, and a tensile failure region. The shapes of the composite yield surface consisting of the three yield surfaces are shown in Fig. 1 in the meridional *p*-*q* space, and in Fig. 2 in the π-plane.

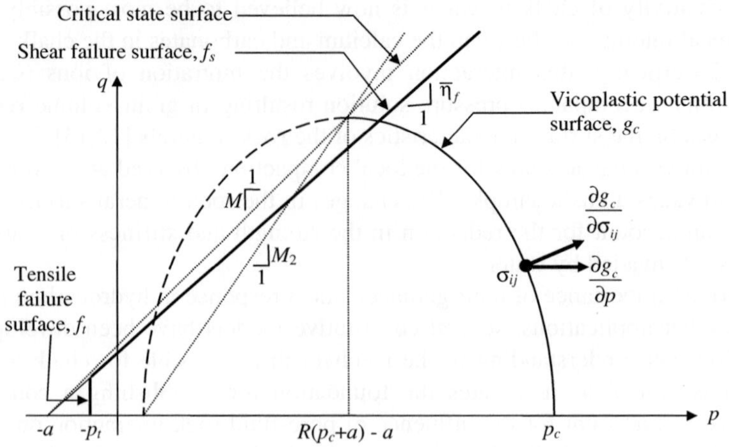

Fig. 1. Shapes of the yield and plastic potential surfaces for pore collapse, shear failure and tensile failure on the *p-q* plane

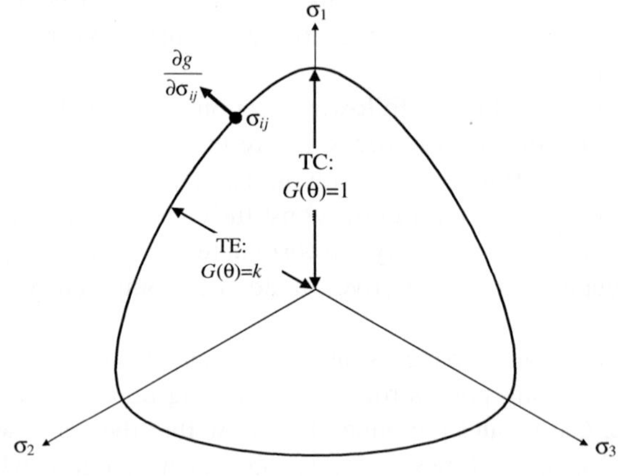

Fig. 2. Shape of the yield and plastic potential surfaces on the π-plane. The shape of the yield and plastic potential surfaces are the same.

Shear failure is represented by a non-hardening linear failure surface f_s and a non-associated plastic flow potential function g_s:

$$f_s = q - G(\theta)\overline{\eta}_f\,(p+a)\ ,\ g_s = q - G(\theta)\overline{\eta}_g\, p \tag{1}$$

where $p = \sigma_{kk}/3$, $q = \sqrt{3J_2}$, and $\sin(3\theta) = \left(-3\sqrt{3}J_3\right)/\left(2J_2^{3/2}\right)$ are the stress invariants ($J_2 = (s_{ij}s_{ij})/2$, $J_3 = s_{ik}s_{kl}s_{li}$, $s_{ij} = \sigma_{ij} - \delta_{ij}p$), $\overline{\eta}_f$ is the shear stress ratio at failure, $\overline{\eta}_g$ is related to the dilation angle, and a is the attraction. The function $G(\theta)$ models the effects of the third stress invariant and three-dimensional stress conditions on yielding, failure and plastic potential. The formulation of $G(\theta)$ is based on the Willam-Warnke [16] failure criterion. The use of the same function $G(\theta)$ in f_s and g_s leads to deviatoric normality.

Elastoplastic deformation due to shear failure follows the additivity postulate, where the total strain increment $\mathbf{d\varepsilon}$ consists of the elastic component $\mathbf{d\varepsilon}^e$ and plastic component $d\boldsymbol{\varepsilon}^p$.

$$d\boldsymbol{\varepsilon} = d\boldsymbol{\varepsilon}^e + d\boldsymbol{\varepsilon}^p = d\boldsymbol{\varepsilon} = [C^e]\,d\boldsymbol{\sigma} + \varphi^p \frac{\partial g_s}{\partial \boldsymbol{\sigma}} \tag{2}$$

where $[C^e]$ is the elastic compliance matrix and φ^p is the plasticity index obtained from the consistency condition, which stipulates that $df_s = 0$.

Tensile failure is controlled by an associated vertical tensile cut-off failure in the p-q space, and calculation of the elastoplastic strain increment follows that of Eq. (2) for the shear failure surface.

Deformation inside the shear and tensile failure surfaces is elasto-viscoplastic, and elasto-viscoplastic strain increments also follow additivity, where the total strain increment $d\boldsymbol{\varepsilon}$ consists of the elastic component $d\boldsymbol{\varepsilon}^e$ and the viscoplastic component $d\boldsymbol{\varepsilon}^{vp}$:

$$d\boldsymbol{\varepsilon} = d\boldsymbol{\varepsilon}^e + d\boldsymbol{\varepsilon}^{vp} = [C^e]d\boldsymbol{\sigma} + \varphi^{vp} \frac{\partial g_c}{\partial \boldsymbol{\sigma}} \;,\; \varphi^{vp} = \gamma\langle \phi(F) \rangle \tag{3}$$

where g_c is the viscoplastic potential, φ^p is the viscoplasticity index, γ is the fluidity parameter, and $\phi(F)$ is a scalar function called the flow function, and F is the overstress function.

The model for viscoplastic deformation is based on an extension of Bjerrum's [17] "reference time lines" approach, which assumes several compressions lines corresponding to different "ages" or "times" in the void ratio e vs. ln(p) as shown in Fig. 3. The compression lines are parallel to each other and have a slope equal to λ. The compression lines with times or age values of $t = t_o, 10t_o, 100t_o, 1000t_o, \ldots,$ where t_o is a reference time, are equidistant from each other. Following Bjerrum's definition, the outermost compression line for $t = t_o$ is called the "instantaneous" compression line.

The time-independent swelling line for overconsolidated materials, which has a slope of κ, intersects the compression lines at different points (Fig. 3). These points may be considered as yield points where compression response changes from normally consolidated to overconsolidated. The yield points depend on the elapsed "time" or the "age" of the material as the mean stress p is increased. These are labelled as $p_c, p_{c1}, p_{c2}, p_{c3}, \ldots$ in Fig. 3 corresponding respectively with the

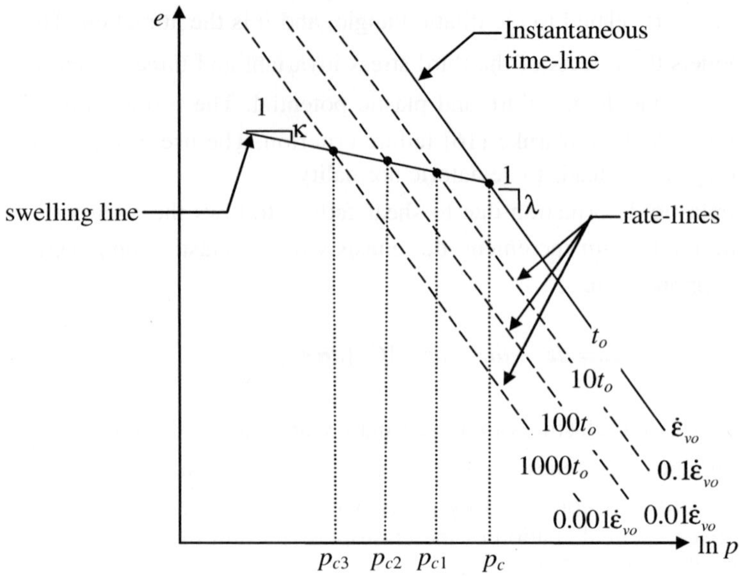

Fig. 3. Bjerrum's "reference time-lines" model. Each time line is also a constant rate- or isotach line. Each isotach has an equivalent time or rate-dependent yield stress $p_c, p_{c1}, p_{c2}, p_{c3},\ldots$ corresponding to $t = t_o, 10t_o, 100t_o, 1000t_o, \ldots.$ or $0.1\dot{\varepsilon}_{vo}, 0.01\dot{\varepsilon}_{vo}, 0.001\dot{\varepsilon}_{vo}, \ldots.$

time values of $t_o, 10t_o, 100t_o, 1000t_o, \ldots.$ The yield point p_c corresponds to the pre-consolidation pressure used in soil mechanics for materials overconsolidated by mechanical loading.

Using the equivalence between volumetric strain increment $d\varepsilon_{vol} = tr(d\boldsymbol{\varepsilon})$ and the change in void ratio de, i.e. $d\varepsilon_{vol} = de/(1+e)$, and the slopes of the swelling line κ in Fig. 3, the following relation between for $d\varepsilon_v^e$ and dp can be obtained:

$$d\varepsilon_{vol}^e = \frac{\kappa}{(1+e)p} dp \tag{4}$$

The time-dependent viscoplastic volumetric strain increment $d\varepsilon_{vol}^{vp}$ resulting from increase in time dt, and can also be obtained by inspection of Fig. 3. Along any vertical line in this figure, the mean stress is constant and only pure viscoplastic or creep deformations will be obtained since $d\varepsilon_v^e$=0 and $dp = 0$. Also, since the compression lines at $t = t_o, 10t_o, 100t_o, 1000t_o, \ldots.$ are equally spaced from each other in the *e*-ln(*p*) plot, it follows that the void ratio during *p*=constant loading changes with log(*t*) as time increases from t_o. Thus, the viscoplastic volumetric strain and its increment can be written as:

$$\varepsilon_{vol}^{vp} = \frac{\psi}{(1+e)} \ln\left(\frac{t}{t_o}\right) \; ; \; d\varepsilon_{vol}^{vp} = \frac{\psi}{(1+e)} \frac{dt}{t} \tag{5}$$

where the parameter ψ can be related to the coefficient of secondary consolidation C_α as $\psi = 0.434 C_\alpha$ (Taylor [18]). Note that Eq. (5) is the same as Taylor's creep law expressed in natural logarithm instead of $\log_{10}$, and that for the above equations to be valid, $t \geq t_o$.

Based on geometric analysis of Fig. 3, the following relation between the corresponding time *t* for a given mean stress *p*, where $p \leq p_c$, can be obtained:

$$t = t_o \left(\frac{p_c}{p}\right)^{\Lambda} \tag{6}$$

where $\Lambda = (\lambda - \kappa)/\psi$. This equation was first derived by Borja and Kavanzanjian [19], and it provides another definition for time *t*, which can now be viewed not only as conventional time but as a measure of the material "age". The farther the current stress *p* is from the pre-consolidation pressure p_c, the older is the material, and the higher is the value of *t*.

The development of aging can be more clearly illustrated for creep loading under constant stress *p* from time *t* to $t + dt$. During the time increment dt the void ratio of the material will decrease because of viscoplastic deformation, and at the same time the pre-consolidation pressure will also increase by dp_c. Following the derivations of [19], the increment dp_c can be obtained from Eq. (6) as:

$$dp_c = \left(\frac{p_c}{\Lambda}\right) \frac{1}{t} \tag{7}$$

The above equation may also be considered as analogous to strain hardening in plasticity, where the hardening variable is now time instead of the plastic strain. Substituting Eq. (6) in (5) completes the viscoplastic formulation, which is now expressed in terms of *p* and p_c:

$$d\varepsilon_{vol}^{vp} = \frac{\psi}{1+e} \left(\frac{p}{p_c}\right)^{\Lambda} dt \tag{8}$$

Extension to two and three-dimensional loading condition can made by introducing a plastic potential surface $g_c = g_c(\boldsymbol{\sigma})$ passing through the current stress point, and using a "volumetric scaling" to give the viscoplastic strain increment tensor $d\boldsymbol{\varepsilon}^{vp}$. To account for the attraction parameter a, p and p_c are replaced by $\overline{p} = p + a$ and $\overline{p_c} = p_c + a$.

$$d\varepsilon^{vp} = \frac{\psi}{1+e}\left(\frac{\overline{p}}{\overline{p}_c}\right)^{\Lambda}\left(\frac{\partial g_c}{\partial p}\right)^{-1}\left(\frac{\partial g_c}{\partial \sigma}\right)dt \tag{9}$$

The similarity between Eqs. (3) and (9) can be noticed if the fluidity parameter γ and the flow function $\phi(F)$ are defined as:

$$\gamma = \frac{\psi}{(1+e)}\left(\frac{\partial g}{\partial p}\right)^{-1}\frac{dt}{t} \quad ; \quad \phi(F) = \left(\frac{\overline{p}}{\overline{p}_c}\right)^{\Lambda} \tag{10}$$

As shown in Fig. 3, the viscoplastic potential surface g_c has an elliptical shape in the meridional *p-q* plane, and has an equation of the form:

$$g_c = q^2 - G(\theta)^2 M^2 R^2 \overline{p}_c^2 + G(\theta)^2 M_2^2 \left(\overline{p} - R\overline{p}_c\right)^2 \tag{11}$$

where M is the slope that controls the position of the center of the elliptical cap surface in *p-q* space, R is the parameter that controls the aspect ratio of the elliptical surface, and M_2 is related to M and R by the equation $M_2 = M \cdot R/(1-R)$. Note that for R=0.5, the yield and plastic potential functions simplifies to that of the Modified Cam Clay equations.

3 Experimental Data on Pore Fluid Effects in Chalk

Experimental results on chalk saturated with a mixture of oil and water, or and water, are used to illustrate the effects of pore fluid composition on the mechanical behavior of rocks. The strength and stiffness of chalk have been shown experimentally to depend on the pore fluid present in the rock. Effects of pore fluid on mechanical behavior of chalk have been noted repeatedly in the literature, and Hickman et al. [21] provide an extensive summary of the test results that have been reported in the literature. Dry or air-saturated chalk has a greater yield and failure strength and greater elastic stiffness than oil-saturated chalk, which is in turn stronger and stiffer than water-saturated chalk. Multiphase fluid saturated chalks exhibit characteristics intermediate to chalks with the end-member fluid compositions. Partially saturated chalks exhibit behavior intermediate to dry and fully water-saturated chalks, and chalks fully saturated with miscible glycol-water pore fluid mixtures exhibit behavior that changes as the pore fluid composition changes. The creep behavior and other rate-dependent characteristics of chalk also depend on pore fluid composition including partial saturation.

The constitutive model presented above requires 12 parameters to fully describe the rate-dependent behavior of chalk. These parameters and how they are affected by water saturation are summarized in Table 1. Based on experimental observations, of the 12 parameters, seven vary regularly as function of pore fluid composition with an oil-water pore fluid composition. The parameters that are dependent on the degree of water saturation are: (1) bulk modulus K; (2) reference time-line anchor N (void ratio for the "instantaneous" compression line for $p=\underline{1}$ stress unit); (3) attraction a; (4) adjusted failure shear stress ratio η_f; (5) cap aspect ratio M; (6) creep parameter ψ; and (7) tensile strength p_t. Of these seven parameters, all decreases with increasing water saturation except for the creep parameter ψ which increases with increasing water saturation.

The effect that pore fluid composition has on each of these parameters was determined from inspection of lab test results. Representative results are shown in Figs. 4 to 6 to illustrate the effects of fluid composition on the mechanical properties of chalk. Fig. 4 shows the differences between the shear failure and "pore collapse" cap surfaces for oil- and water-saturated chalks. As can be seen, despite the scatter in data, water-saturated chalk has statistically much lower shear strength and pore collapse yield strengths than oil-saturated chalk. Fig. 5 shows that the lower pore collapse yield strength for oil-saturated chalk is observed for all chalk porosities. Fig. 6 shows similar results that for any given porosity, the bulk modulus of water saturated chalk is also generally lower than that of oil saturated chalk. Similar decreasing trends (again except for ψ) for all porosity values are observed for the other constitutive parameters that are affected by the pore fluid composition.

Table 1. Required parameters for the soft rock constitutive model

Parameter	Effect of increasing water saturation
Bulk modulus, K	Decreases
Poisson's ratio, ν	Constant
Compression coefficient, λ	Constant
Reference time-line void ratio, N	Decreases
Critical state slope, M	Decreases
Eccentricity parameter, R	Constant
Attraction, a*	Decreases
Creep parameter, ψ	Increases
Minimum volumetric age, $t_{v,min}$	Constant
Adjusted failure shear stress ratio, $\overline{\eta}_f$	Decreases
Tensile strength, p_t	Decreases
William-Warnke scaling parameter, k	Constant

*Attraction describes both rate-dependent pore collapse behavior and shear failure surface.

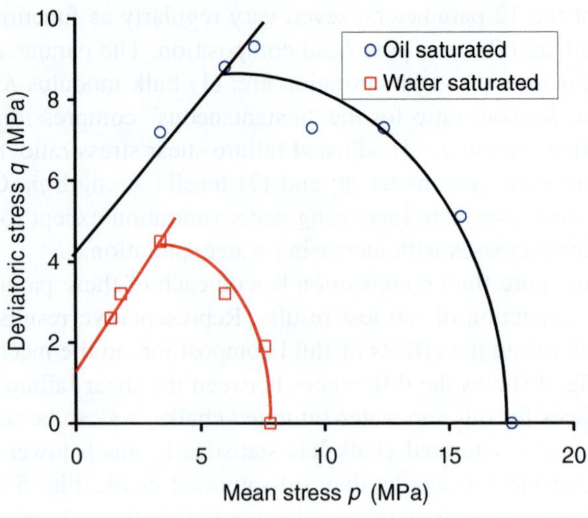

Fig. 4. Shear failure and "pore collapse" volumetric yield surfaces for oil- and water-saturated chalk

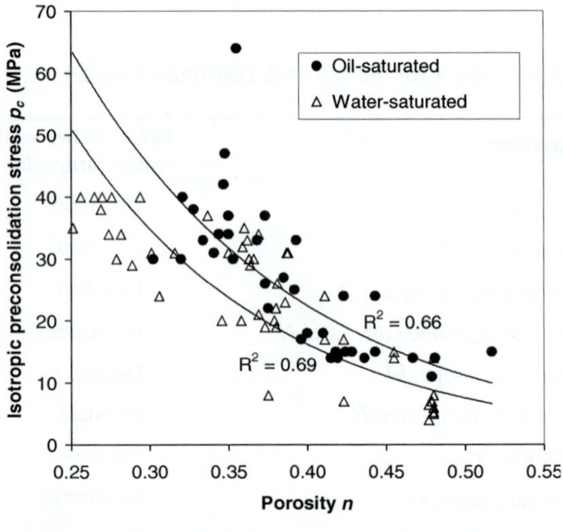

Fig. 5. Pore collapse yield strength for oil- and water-saturated chalk as function of porosity

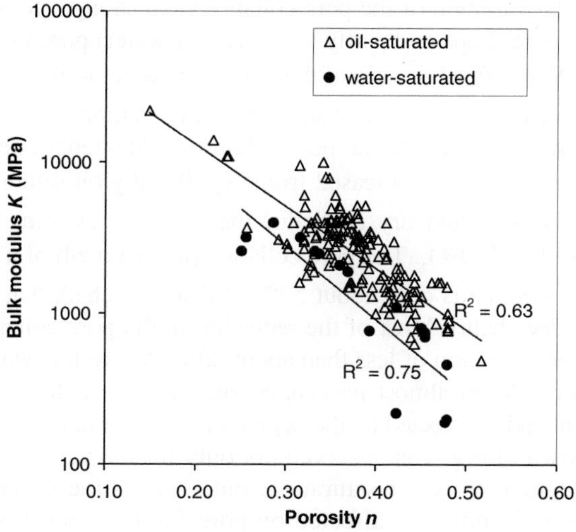

Fig. 6. Bulk modulus for oil- and water-saturated chalk as function of porosity

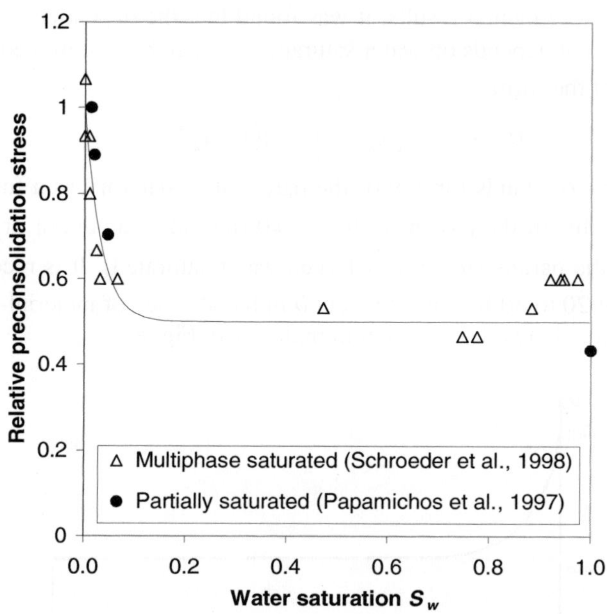

Fig. 7. Pore collapse yield stress as function of water saturation in chalk with oil-water pore fluid, and air-water pore fluid

Fig. 7 shows the variation of the pore collapse yield strength for chalk with multiphase (oil-water) pore fluids and partly-saturated (air-water) pore fluids as function of water saturation S_w (defined as ratio of the volume of water to the volume of voids for both multiphase oil-water pore fluid system and partly saturated air-water system). As can be seen, the relative value of the pore collapse yield strength decreases very rapidly as the water saturation is increased from S_w=0 (fully oil saturated for the multiphase pore fluid system and dry chalk for the air-water system) to a fully water-saturated system (S_w=100%). The pore collapse yield strength of a chalk with 100% water in the pore volume is only about 50% of that of a chalk that is fully saturated with oil (or of a dry chalk). Most of the reduction in the pore collapse yield strength occurred for water saturation of less than about 20%. Above this value, an increase in water saturation results in almost no change and the pore collapse yield strength is relatively constant and unaffected by the degree of water saturation.

The trend shown in Fig. 7 is observed not only for the variation of the pore collapse yield strength with water saturation, but also for the other constitutive parameters that were found to be affected by pore fluid composition (i.e., as shown in Table 1, these parameters are *K*, *N* $a, \bar{\eta}_f$, M and p_t). Similar to the pore collapse yield strength, the values of these parameters also rapidly decrease as S_w is increased from 0 to about 20%, after which the values remain almost constant.

Based on experimental results, it was found that the dependency of any model parameter α that depends on water saturation S_w can be represented by a uniform relationship of the form:

$$\alpha = \alpha_{min} + (\alpha_{max} - \alpha_{min})(1 - S_w)^b \tag{12}$$

In Eq. (12), α stands for any of the parameters that vary as a function of S_w. α_{max} is the value of the parameter for S_w=0 (i.e. oil saturated or dry) and α_{min} is the value of the parameter for S_w=1 (i.e., water saturated). It is recommended to use a value of 20 to 30 for the exponent *b* in the absence of material-specific laboratory data. Eq. (12) is shown diagrammatically in Fig. 8.

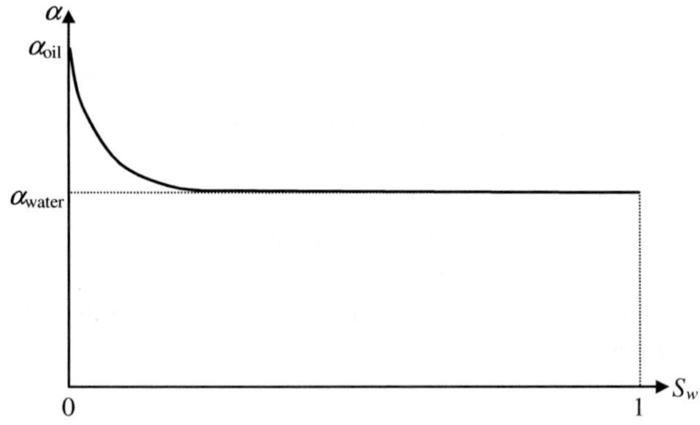

Fig. 8. Change in water-dependent model parameters with water saturation

4 Model for Pore Fluid Dependent Mechanical Behavior

To implement the experimentally observed dependency of chalk to pore fluid composition as described above, the conventional elastic, elastoplastic and elasto-viscoplastic formulations for constitutive modeling are modified to allow for dependency of the constitutive parameters to fluid composition. The modeling follows the chemo-plasticity framework developed by Pietruszczak [22] for alkali-aggregate reaction in concrete, and Pietruszczak et al. [23] for chemo-mechanical coupling in chalk. The chemo-plasticity framework is extended below for the case of elasto-viscoplastic constitutive models. Also, instead of using a scalar parameter, which measures the water induced chemical dissolution in chalk in the constitutive relation, as was done by Pietruszczak et al. [23], the approach followed is to use direct observations of relationships between constitutive parameter and water saturation.

Experimental results indicate that the elastic compliance matrix $[\mathrm{C}^{\mathrm{e}}]$, shear yield function f_s and viscoplastic parameter $\varphi^{vp}=\gamma\langle\phi(F)\rangle$ are all dependent on water saturation S_w in both air-water and oil-water pore fluid systems. These dependencies can be represented by the following mathematical expressions:

$$[\mathrm{C}^{e}]=[\mathrm{C}^{e}(S_w)]\ ,\quad f_s=f_s(\boldsymbol{\sigma},\boldsymbol{\varepsilon}^{\mathrm{p}},S_w)\quad ,\quad \varphi^{vp}=\varphi^{vp}\left(\gamma,\phi(F),S_w\right) \tag{13}$$

The elastic compliance $[\mathrm{C}^{\mathrm{e}}]$ matrix is dependent on S_w since K varies with S_w. Shear failure is dependent on S_w since the parameters $\overline{\eta}_f$ and a in the failure criterion f_s vary with S_w. Viscoplastic deformation is dependent on S_w since the parameters N, M, a and ψ in the viscoplastic parameter φ^{vp} vary with S_w.

Elastic response from the conventional elasticity relation $\boldsymbol{\varepsilon}=[C^{e}]\boldsymbol{\sigma}$ is modified to account for the dependency of the elastic compliance matrix $[\mathrm{C}^{\mathrm{e}}]$ to water saturation S_w:

$$d\boldsymbol{\varepsilon}=[C^{e}]d\boldsymbol{\sigma}+\mathbf{B}^{e}dS_w \tag{14}$$

where $\mathbf{B}^{e}$ is the elastic degradation matrix obtained from the rate of change of $[\mathrm{C}^{\mathrm{e}}]$ with respect to the change in water saturation dS_w:

$$\mathbf{B}^{e}=\frac{\partial[C^{e}]}{\partial S_w}\boldsymbol{\sigma} \tag{15}$$

Elasto-plastic response is obtained by adding plastic strain increments $d\boldsymbol{\varepsilon}^{p}$ to Eq. (14):

$$d\boldsymbol{\varepsilon}=[C^{e}]d\boldsymbol{\sigma}+\mathbf{B}^{e}dS_w+d\boldsymbol{\varepsilon}^{p} \tag{16}$$

The plastic strain increments $d\boldsymbol{\varepsilon}^p$ are obtained when the failure criterion is satisfied and from the flow rule:

$$f_s = f_s(\boldsymbol{\sigma}, \boldsymbol{\varepsilon}^p, S_w) = 0 \text{ and } d\boldsymbol{\varepsilon}^p = \varphi^p \frac{\partial g_s}{\partial \boldsymbol{\sigma}} \tag{17}$$

Applying the consistency condition:

$$df_s = \left(\frac{\partial f_s}{\partial \boldsymbol{\sigma}}\right)^T d\boldsymbol{\sigma} + \left(\frac{\partial f_s}{\partial \boldsymbol{\varepsilon}^p}\right)^T \varphi^p \left(\frac{\partial g_s}{\partial \boldsymbol{\sigma}}\right) + \frac{\partial f_s}{\partial S_w} dS_w = 0 \tag{18}$$

$$\varphi^p = \frac{1}{H}\left[\left(\frac{\partial f_s}{\partial \boldsymbol{\sigma}}\right)^T d\boldsymbol{\sigma} + \frac{\partial f_s}{\partial S_w} dS_w\right] \text{ and } H = -\left(\frac{\partial f_s}{\partial \boldsymbol{\varepsilon}^p}\right)^T \left(\frac{\partial g}{\partial \boldsymbol{\sigma}}\right) \tag{19}$$

Substituting Eqs. (19) in (17), gives the follow elasto-plastic increment relation which also accounts for the change in water saturation:

$$d\boldsymbol{\varepsilon} = [C^{ep}]d\boldsymbol{\sigma} + \mathbf{B}^{ep} dS_w \tag{20}$$

where $[C^{ep}]$ is the conventional elasto-plastic compliance matrix, and $\mathbf{B}^{ep}$ is the elasto-plastic degradation matrix:

$$[C^{ep}] = [C^e] + \frac{1}{H}\frac{\partial g_s}{\partial \boldsymbol{\sigma}}\left(\frac{\partial f_s}{\partial \boldsymbol{\sigma}}\right)^T d\boldsymbol{\sigma} \quad , \quad \mathbf{B}^{ep} = \mathbf{B}^e + \frac{1}{H}\frac{\partial f_s}{\partial S_w}\frac{\partial g_s}{\partial \boldsymbol{\sigma}} \tag{21}$$

For elasto-viscoplastic response with both the elastic compliance matrix $[C^e]$ and the viscoplasticity index φ^{vp} dependent of saturation, superposition of elastic and viscoplastic strain increment yields:

$$d\boldsymbol{\varepsilon} = [C^e]d\boldsymbol{\sigma} + \mathbf{B}^e dS_w + d\boldsymbol{\varepsilon}^{vp} \tag{22}$$

$$d\boldsymbol{\varepsilon} = \left\{[C^e]d\boldsymbol{\sigma} + \varphi \frac{\partial g_c}{\partial \boldsymbol{\sigma}}\right\} + \mathbf{B}^{vp} dS_w \tag{23}$$

where $\mathbf{B}^{vp}$ is the viscoplastic degradation matrix:

$$\mathbf{B}^{vp} = \mathbf{B}^e + \frac{\partial \varphi}{\partial S_w}\frac{\partial g_c}{\partial \boldsymbol{\sigma}} \tag{24}$$

The derivatives $\partial[C^e]/\partial S_w$, $\partial f_s/\partial S_w$ and $\partial\varphi/\partial S_w$ required to formulate the degradation matrices $\mathbf{B}^e$, $\mathbf{B}^{ep}$ and $\mathbf{B}^{vp}$ can be obtained from chain rule:

$$\frac{\partial[C^e]}{\partial S_w} = \sum_i \frac{\partial[C^e]}{\partial \alpha_i}\frac{\partial \alpha_i}{\partial S_w} \quad ; \quad \frac{\partial f_s}{\partial S_w} = \sum_i \frac{\partial f_s}{\partial \alpha_i}\frac{\partial \alpha_i}{\partial S_w} \quad ; \quad \frac{\partial \varphi^{vp}}{\partial S_w} = \sum_i \frac{\partial \varphi^{vp}}{\partial \alpha_i}\frac{\partial \alpha_i}{\partial S_w} \tag{25}$$

where α_i are the set of model parameters that are dependent on S_w (i.e., $\alpha_i = K$ for $[C^e]$, $\alpha_i = (\overline{\eta}_f, a)$ for f_s, and $\alpha_i = (N, M, a, \psi)$ for φ^{vp}). For the three degradation matrices, the change of the pore fluid dependent model parameter α_i with respect to S_w can all be obtained from Eq. (12) as:

$$\frac{\partial \alpha}{\partial S_w} = -b\left(\alpha_{2\max} - \alpha_{2\min}\right)\left(1 - S_w\right)^{b-1} \tag{26}$$

Fig. 9 gives an illustration of the pore-fluid dependent constitutive model for chalk. The figure schematically interprets the different equations presented above on how fluid content affects the size of the three different failure and yield surfaces involved in the model. As the water saturation is increased and the chalk changes from being oil saturated (or dry in partially saturated chalk) to fully water saturated, the shear and tensile failure, and pore collapse cap surfaces shrink. As a result, the chalk weakens and undergoes additional deformation as it gets invaded with water.

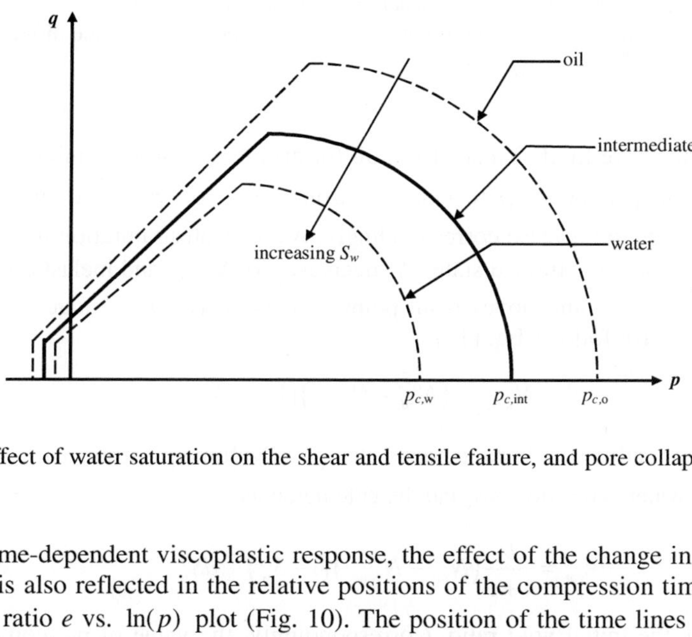

Fig. 9. Effect of water saturation on the shear and tensile failure, and pore collapse surfaces for chalk

For time-dependent viscoplastic response, the effect of the change in water saturation is also reflected in the relative positions of the compression time lines in the void ratio e vs. $\ln(p)$ plot (Fig. 10). The position of the time lines (given by N) changes from that of N_{oil} (or N_{dry}) to that of N_{water} as S_w is increased. Although the time line positions are different, limited experimental data indicate that the slope of the time lines given by λ is independent of water saturation. Fig. 10 reflects the behavior of chalk during invasion of water. Volumetric age decreases and creep rate increases as water saturation increases under constant stress. At the constant stress state of point A, the reference void ratio decreases from N_{oil}

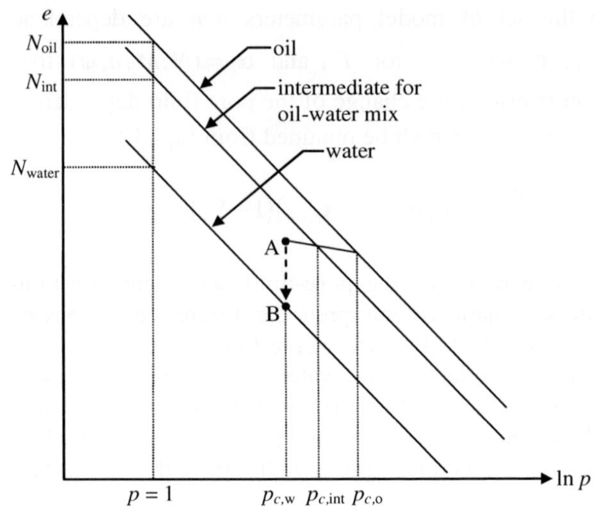

Fig. 10. Relative positions of the compaction time lines in the void ratio e vs. $\ln(p)$ axis for various pore fluids. The void ratio at p=1 unit, N, is the anchor for the different curves. $N_{\text{oil}} > N_{\text{water}}$ but $\lambda_{oil} = \lambda_{water}$.

to N_{int} as the pore fluid changes from oil (or dry) to an oil-water mixture, causing the pore collapse stress p_c to decrease from $p_{c,oil}$ (or $p_{c,dry}$) to $p_{c,\text{int}}$ and the volumetric age to decrease correspondingly. As the water saturation increases further to a full water-saturated state, N decreases to N_{water}, and inelastic strains occur as the stress point moves from point A to point B. The change in reference void ratio N also follows Eq. (12):

$$N = N_{water} + (N_{oil} - N_{water})(1 - S_w)^b \tag{27}$$

From this equation, the instantaneous change in volumetric strain $d\varepsilon^s_{vol}$ due to the change in water saturation dS_w can be calculated as:

$$d\varepsilon^s_{vol} = \frac{b}{1+e}(N_{oil} - N_{\text{water}})(1 - S_w)^{b-1}\, dS_w \tag{28}$$

where e is the initial void ratio. Correspondingly, the value of pre-consolidation stress $\bar{p}_c$ is adjusted according to the equation of the time line:

$$\bar{p}_c = \bar{p}_{c,oil} \exp\left(\frac{N_{oil} - N}{\lambda}\right) \tag{29}$$

Due to the change in p_c, the volumetric age also changes according to Eq. (6). Since ψ also changes with S_w (following Eq. 12), the viscoplastic strain rate

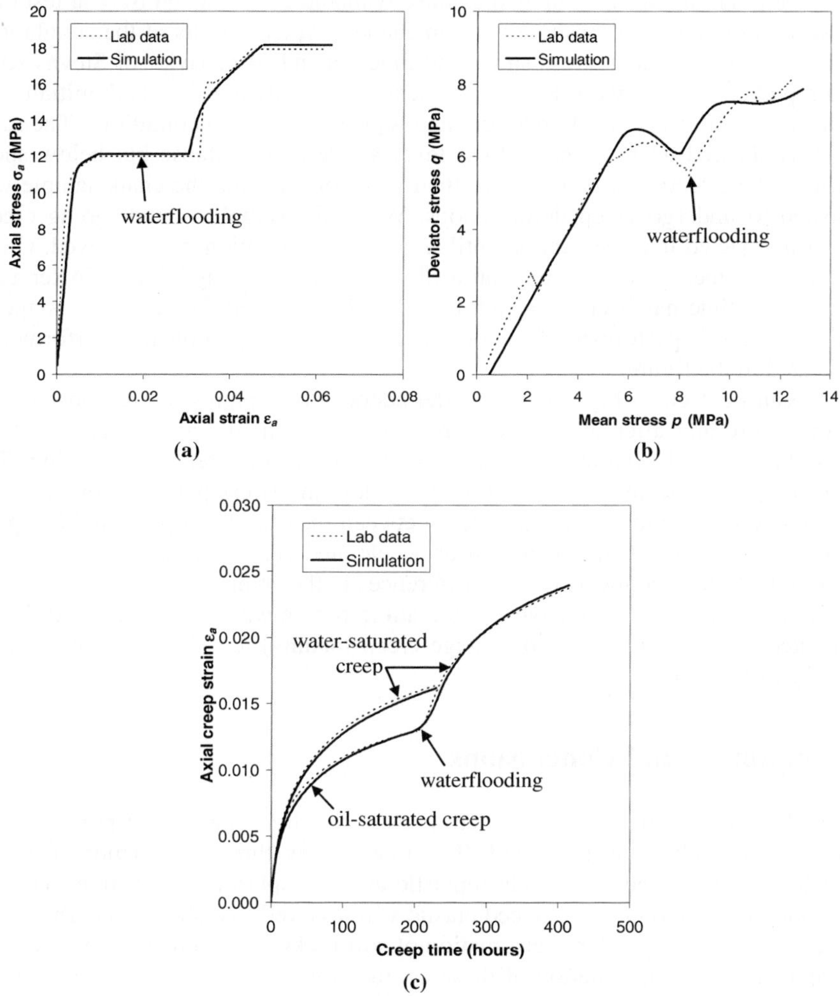

Fig. 11. Results of simulated waterflooding test on Stevns Klint outcrop chalk. Results include (a) stress-strain curve, (b) stress path and (c) creep curves.

isalso affected by S_w. Thus, when water saturation is increased, the result is to accelerate creep response.

5 Comparison with Experimental Results

Extensive laboratory tests have been performed in many laboratories to test the behavior of chalk under various stress- and strain-controlled, and pore fluid conditions. To validate the pore fluid dependent constitutive model proposed here, model simulations were compared against available experimental triaxial test data on several offshore and onshore outcrop chalks. Typical results of the simulations of pore fluid dependent stress-strain and time-dependent behavior of Stevns Klint outcrop chalk are shown in Fig. 11. A specimen of Stevns Klint chalk initially saturated with oil was loaded under drained K_0-triaxial loading conditions. The sample was sheared up to a shear stress level above pore collapse but below shear failure. The effective stresses where then kept constant and the chalk sample was allowed to undergo creep deformations. While the sample is undergoing creep, water is injected into the sample until residual oil saturation was achieved. Creep deformation measurements are continued while water is being injected. After a desired creep time has been achieved, the shear is increased then a second stage of creep loading is performed. The point at which water injection was performed is indicated in the figure.

As can be seen in Fig. 11, the model adequately captures the response of the chalk before and after water injection in terms of: (a) the axial strain vs. axial stress behavior, (b) the effective stress path, and (c) the creep deformation. The simulated creep behavior, stress-strain behavior, and stress path all closely match the observed behavior. The agreement between measured and predicted creep response, particularly in the increased or accelerated creep deformations right after water injection, is noteworthy. The differences in the creep behavior of water saturated chalk (from the first stage creep loading before water injection), and the oil saturated chalk (from the second stage creep loading) are also adequately replicated by the model.

6 Summary and Conclusions

A model for the pore fluid dependent constitutive behavior of soft porous rocks, particularly chalk, was presented. The model uses three degradation matrices, which are namely the elastic, elastoplastic and viscoplastic degradation matrices, to model, respectively, the reduced elastic stiffness, reduced shear strength and the accelerated time-dependent deformation of soft rocks due to changes in pore-fluid composition. The formulation of these degradation matrices were based on direct experimental observations of the effects of pore fluid content on the parameters required in the constitutive model. Comparisons with published experimental data indicate that the model is capable of reproducing observed behavior of chalk under a variety of loading conditions and changes in pore fluid composition.

References

1. Horne, H.M., Deere, D.W.: Frictional characteristics of minerals. Geotechnique 12, 319–335 (1962)
2. Colback, P.S.B., Wiid, B.L.: The influence of moisture content on the compressive strength of rocks. In: Proc. Rock Mech. Symp. Univ. Toronto, Canada (1965)
3. Street, N., Wang, F.D.: Surface potentials and rock strength. In: Proc. 1st Intl. Cong. Intl. Soc. Rock Mech., vol. 1, pp. 451–456 (1966)

4. Simpson, D.R., Fergus, J.H.: The effect of water on compressive strength of diabase. J. Geophys. Res. 73, 6591–6594 (1968)
5. Forsans, T.M., Schmitts, L.: Capillary force: The neglected factor in shale stability. In: Proc. Eurock 1994 Conf. Delft, pp. 71–74 (1994)
6. Gutierrez, M., Øino, L.E., Høeg, K.: The effect of fluid content on the mechanical behaviour of fractures in chalk. Rock Mech. Rock Eng. 33(2), 93–117 (2000)
7. Alonso, E.E., Gens, A., Josa, A.: A constitutive model for partially saturated soils. Geotechnique 40(3), 405–430 (1990)
8. Delage, P., Schroeder, C., Cui, Y.: Subsidence and capillary effects in chalks. In: Proc. Eurock 1996 Conf., vol. 2, pp. 1291–1298 (1996)
9. Papamichos, E., Brignoli, M., Santarelli, F.: An experimental and theoretical study of a partially saturated collapsible rock. Mech. Cohesive-Frictional Matls. 2, 251–278 (1997)
10. Collin, F., Cui, Y.J., Schroeder, C., Charlier, R.: Mechanical behavior of Lixhe chalk partly saturated by oil and water: Experiment and modeling. Intl. J. Num. Analy. Meth. Geomech. 26, 897–924 (2002)
11. De Gennaro, V., Delage, P., Cui, Y.J., Schroeder, C., Collin, F.: Time-dependent behaviour of oil reservoir chalk: A multiphase approach. Soils Found. 43(4), 131–147 (2003)
12. Rehbinder, P.A., Likthman, V.: The effect of surface active media on strain and rupture in solids. In: Proc. 2nd Intl. Cong. Surf. Act., vol. 3, pp. 563–580 (1957)
13. Butenuth, C., DeFreitas, M.H.: Studies of the influence of water on calcite. In: Proc. Intl. Chalk Symp., Brighton Polytech., pp. 103–108 (1989)
14. Carles, P., Lapointe, P.: Water-weakening of under stress carbonates: New insights on pore volume compressibility measurements. In: Proc. Intl. Soc. Core Analysts, Abu Dhabi, p. 12 (2004)
15. Papamichos, E., Tronvoll, J., Vardoulakis, I., Labuz, J.F., Skjaerstein, A., Unander, T.E., Sulem, J.: Constitutive testing of Red Wildmoor sandstone. Mech. Cohesive-Frictional Matls. 5, 1–40 (2000)
16. Willam, K.J., Warnke, E.P.: Constitutive model for the triaxial behavior of concrete. In: ISMES Seminar on Concrete Structures Subjected to Triaxial Stress, Bergamo, Italy, pp. 1–30 (1975)
17. Bjerrum, L.: Engineering geology of Norwegian normally consolidated marine clays as related to settlements of buildings. Geotechnique 17(2), 83–117 (1967)
18. Taylor, D.W.: Fundamentals of soil mechanics, p. 700. John Wiley and Sons, New York (1978)
19. Borja, R.I., Kavazanjian, E.: A constitutive model for the stress-strain-time behavior of wet clays. Geotechnique 35(3), 283–298 (1985)
20. Hickman, R.J., Gutierrez, M.S.: Formulation of a three-dimensional rate-dependent constitutive model for chalk and porous rocks. Intl. J. Num. Analy. Meth. Geomech. 31(4), 583–605 (2007)
21. Hickman, R.J., Gutierrez, M., DeGennaro, V., Delage, P.: Modeling of pore fluid-rock interaction as a weathering process. Intl. J. Num. Analy. Meth. Geomech. 32, 1927–1953 (2008)
22. Pietruszczak, S.: On the mechanical behavior of concrete subjected to alkali-aggregate reaction. Comp. Struct. 58, 1093–1099 (1996)
23. Pietruszczak, S., Lydzba, D., Shao, J.F.: Modeling of deformation response and chemo-mechanical coupling in chalk. Intl. J. Num. Analy. Meth. Geomech. 30, 997–1018 (2006)

Natural Processes and Strength Degradation

Jim Graham, Marolo Alfaro and James Blatz

Department of Civil Engineering, University of Manitoba,
Winnipeg, Manitoba, Canada, R3T 2N2
e-mail: jgraham@cc.umanitoba.ca, alfarom@cc.umanitoba.ca,
blatzja@cc.umanitoba.ca

Abstract. Many engineering projects are designed on the basis of laboratory tests using so-called 'undisturbed' samples of clay taken from the field. There is a tendency to test only intact specimens and discard specimens that appear disturbed, fissured or otherwise weaker. It is known, however, that natural processes such as wetting-drying, freezing-thawing, desiccation, heating-cooling, and alterations in chemistry can affect the structure of clays and significantly change their compressibilities, hydraulic conductivities and strengths. For example, plastic clays that have been fissured by desiccation or freezing cannot reliably provide peak strength resistance in slopes and under engineered embankments. The paper shows examples of projects where natural processes degraded the strengths of natural and reconstituted clays. The case histories in the paper provide a reminder of the importance of recognizing natural processes and the limitations of laboratory measurements when selecting appropriate parameters for numerical modeling.

1 Introduction

Much of the testing done in geotechnical laboratories examines the peak strengths of clays. Additional information needed for numerical modeling often includes deformation characteristics, volume changes, hydraulic conductivities, and pore fluid pressures, including gas pressures. We note too, that there are many applications in geotechnical practice in which shear strengths may decrease with time or in response to various physical processes. Projects that have functioned safely for many years may become unsafe. Our objective is to show the high frequency of process-induced changes in clay behaviour, and by implication, to draw attention to the potentially broad significance of climate change. The physical processes include wetting-drying; freezing-thawing; viscosity, strain-rate and stress-rate effects; rainfall and consequent reduction of soil suctions; heating-cooling; swelling in expansive clays; and changes in pore fluid chemistry.

Reductions in strength related to these processes will be considered as 'degradation'. Some involve physical disturbance of the macrostructure or microstructure of the clay, while others involve electro-chemical changes in 'bound' water attached to the clay particles. The processes can alter the constitutive behaviour of clays in the following ways:

R. Wan et al. (Eds.): Bifurcations, Instabilities and Degradations in Geomaterials, SSGG, pp. 187–210.
springerlink.com

- wetting and weakening in swelling clays,
- frost-induced fracturing and weakening in soft clays and weak rocks,
- changes in microstructure, with resulting changes in hydraulic conductivity,
- rainfall-induced reductions in suction and therefore in shear strength,
- reduction of the region of elastic behaviour due to leaching of cementation,
- consolidation, strengthening and sometimes weakening caused by heating, and
- changes in compressibility and strength due to creep (viscoplasticity).

In terms of impact on engineered structures, the processes can cause:

- damage to buildings and earth-retaining structures on swelling clays,
- damage to highway pavements, bases and sub-bases through the development of ice lenses and frost boils, with subsequent settlements and softening when the ice melts,
- reduced protection provided by frost-damaged riprap,
- instabilities in natural hillsides, riverbanks, and highway cuts resulting from increases in pore water pressures or decreases in soil suctions caused by sustained and/or heavy rainfall,
- failure of water retention dykes through seepage of reservoir water and leaching of cementation,
- thaw settlements and other movements of cold storage facilities, industrial, furnaces, nuclear waste disposal containers, etc, and
- large, ongoing settlements of embankment fills on soft clays and organic silts.

The following paragraphs outline a number of the authors' field and laboratory projects in which strengths have degraded considerably as a result of natural physical processes. We have chosen to emphasize the frequency of occurrence of these processes and the impact they have on material properties. We have not dealt here with the associated mathematics of analysis or design. Mathematical treatment of the various processes can be found in related references given in the text.

2 Freezing-Thawing and Wetting-Drying

Figure 1 shows results from carefully sampled specimens of plastic (CH) proglacial Lake Agassiz clay from Winnipeg, Canada. The liquid limit was 75% – 80%, liquidity index about 0.6, and undrained shear strength 50 – 60 kPa. The clay contains montmorillonite and gypsum. The montmor-illonite permits a pressure-swelling relationship when the clay has access to water, while gypsum cementation means it can strain-soften on shearing.

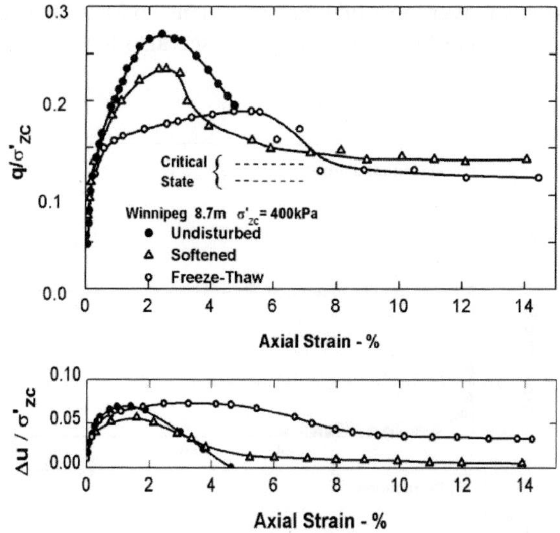

Fig. 1. Degradation of peak undrained shear strength in plastic clay with wetting (softening) and freezing-thawing. Lake Agassiz clay from Winnipeg. (Graham and Au, 1984)

Some of the specimens were tested in drained (constant-p') and undrained triaxial compression tests (Graham and Au 1985). Other specimens were subjected to five freeze-thaw cycles before shearing. Freezing was at -5ºC or -25ºC, with each cycle lasting up to 4 days. Compressive strains of the order +3% to +6% were observed. The original 'intact' structure became 'nuggety', with clearly-developed small fissures forming a macrostructure of about 5 mm size. A third group was permitted to swell freely under low confining pressures. Average volume and height expansions during this period were -6% and -3% respectively.

The structural disturbance caused by wetting-induced swelling in Fig.1 reduced the shear strength in this gypsum-cemented clay by about 15%, while freezing-thawing reduced the strength by about 30%. 'Large strain' (Critical State) strengths are independent of the pre-existing microstructure and were essentially unaffected.

A second series of tests (Fig.2) showed significant freezing-generated compressions and increases in hydraulic conductivity. Mercury intrusion porosimetry on these specimens showed that the clay microstructure changed from a single peak at about 0.02 μm to a bimodal distribution with peaks at about 0.06 and 60 μm respectively (Yuen et al. 1998).

Comment: These results show that seasonal wetting-drying and freezing-thawing can produce significant changes in inter-particle structures of clays. These changes will alter strength, stiffness, brittleness, pore water pressure generation and hydraulic conductivity. While laboratory tests will often measure 'peak' strength envelopes in overconsolidated specimens, particularly in plastic clays, these should generally not be used in design – see also Rivard and Lu (1978) and Schofield (2005).

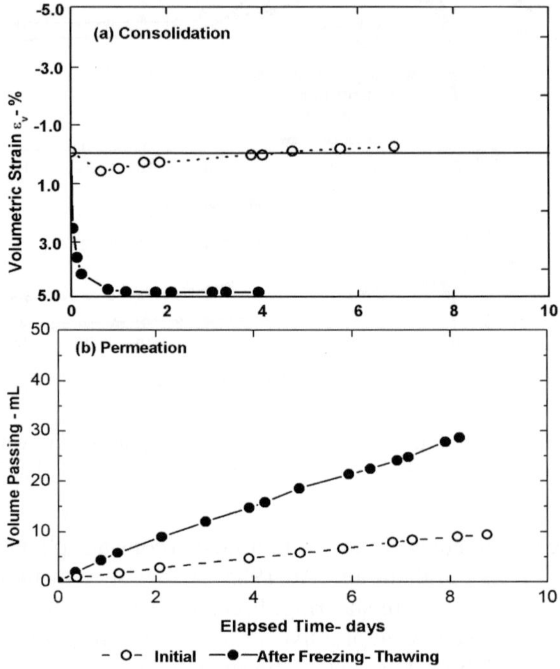

Fig. 2. Consolidation and permeation results after 1-D open drainage freezing thawing (Yuen et al., 1998)

3 Frost-Damage to Rockfill

Figure 3 shows another example of material degradation caused by freezing. In this case, limestone was used for riprap protection on 24 km of water retention dikes at a hydroelectric generating station in N. Manitoba. The owners, Manitoba Hydro, used engineering judgment to select cheaper limestone with higher maintenance costs in preference to superior granite with higher initial transportation costs. The rock consisted of variable argillaceous (clayey or shaley) limestone from local quarries.

Riprap is required to have a 'particle size' (D_{50}) that will withstand wave action from a 'significant' storm event with a 1:100 year return time (ASTM 1994). Here, D_{50} is the particle size where 50% of the material by mass is smaller in size. Evaluating D_{50} is not easy when rock pieces are large and sieve analysis is not possible. The approach recommended by ASTM is to measure an 'equivalent' particle size and then convert this to mass using a mass-volume relationship obtained from representative samples. Size measurements were taken at 80 sites, each 3 m square, along the dikes during three summers between 1998 and 2002. For safety reasons, measurements had to be made on site without damaging the integrity of the dikes. In all, 3324 pieces of riprap were measured and weighed.

Fig. 3. Limestone riprap showing degradation caused by natural freezing-thawing (Graham et al. 2007)

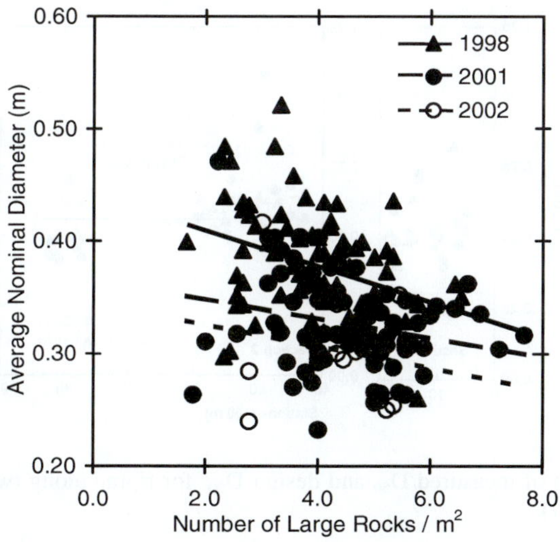

Fig. 4. Average nominal diameter versus number of large rocks (Graham et al. 2007)

Figure 4 compares average nominal diameters and the numbers of 'large rocks' (here taken as larger than 0.2 m) on each of the sampling areas. Clearly the size of the available riprap was degrading with time over the five years of the study period (Graham et al. 2007).

The question then arises whether degradation of the riprap was putting protection of the dikes at unacceptable risk of failure during storms. It will be appreciated that the 'demand' D_{50} varies with factors such as the design storm, the fetch distance, the water depth, and the orientation of the section relative to the prevailing wind direction. The 'capacity' D_{50} varies with the quality of the riprap and with time, if freezing-thawing causes degradation.

Figure 5 shows results from two different sections of dike. The fetch distance is short in Section 1 and the available capacity is comfortably greater than the demand (design) D_{50}. At Section 2, demand exceeds capacity and the deficit appears to be getting worse with time. The owner has been proactive in providing improvements to the riprap on these dikes and no breaching has occurred.

Comment. Frost action is again seen to be an important natural process in degrading the performance of geomaterials. Tests showed that while the limestone was not significantly affected by wetting-drying from rainfall or blowing spray from waves, it was adversely affected by freezing-thawing cycles. Standard ASTM freezing-thawing tests are lengthy and not suitable for on-site assessment of rock quality. Good correlations were found between results from standard ASTM tests and Iowa Pore Index Tests (Iowa Dept. of Transportation 1980), which were originally developed for frost-susceptible concrete aggregates (Graham et al. 2007).

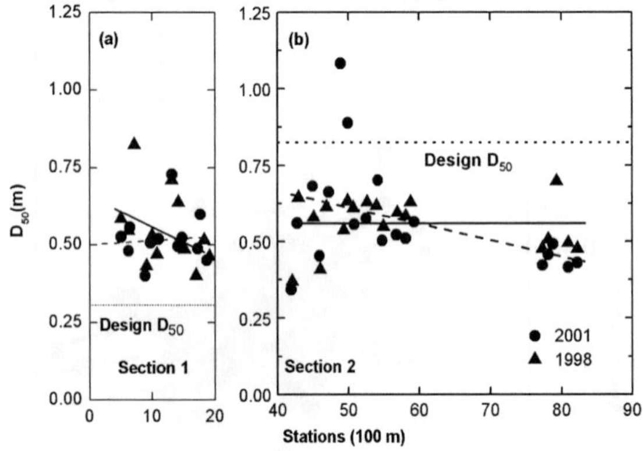

Fig. 5. Comparison of measured D_{50} and design D_{50} for riprap along two sections of dyke (Graham et al. 2007)

4 Rainfall, Degrading Soil Suctions, and Triggering of Slope Failure

In dry climates, slopes may be stable at inclinations steeper than would be expected if the analysis used strength parameters from saturated samples. In unsaturated soils, suctions, that is, pore water pressures below atmospheric pressure, provide increased forces between soil particles, higher strengths and increased stability. Unfortunately, if the degree of saturation increases through infiltration after rainfall, suctions will decrease, strengths will degrade, and the slope may become unstable.

Infiltration is the most common process leading to decreases in suction and degradation of strength in unsaturated soils. Due to the relationship between hydraulic conductivity and suction, infiltration is faster in 'moist' soils than in 'dry' soils. As a result, failures are more likely if heavy rainfall comes after a period of moderate rainfall and less likely if it follows a dry period (Blatz et al. 2004).

Figure 6 shows failure of a cut slope in a colluvial deposit on a provincial highway in W. Manitoba. The slope had been stable for about thirty years before becoming unstable following an unusually long period of rainfall in 1999.

Fig. 6. Failure in cut slope near Virden , Manitoba (Blatz et al. 2004)

Preliminary calculations showed that the slide mass would have been unstable without the additional strength that comes from the suctions in unsaturated soils. The difficulty with confirming that dissipation of suctions triggered the failures lies in assessing the *in situ* suction conditions at the time of failure and the time-dependent mechanism of infiltration of rainfall.

The site was assessed using boreholes and test pits (Fig.7) during a relatively dry summer in 2000. Matric suctions were measured using a tensiometer in the side walls of two test pits at intervals of 0.5 m below ground surface until a zero reading of suction was obtained. Suctions decreased from about 75 kPa at

0.5 – 1.0 m depth to 0 kPa at 2.0 – 2.5 m depth. A distinct shift in the soil suction profile was noted in both pits at a depth of 1.0 – 1.5 m, indicating a pre-existing suction gradient that was probably due to rainfall events in preceding months.

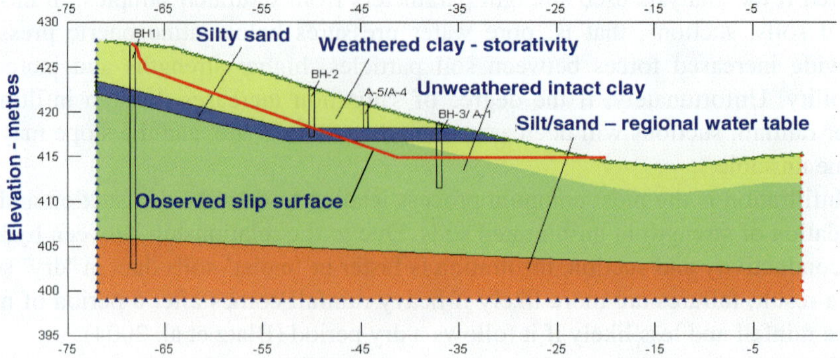

Fig. 7. Slope profile and slide surface. (Blatz et al. 2004)

Details of the choices made in the modeling program have been given by Blatz et al. (2004). They can be summarized as follows. Modeling the unsaturated soil component of suctions and flow in the upper weathered zone required soil-water characteristic curves (SWCC) and hydraulic conductivity functions. Based on measured grain-size distributions, the SWCCs for the soil types in the study area were determined using the modified Kovács method (Aubertin et al. 1998). The associated hydraulic conductivity functions were estimated using the van Genuchten method (1980). To account for the presence of secondary structures caused by weathering, the calculated SWCC and hydraulic conductivity function were calibrated against the measured suctions. The hydraulic conductivity used for the lower unweathered clay layer was taken as the average measured value from six flexible-walled permeameter tests without further adjustments, namely 1.8×10^{-5} m/day.

Following calibration, local rainfall records were converted into environmental flux boundary conditions. Figure 8 shows the variation of boundary flux and suction with time at a typical element near the toe of the slope starting, from 01 April 1999. Spring melt and a wet spring (low suctions) are followed by a dry summer (higher suctions) and a return to wetter conditions (lower suctions) in the fall.

Upon completion of the transient seepage analyses for 1999 and 2000, the corresponding stability of the slope was examined using the Morgenstern–Price method with a constant interslice force function. Figure 9 shows the variation of the safety factor (FS) with time for the years 1999 and 2000.

The calculated factors of safety remain significantly above 1.0 during the dry summer of 2000, with a minimum factor of safety of 1.72. Variations in the factor of safety correspond to changes in the seasonal wetness but were relatively small, in the range 1.7–2.2.

In contrast, the modeled factors of safety for the wet year in 1999, when the slope failed, changed considerably in short periods of only a few days. For the most part, significant fluctuations in the factor of safety occurred in May and June (days 30–60), corresponding to the extended periods of rainfall represented by the environmental flux boundary function in Fig. 8. Figure 9 shows several occasions during days 30–60 in 1999 when the calculated factors of safety were only marginally above unity (FS = 1.04). These factors of safety suggest that slope movements would have occurred. The effect of long-duration, low-intensity events is reflected in decreases in the modeled factor of safety from higher values (around 2.0) to values just above unity.

Comment: In slopes where a potential failure is relatively shallow, stability can be degraded by rainfall and infiltration. Groundwater pressures, including suctions, can rarely be measured at exactly the time landslides occur. Detailed modeling with time is therefore required to calibrate flux conditions at the boundaries and calibrate them against field measurements. Once this has been done, it is possible to simulate pore water pressure distributions and stress-deformation conditions in the slope. The challenge in this work is the selection of material properties and boundary conditions, and not the development of new analytical tools. Suitable algorithms and programs already exist.

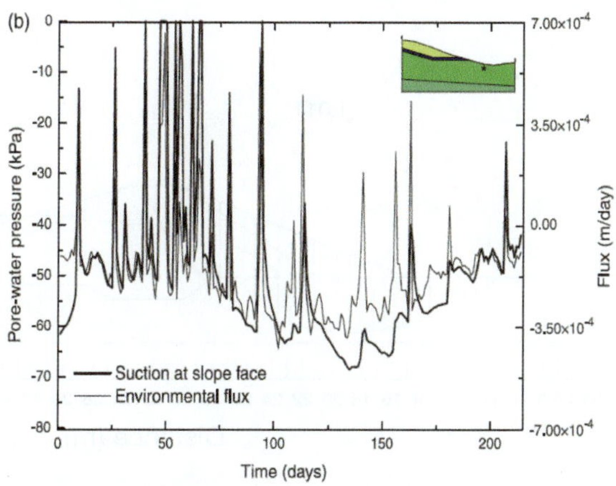

Fig. 8. Suction and surface environmental flux with time, 1999 (Blatz et al. 2004)

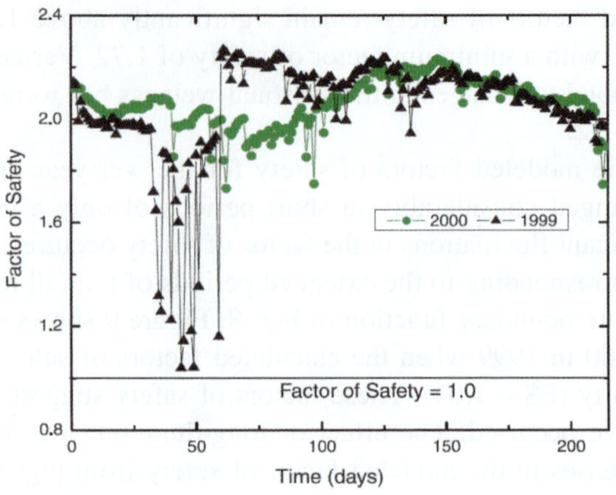

Fig. 9. Variation of safety factor with time, years 1999, 2000 (Blatz et al. 2004)

5 Leaching of Cementation

Garinger et al. (2004) examined a problem of intermittent instabilities in water retention dikes at Seven Sisters Generating Station in S.E. Manitoba. The dikes at Seven Sisters are relatively low (Fig.10) but extensive, with a total length of about 12.8 km and a crest width of approximately 4.3 m. They were initially constructed in 1929 on soft, highly plastic clay with some challenging properties. They performed satisfactorily until they were heightened and lengthened in the late 1940s to gain additional head and capacity.

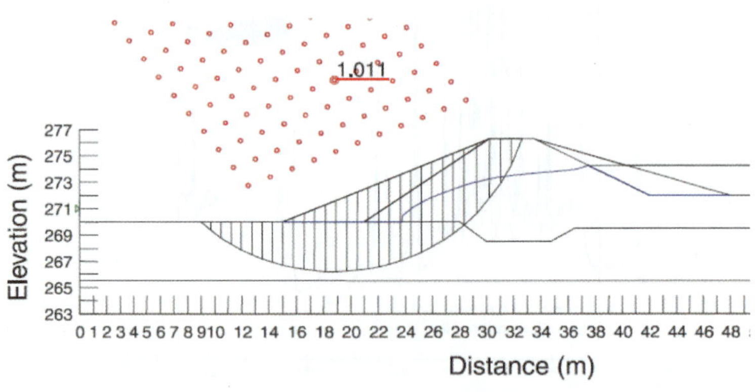

Fig. 10. Cross-section of Seven Sisters dike, SE Manitoba (Garinger et al. 2004)

After they were heightened, 13 separate instabilities occurred intermittently on the dry side of the dykes. The instabilities occurred at apparently random locations and times, despite generally similar soil conditions and geometries. Engineers who worked on the instabilities, including Casagrande and Bjerrum, did not examine the question raised by Garinger et al. (2004), namely the potential effects of changes in pore fluid chemistry and cementation on the stress-strain behaviour of the foundation clay.

The local clay (also from Lake Agassiz) contains nodules and streaks of gypsum (calcium sulphate) that provides cementation between the clay particles and additional strength. Gypsum is strongly soluble in water.

Figure 11 shows concentrations of calcium cations in pore fluid taken from the foundation clay below a stable section of dike, an unstable section, and a location outside the dikes. Clearly, there has been significant reduction of calcium during the lifetime of the dikes. Similar results were obtained for sulphate anions. Little change was observed for other cations and anions (Garinger et al. 2004). The results suggest that seepage produced by the head difference across the dikes dissolved gypsum cementation in the foundation clay. Consequent weakening of the clay led to the observed instabilities. The irregularity of the instabilities can be explained by differences in advection and diffusion resulting from localized sand partings and seams that were produced in the foundation clay by flood events during deposition (Man et al. 2006).

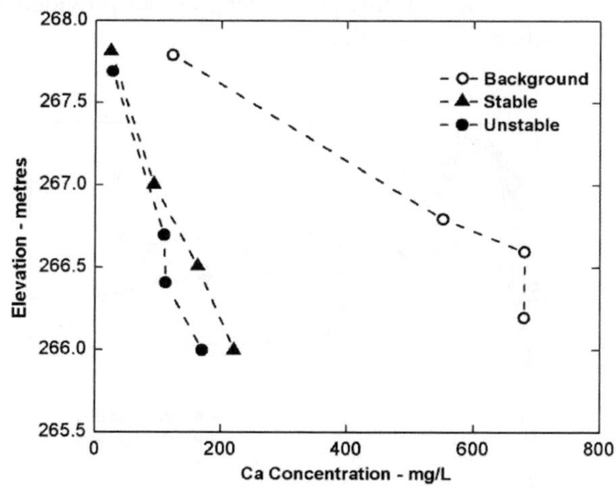

Fig. 11. Calcium concentrations in the foundation clay below the dikes (Garinger et al. 2004)

The question of how pore fluid chemistry affects stress-deformation performance was examined using specimens reconstituted with different pore water chemistries (Man and Graham 2008). Figure 12 shows drained triaxial compression results from specimens that were a) washed with 6 pore volumes of distilled water to remove gypsum cementation; b) gypsum rich; c) enriched with sodium chloride to alter the monovalent/bivalent ratio; and d) buffered with sulphuric acid to reduce possible cementation by carbonates. The following conclusions can be drawn. The washed specimen with depleted gypsum was much weaker than the others that retained gypsum cementation. The strength of the gypsum-rich natural clay can be degraded by seepage beneath the dikes. Changes in chemistry affect brittleness and the deformations at which strain-softening takes place.

Additional drained tests were loaded incrementally along a variety of stress paths with constant values of $\Delta q/\Delta p'$ to examine yielding behaviour. Figure 13 shows yield loci for three different levels of gypsum concentration ranging from 0 to about 2500 mg/litre. The region of stiff, largely-recoverable behaviour (inside the state boundary surface) decreases as gypsum is removed.

Comment. These figures show that strength, brittleness, strain-softening, and compressibility are all affected by natural leaching processes that remove gypsum from the foundation soil. When this understanding is fed into stability analyses of the Seven Sisters dikes, it becomes possible to separate stable and unstable sections of dike. This research supports the conclusion of Rivard and Lu (1978), who argued that peak strengths are unreliable and should not be used for design. Their assessment, mentioned earlier in this paper, incorporated analysis for the Seven Sisters site using both peak and post-peak strengths but provided no explanation for the need to use only post-peak values.

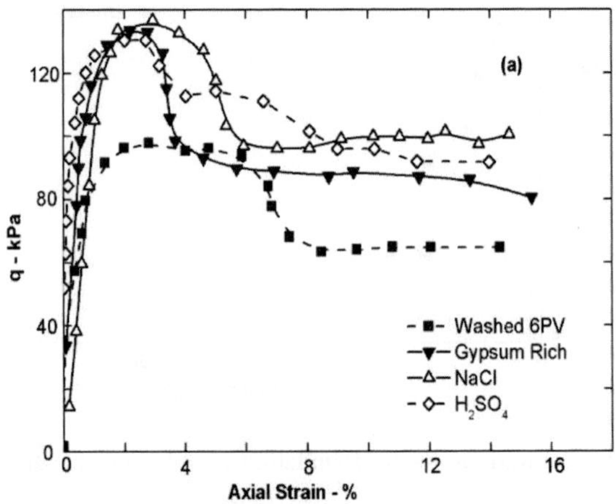

Fig. 12. Drained triaxial compression test results on specimens with different pore fluid chemistries (Man and Graham 2008)

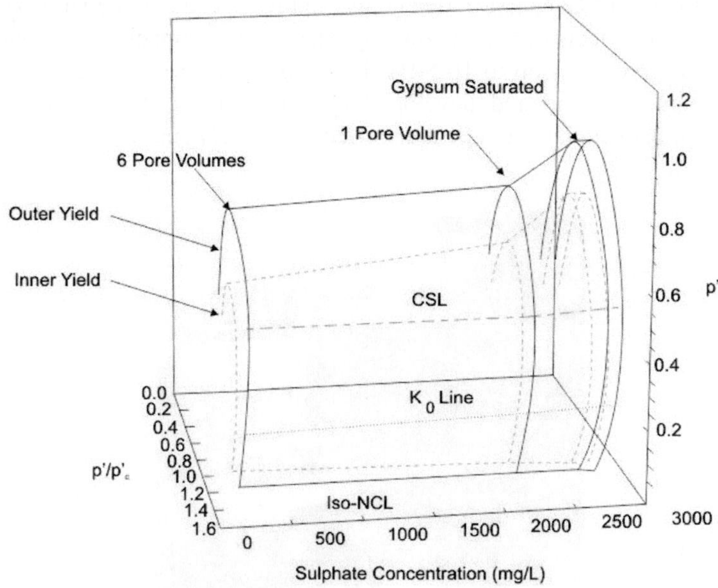

Fig. 13. Yielding in q/p'_c versus p'/p'_c –space as a function of sulphate concentration (mg/litre) (Man and Graham 2008)

6 Heating-Cooling; Changes in Temperature

While there has been for many years an interest in temperature redistributions produced by heat transfer in soil, especially in colder climates, little work has been done on constitutive modeling of related stress-strain-temperature effects. Applications include the growth of frozen regions under cold stores and sports arenas, 'heat syphons' for stabilizing roads and pipelines on permafrost, foundations for transmission towers, and ground freezing for constructing excavations and tunnels. Alternatively, heating may cause thermal degradation of permafrost and thermal consolidation in unfrozen ground. We are currently working on projects that involve degraded (thawed) permafrost in N. Manitoba.

International interest in the possible disposal of nuclear fuel waste in deep underground vaults has provided an important stimulus to research on the thermomechanical effects of changes in temperature in soils. Spent nuclear fuel is radioactive, highly toxic and an ongoing heat source. A common approach to disposing of radioactive waste is to surround canisters of waste with compacted mixtures of bentonite and sand, known commonly as 'buffer' or 'BSB". The buffer supports the canister, transfers heat into the surrounding rock, and provides a hydraulic barrier to possible leakage of radionuclides to the biosphere. Because of heat loading from the fuel waste, it is important to understand how heating will affect the strength and compressibility of bentonite-rich buffers.

Atomic Energy of Canada Limited contracted researchers at the Univ. of Manitoba to study the effects of heating on clay behaviour at temperatures from +25°C to +100°C and pressures up to 10 MPa. The research (Fig.14) studied compacted illite and bentonite-sand and developed a semi-empirical thermoelastic-plastic (TEP) model that simulated temperature effects relatively successfully.

Fig. 14. High-temperature, high-pressure triaxial equipment at the University of Manitoba

Descriptions of the testing program and development of the TEP model were given by Tanaka et al. (1995) and Graham et al. (2001). The TEP model is an extension of Modified Cam Clay and is similar in some ways to the more mathematical approach proposed by Hueckel and Baldi (1990). It assumes:

- the Critical State Line (CSL) is unaffected by temperature in the q,p'-plane,
- in the *ln*(p'),V compression plane, CSLs, have constant slope λ but different values of specific volume V,
- normal consolidation lines (NCLs), also have slope λ and vary with V,
- yield loci are elliptical and the flow rule is Associated,
- the slopes of κ-lines vary with temperature.

This latter assumption implies thermoelasticity. In view of some recent work on elastic-viscoplasticity and our research on highways on degrading permafrost, we intend to improve the TEP model so that temperature effects are included with the plastic strains. That is, it will become an elastic-thermoplastic (ETP) model.

Some of the complexity of temperature effects can be seen in Figs.15 and 16. Figure 15 shows results from undrained triaxial compression tests on three specimens sheared at the same confining pressure but at different temperatures. In this case, the hotter specimen had the highest strength. In Figure 16, the diagram on the left shows measured volume strains during drained heating at constant effective stress. Normally consolidated specimens compress on heating, whereas overconsolidated specimens expand. The data are similar to those reported by Hueckel and Baldi (1990) and explain some of the confusing reports of volume changes on heating in the research literature. The diagram on the right in Fig.16 shows similar behaviour predicted by the TEP model.

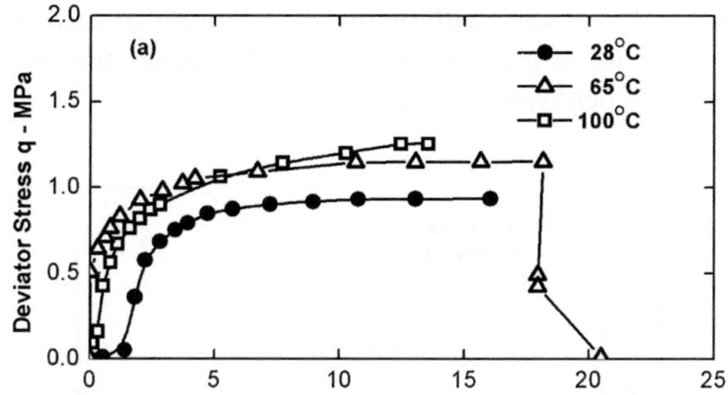

Fig. 15. Shearing at different temperatures produces different undrained shear strengths (Graham et al. 2001)

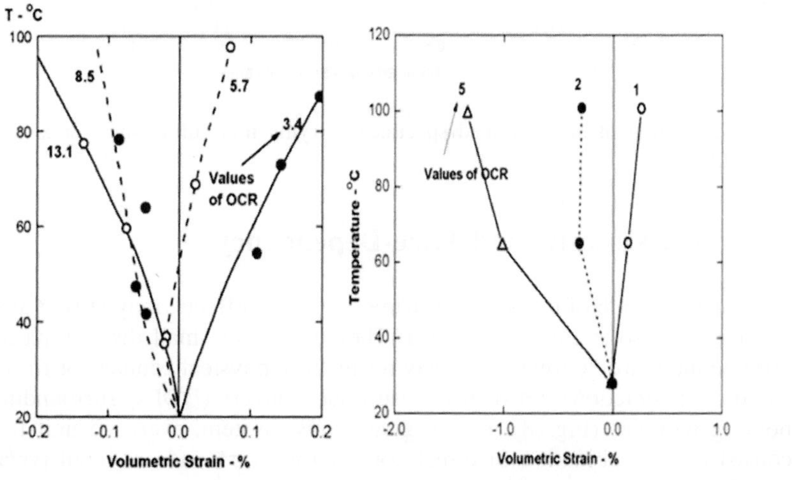

Fig. 16. Volume changes resulting from drained heating at constant pressure (Graham et al. 2001)

Perhaps more significant in the current discussion of degradation, is the case shown in Fig.17. Here the specimen was consolidated isotropically to just over 1.5 MPa at 26°C (Point 1). Axial loading was then applied until a (non-failing) deviator stress of 0.5 MPa was reached at Point 2. At that stage, no further shear stress was added but the specimen was heated, still undrained, to 100ºC. Pore water pressures increased and mean effective stresses decreased along the stress path 2-3. With unchanged external loading, the specimen that was previously stable at 26ºC has now become unstable at 100ºC. The TEP model simulates this behaviour along the line OAB. The conclusion here is that rapid heating of clay with low hydraulic conductivity can produce instability.

Comment. Most laboratory tests for compressibility, strength, and hydraulic conductivity are done at room temperatures, whereas in-ground temperatures are typically much lower and may vary with the nature of future construction and service. More attention needs to be given to the effects of temperature on laboratory, in-ground, and in-service behaviour of clays.

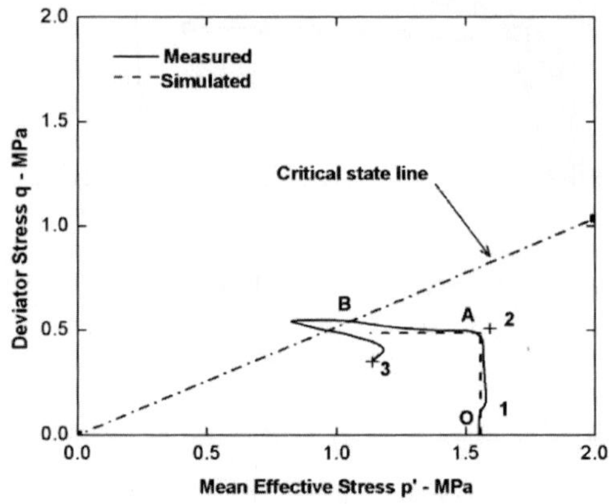

Fig. 17. Heating of an undrained specimen carrying non-failing shear stresses (Graham et al. 2001)

7 Creep, Viscosity and Time-Dependency

Our final example of degradation, namely viscous effects, may at first appear surprising. Clay soils are not continua. They consist of small discrete particles that provide interparticle forces that may result from physical contacts or from electrochemical interactions between diffuse double layers (DDLs) surrounding neighbouring particles (Fig.18). In any given stress system, averaged unit forces will consist of a combination of direct contact forces σ* and net repulsive/attractive unit forces $|R - A|$. Effective stresses σ', which control stress-deformation behaviour, can then be written

$$\{\sigma'\} = \{\sigma\} - u \times \{I\}$$
$$= \{\sigma^*\} + \{\,|R - A|\,\}$$

where {σ} is the total externally applied stress, u is the pore water pressure and {I} is the unit tensor.

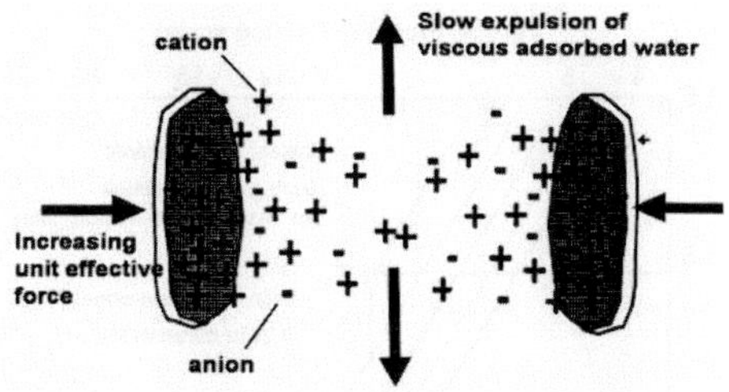

Fig. 18. Slow 'creep' movement of viscous adsorbed water from DDLs in active clays (Graham et al.1992)

In such systems, some inter-particle contacts carry larger forces than others. With time, there will be a general relaxation of high forces on to other contacts. Adsorbed water is more viscous than ordinary water and the 'easing' of highly stressed contacts produces time-dependent volume changes, even at constant (averaged) effective stresses. This creep degradation of localized highly stressed contacts is commonly known in 1-D compression as 'secondary compression'.

It is becoming more widely appreciated that viscosity can significantly influence the stress-strain behaviour of clay soils in all stress states (particularly in shear), and hence the behaviour of engineered structures on clay (Kelln et al. 2008a). Commonly, 'secondary compression' is assumed to begin only after 'primary consolidation' has finished (for example, Mesri and Castro 1987). More recently, researchers like Yin et al. (2002) and Kelln et al. (2008) have accepted that viscous effects are present at all stages of clay behaviour, including the period of primary consolidation. They affect not only compression, but also the dependency of strength on strain rate and loading rate. Viscosity effects can be easily described using the coefficient of secondary compression $C_\alpha = (e_1 - e_2)/\log_{10}(t_2/t_1)$, or as Yin and Kelln have written, $\psi = (e_1 - e_2)/\ln(t_2/t_1)$. An important feature of this work, as in Critical State Soil Mechanics, is the mapping of behaviour between stress planes and compression planes in terms of specific volumes V.

Perhaps the most familiar introduction to the effects of viscosity is Bjerrum's (1967) treatment of 'aging' (Fig.19). He idealized parallel lines of constant duration of loading for normal consolidation, that is, for first time loading. At a given stress level, the clay continues to compress after primary consolidation is complete. Subsequent increases in loading identify apparent preconsolidation pressures, even though the clay has never been loaded beyond this level.

Viscous effects are seen in all stress-strain behaviour and not only in 1-D compression. For example, Fig.20 shows viscous effects in undrained triaxial compression. Here, increasing the straining rate by two orders of magnitude increases the undrained shear strength, decreases pore water pressure generation, but leaves the Critical State Line (CSL) unchanged.

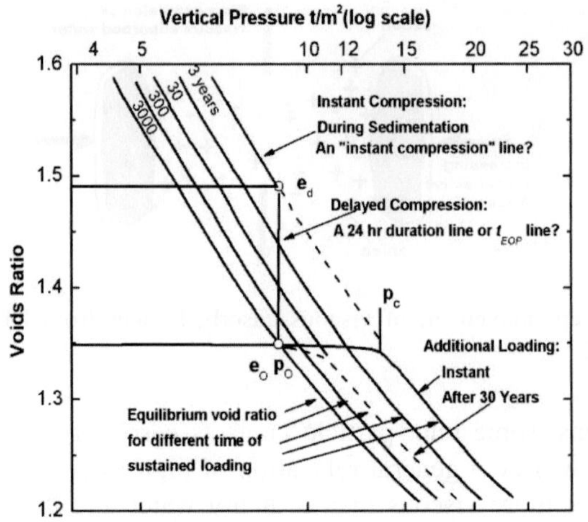

Fig. 19. Delayed compression (creep), apparent overconsolidation,and time dependency (Bjerrum 1967)

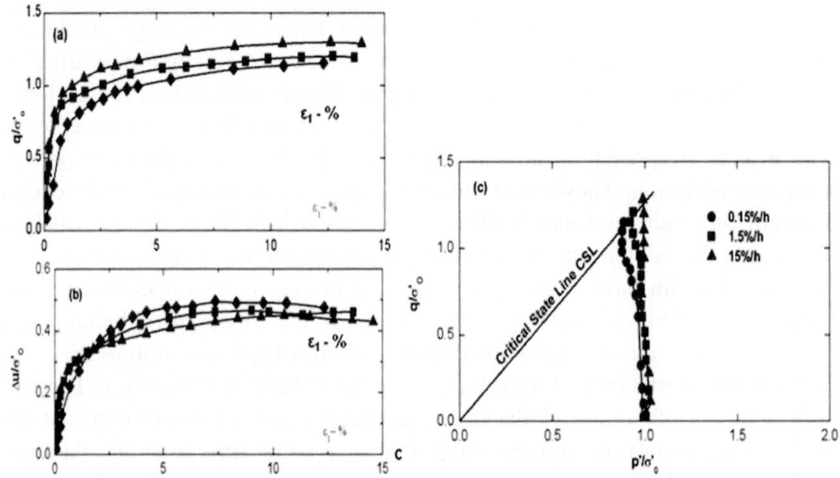

Fig. 20. Variation of shearing resistance, pore water pressure and effective stress paths with strain rate in triaxial compression (Yin et al. 2002)

Modeling clay behaviour and interpreting field behaviour without taking account of viscosity must now be considered oversimplification. One of the difficulties of incorporating viscoplasticity is the handling of the transition from an initial set of stresses where creep rates are low, to a second set of stresses where creep rates will be much higher. The first set may correspond, for example, to *in situ* stresses, while the second set might result from construction loading. This has been handled in several ways. Perzyna (1963), in the 'overstress plasticity' model relates viscoplastic strain rate to the extent by which an elastic prediction of the effective stress lies outside the current static yield locus. This approach appears kinematically inadmissible.

Starting from earlier work by Yin et al. (2002) based on Critical State Soil Mechanics, Kelln et al. (2008a) have clarified this issue and produced the elastic viscoplastic (EVP) model shown in Fig. 21. The model incorporates a viscoplastic limit line (vpl) and viscoplastic strains at all stages of straining. Naturally, when overconsolidation ratios are high, viscoplastic strain rates are low.

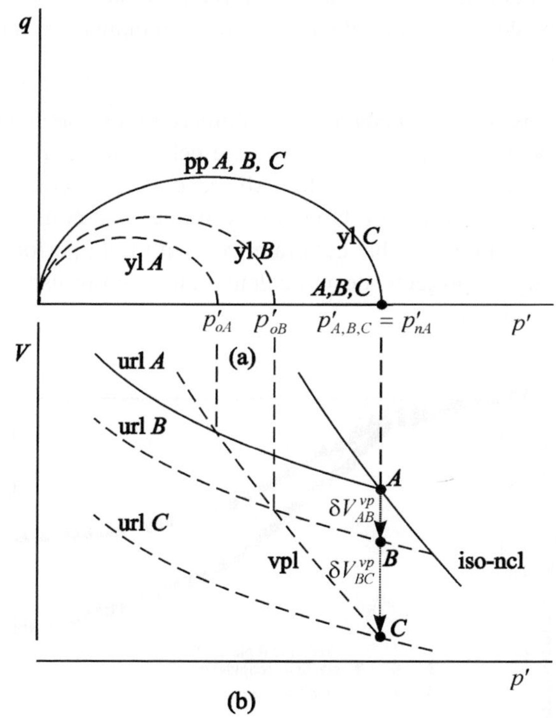

Fig. 21. An upgraded Modified Cam Clay model that takes account of elastic viscoplasticity (Kelln et al.2008)

Kelln et al. (2008b) provide a detailed description of implementation of the EVP model in a finite element program. Later, (Kelln et al. 2009) they used the program to simulate behaviour of a highway embankment over soft organic silty clay in N. Ireland. Settlements of up to 1.3 m were observed under the 4 m high embankment, which was supported on filter drains and a geotextile mat.

Figure 22 shows measured settlements (solid line) under the south shoulder of the embankment. It also shows settlements calculated using the time-independent Modified Cam Clay (MCC) model and the new time-dependent elastic-viscoplastic (EVP) model. The material parameters used in both simulations were the same, except that a creep coefficient was added for the EVP modeling.

With the organic content at this site, it is clear that viscoplastic behaviour would be present. This is confirmed by the good agreement between the observed settlements and those simulated by the EVP model. The MCC model underestimates the settlements.

Simulating lateral displacements is usually difficult – most finite element solutions underestimate lateral movements and show them extending too far away from the embankment. In Fig. 23, the EVP model produces a better simulation of inclinometer displacements under the toe of the embankment compared with those obtained from MCC.

Comment. All stress-strain behaviours – compressions, preconsolidation pressures, strengths, yield loci, pore water pressure behaviour, and volume changes – vary with strain rate or loading rate. All have components that depend on viscosity. This implies two things. One, it is fortuitous that common testing procedures produce soil parameters that calibrate well with field performance. Two, especially in challenging projects, design calculations need to use newly developed EVP models for analysis.

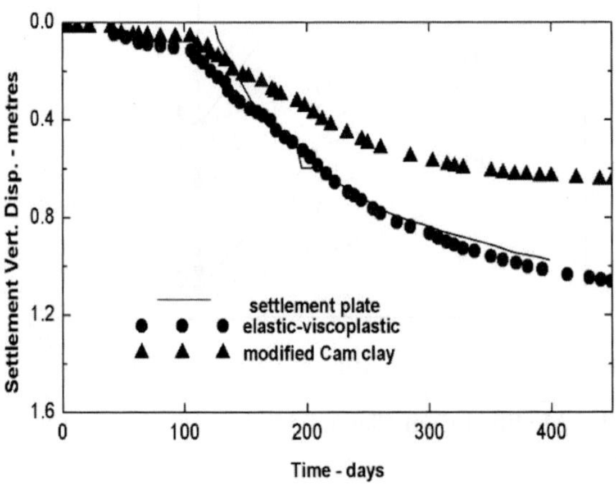

Fig. 22. Measured and simulated settlements under shoulder of embankment (Kelln et al. 2009)

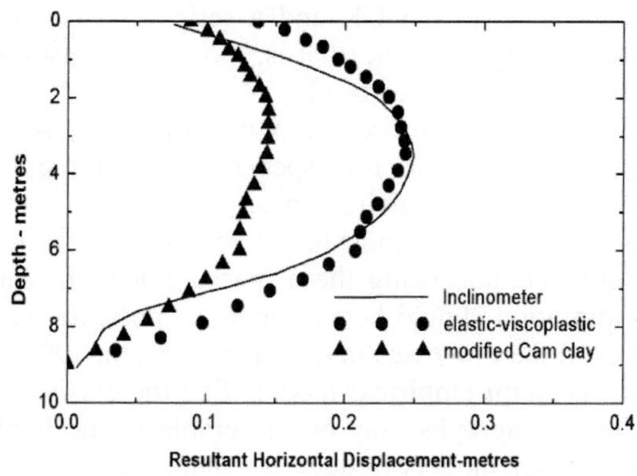

Fig. 23. Computed and measured inclinometer profile at Day 218 under South toe of the embankment, (Kelln et al. 2009)

8 Concluding Remarks

We have chosen to outline a series of projects that reflect the high frequency of naturally-occurring and engineered processes that reduce the strength and compressibility of (mostly) clay soils. Detailed information, including mathematical treatments where appropriate, can be found in the attached list of references.

One of the challenges in designing engineering projects is to take short-term or peak values measured in the laboratory and assess how they will degrade during the operating life of the project. Potential degradation should be an integral feature of the design process. It should not simply be assigned to 'uncertainty' and incorporated into 'safety factor' in limit equilibrium analysis, or 'capacity factors' in limit state design.

In some cases, for example the effects of viscosity, the process has been studied widely and good mathematical modeling is available. We have chosen not to recount this work here, but have left interested readers to recover it separately. In other cases – swelling, unsaturated soils, and soil chemistry effects for example – some work has been done in developing numerical modeling but further work is needed. In other cases such as fracturing caused by freezing-thawing, it appears that little has been done. These problems may simply be intractable to analysis, so that empirical procedures may be the best that can be expected. We note that good modeling exists for changes in temperature and the development and thawing of frozen soils.

One degradation issue remains surprisingly contentious – the question of whether peak strengths can be used for design in overconsolidated clays. It has been known in Europe for at least sixty years (Schofield 2005) and in Canada for at least thirty years (Rivard and Lu 1978) that earth structures on plastic fissured clays should not be designed using overconsolidated peak strengths. Field evidence from railway cuts in the UK and a series of small embankments and weirs in Canada shows that post-peak (that is, 'normally consolidated' or 'Critical State') strength parameters should be used.

Practical projects are often constrained as much by time as by cost. Provided that site conditions and loadings are not especially challenging, it should be possible to introduce good quality modeling in an acceptable time frame. The constraining principle here is the relationship between measuring material properties in laboratory or field tests and using these measurements to calibrate sound numerical models. Calibration should be as simple as possible while still capturing the essential features of the processes in question (Graham 2006).

This may mean that semi-empirical models like the elastic-viscoplastic model described in previous paragraphs may be preferable to mechanics- and physics-based models. These latter models frequently contain relatively large numbers of inter-dependent functionals that are difficult to identify successfully in the laboratory. Separating them depends on algorithms that may or may not be valid. Uniqueness becomes questionable.

An excellent example of this dilemma can be found in work on the bentonitic sealing materials that have been proposed for nuclear waste disposal in various countries. The bentonites will experience (1) physical loading from the weight of canisters of fuel waste, (2) temperature gradients from the hot waste to surrounding cooler rock, and (3) water pressure gradients between positive water potentials in the surrounding ground and negative potentials (suctions) in the sealing materials. While the waste is hot, water will move down the temperature gradient towards the rock. Later, after cooling, water will be drawn back into the bentonite through a combination of swelling, suction, and inflow from the ground water. There may also be chemistry differences between ground water and the water in the bentonite.

Material characterization for this application falls into the category of HTM (hygro-thermomechanical) modeling. We have worked on both physics-based soil models (Zhou et al. 1998) and semi-empirical soil models that were mostly based on developments of volumetric compressible elasto-plasticity (Graham et al. 2001, Blatz and Graham, 2003). Mechanics-based modeling provides better insights into the physics of what is happening but semi-empirical models are much easier to calibrate.

Acknowledgments. Funding has been provided by the Natural Sciences and Engineering Research Council of Canada, Manitoba Infrastructure and Transportation, Atomic Energy of Canada Limited, and Manitoba Hydro.

References

Aubertin, M., Ricard, J.-F., Chapuis, R.P.: A predictive model for the water retention curve: application to tailings from hard-rock mines. Can Geotech. J. 35, 55–69 (1998)

ASTM, Standard Test Method for Particle Size Analysis of Natural and Man-Made Riprap Materials. Standard ASTMD 5519 – 94, Am. Soc. Testing Mater, Philadelphia PA (1994)

Bjerrum, L.: Engineering geology of normally-consolidated marine clays as related to the settlements of buildings. Géotech. 17, 81–118 (1967)

Blatz, J.A., Graham, J.: Elastic-plastic modeling of unsaturated soil using results from a new triaxial test with controlled suction. Symposium in Print on Suction in Unsaturated Soils, Géotechnique 53, 113–122 (2003)

Blatz, J.A., Ferreira, N.J., Graham, J.: Effects of near-surface environmental conditions on instability of an unsaturated soil slope. Can Geotech. J. 41, 1111–1126 (2004)

Garinger, B., Alfaro, M., Graham, J., et al.: Instability of dykes at Seven Sisters Generating Station. Can Geotech. J. 41, 959–971 (2004)

Graham, J.: The 2003 R. M. Hardy Lecture: Soil parameters for numerical analysis in clay. Can Geotech. J. 43, 187–200 (2006)

Graham, J., Au, V.C.S.: Influence of freeze thaw and softening effects on stress-strain behaviour of natural plastic clay at low stresses. Can Geotech. J. 22, 69–78 (1985)

Graham, J., Oswell, J.M., Gray, M.N.: The effective stress concept in saturated sand-clay buffer. Can Geotech. J. 29, 1033–1043 (1992)

Graham, J., Tanaka, N., Crilly, T., et al.: Modified Cam-Clay modelling of temperature effects in clays. Can Geotech. J. 38, 608–621 (2001)

Graham, J., Franklin, K., Alfaro, M., et al.: Degradation of shaley limestone rip-rap. Can Geotech. J. 44, 1265–1272 (2007)

Hueckel, T., Baldi, G.: Thermoplasticity of saturated clays: experimental constitutive study. Am. Soc. Civ. Eng., J. Geotech. Eng. 116, 1778–1796 (1990)

Iowa Department of Transportation (IDOT), Iowa Pore Index Test. Interim Report, Iowa Dept Transport, Ames, Iowa (1980)

Kelln, K., Sharma, J., Hughes, D., et al.: An improved framework for an elastic-viscoplastic soil model. Can Geotech. J. 45, 1356–1376 (2008a)

Kelln, C., Sharma, J., Hughes, D.: A finite element solution scheme for an elastic–viscoplastic soil model. Comp. and Geotech. 35, 524–536 (2008b)

Kelln, C., Sharma, J., Hughes, D., et al.: Finite element analysis of an embankment on soft estuarine deposit using an elastic-viscoplastic soil model. Can Geotech. J. (2009) (in press)

Man, A., Graham, J., Van Gulck, J.: Effect of pore fluid chemistry on strain-softening behaviour of reconstituted plastic clay. In: Proc. 5th Int. Conf. Geoenvir. Eng., Cardiff, Wales (June 2006)

Man, A., Graham, J.: Pore fluid chemistry and the stress-strain behaviour of a reconstituted highly plastic clay. Geotechnique (2008) (in review)

Mesri, G., Castro, A.: The $C\alpha$/Cc concept and Ko during secondary compression. Amer. Soc. Civ. Eng., J. Geotech. Eng. 119, 230–247 (1987)

Perzyna, P.: The constitutive equations for rate sensitive plastic materials. Q. J. Appl. Math. 20, 321–332 (1963)

Rivard, P.J., Lu, Y.: Shear strength of soft fissured clay. Can Geotech. J. 15, 382–390 (1978)

Schofield, A.N.: Disturbed Soil Properties and Geotechnical Design. Thomas Telford Limited, London (2005)

Tanaka, N., Graham, J., Lingnau, B.E.: A thermal elastic plastic model based on Modified Cam Clay. In: 10th Pan. Am. Conf. Soil Mech. Found. Eng., Guadalajara Mex, vol. 1, pp. 534–546 (October 1995) van Genuchten, M.T.: A closed-form equation for predicting the hydraulic conductivity of unsaturated soils. J. Soil Sci. Soc. Am. 44, 892–898 (1980)

Yin, J.-H., Zhu, J.-G., Graham, J.: A new elastic viscoplastic model for time-dependent behaviour of normally and overconsolidated clays: theory and verification. Can Geotech. J. 39, 157–173 (2002)

Yuen, K., Graham, J., Janzen, P.: Weathering-induced fissuring and hydraulic conductivity in a natural plastic clay. Can Geotech. J. 35, 1101–1108 (1998)

Zhou, Y.: Rajapakse RKND and Graham J A coupled thermoporoelastic model with thermo-osmosis and thermal filtration. Int. J. Solids Struct. 35, 4659–4683 (1998)

Local Behavior of Pore Water Pressure During Plane-Strain Compression of Soft Rock

M. Iwata, A. Yashima, and K. Sawada

Gifu University, Gifu, Japan

Abstract. In order to understand the mechanical behavior of soft rock, many laboratory tests on sedimentary soft rock have been conducted in which the influence of the intermediate principal stress was not taken into account. However, in predicting the precise behavior of soft rock, the influence of intermediate principal stress always plays an important role. Therefore, in this study, a series of plane-strain compression tests on sedimentary soft rock were carried out. Fully saturated specimens were used for all tests. The plane-strain apparatus used in this research has a special feature in which the pore water pressure in the specimen can be measured. From the laboratory tests on soft rock, it was found that the pore water pressure distribution in the specimen was not uniform. In this research, the plane-strain compression tests were regarded as non-homogeneous model tests and the behavior of the local pore water pressure in the specimen was investigated.

1 Introduction

Soft rock ground is widely distributed throughout Japan and failures in soft rock slope occur frequently. Slope failures in soft rock are particularly prevalent in Gifu, Japan. For example, a large slope failure occurred in September, 1999 at a site located between Mino Interchange and Minami Interchange of Tokai-Hokuriku Expressway. The failed slope involved a cut slope and the failed area was 120 m in width and 125 m in length. Another slope case occurred in May, 2006 when the left bank of the Ibi River failed. The scale of the failed area was 150 m in width and 100 m in length. Both cases were large-scale slope failures, which did not cause heavy human casualties because a small failure was found before the occurrence of the larger failure.

In general, a numerical analysis is considered to be an effective tool to predict this kind of slope failure. However, in order to numerically predict a slope failure, it is necessary to understand the mechanical behavior of soft rock and to develop a constitutive model which can appropriately describe the mechanical behavior of soft rock. In the past several decades, many experimental researches on the triaxial testing of soft rock were conducted. Based on the results of a series of triaxial tests, a constitutive model for soft rock was proposed [1] and [2]. However, it is known

R. Wan et al. (Eds.): Bifurcations, Instabilities and Degradations in Geomaterials, SSGG, pp. 211–227.
springerlink.com

that the mechanical behavior of sedimentary soft rock is largely dependent on stress conditions. Plane-strain tests which can consider the intermediate principal stress should be conducted in order to understand the mechanical behavior of soft rock.

As in most slope failures, the main triggering mechanism is normally related to a change in groundwater. In the first slope failure case reported in the beginning of this introduction, a heavy rain was observed before the failure event. In addition, there was a possibility that drainage in the slope had been impaired and thus caused the groundwater to rise. In contrast, limited rainfall occurred in the second case, but a heavy snow fall had been observed in the preceding winter season. Therefore, failure was possibly induced by snowmelts infiltrating into the ground. At any rate, it was eventually considered that one of the triggers of the soft rock failure in both cases was the change in groundwater conditions.

Therefore, in order to understand the mechanical behavior of soft rock in detail, it is necessary to take into account not only the effect of the intermediate principal stress, but also that of pore water pressures in a lab testing experiment. In this research, a series of plane-strain compression tests were carried out in order to consider the influence of the intermediate principal stress. In addition, in order to verify the change of the pore water pressure inside the specimen, a tube which is connected to a piezometer was installed at the center of the specimen and pore water pressure within/near a shear band was measured during shearing.

2 Plane-Strain Test Apparatus and Test Specimen

2.1 Plane-Strain Test Apparatus

Some features of the plane-strain test apparatus for soft rock used in this research are herein explained. A detailed description of test system, such as the precision gap sensor for measuring the axial strain, can be found in the corresponding reference [3]. Figure 1 shows the schematic illustration of the hydraulic pressure cell.

In order to satisfy an isotropic consolidation before shearing and a plane-strain condition during shearing, several new treatments were made. The first feature of the plane-strain test apparatus used in this study is the device that keeps the specimen in plane-strain condition. In a classic plane-strain compression apparatus, the fixed out-of-plane confining platens are usually installed before the isotropic consolidation process, which makes it impossible to keep the consolidation to be isotropic because the specimen will contract during the consolidation process. In order to avoid this shortcoming, a new type of plane-strain platens was designed. Figure 2 shows the schematic illustration of the plane-strain constraint device. Two confining steel platens linked together by four rigid stainless rods form the plane-strain frame.

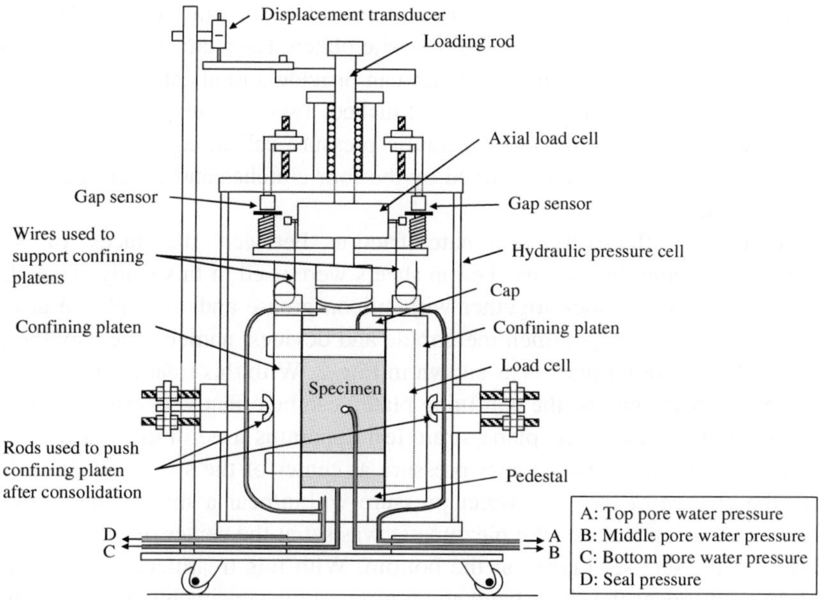

Fig. 1. Schematic illustration of plane-strain constraint device

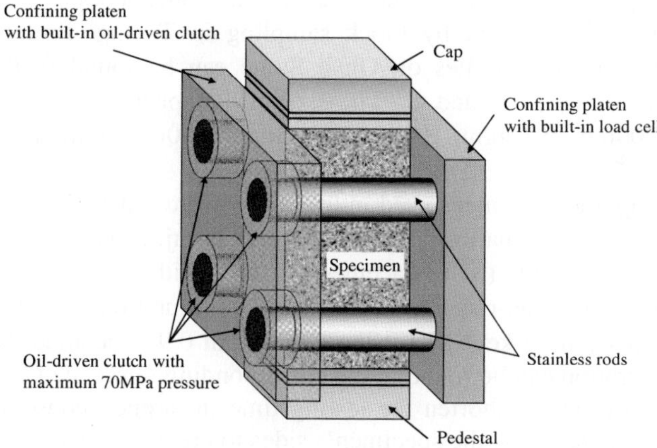

Fig. 2. Schematic illustration of plane-strain constraint device

One end of the rods is fixed onto one platen while the other end is clamped by an oil-driven clutch embedded in the opposite platen. Each clutch is actuated by an oil pressure of 70 MPa. All four clutches can provide a total catching force of 100 kN. After the consolidation process is finished, two confining steel platens can then be pushed from outside the hydraulic pressure cell to contact the specimen. The four rigid stainless rods are fixed by the four clutches and a rigid plane-strain frame is formed.

In order to effectively eliminate friction between the faces of specimen/membrane and the platens, Teflon sheets were used in this study. Two sheets of Teflon sheet were stuck together with silicon grease and were placed at every contact faces between specimen/membrane and devices, namely, the cap, the pedestal and the confining platens as shown in Fig. 2. With this treatment, the friction between the specimen and the confining platen can be effectively eliminated.

The second feature of the plane-strain test apparatus used in this study is a piezometer measuring the pore water pressure at center of the specimen. In order to verify the change of the pore water pressure within/near a shear band formed in post-peak shearing process, the piezometer was set at the center of the specimen in addition to the one set placed at the bottom. With this treatment, the pore water pressure within/near the shear band and bottom of the specimen can be measured.

2.2 Test Specimen

The rock sample used in this study was Ohya stone, a kind of sedimentary soft rock, which was mined by block sampling in Tochigi Prefecture, Japan. Numerous experimental studies on Ohya Stone can be found in the literature, such as the works by [4, 5 and 6]. The size of the specimen used in the plane-strain test is 200mm in height, 100mm in width and 80mm in thickness as shown in Fig. 3.

Before testing, the specimens used in this study were saturated according to the following steps; (1) evacuation in a desiccator, (2) injection of the sample with carbon dioxide gas, CO_2 (3) replacement of CO_2 with gas-free water, and (4) evacuation in the desiccator again. Using the above saturation process, the B value of the specimens can be kept to a value larger than 0.95. Detailed description of method for saturation can be found in the corresponding reference [7].

Moreover, in order to shorten the testing time in drained consolidation, filter paper strips were placed on the specimen's sides to create drainage paths as shown in Fig. 4.

In order to verify the change of the pore water pressure within/near the shear band, an inner tube, which is connected to a piezometer set outside cell, was placed at the center of the specimen, as shown in Fig. 5. The detailed process for preparing the inner piezometer in a cubic specimen can be found in the corresponding reference [7]. At first, a hole with a diameter of 2 mm was drilled at the center of the specimen; followed by a further drilling with cooling water down to a depth of 40 mm, and finally an expanding drill of the hole to enlarge the diameter from 2 mm to 3 mm with cooling water. The inner tube was inserted into the hole and then sealed with bonding and was stuck to the membrane, as shown in Fig. 5.

Fig. 3. Specimen of sedimentary soft rock used in plane-strain tests

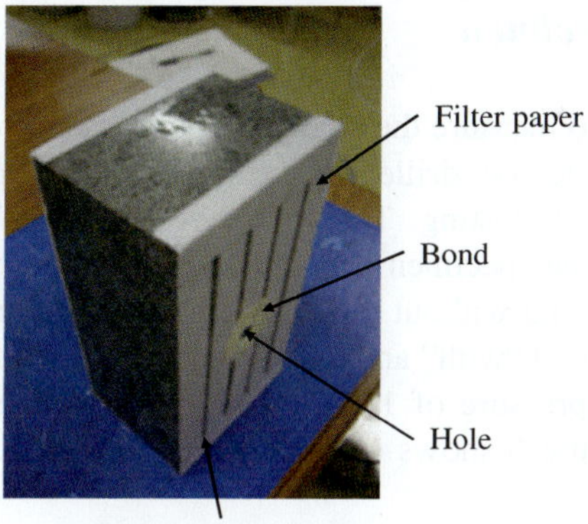

Fig. 4. Specimen covered with filter paper strips

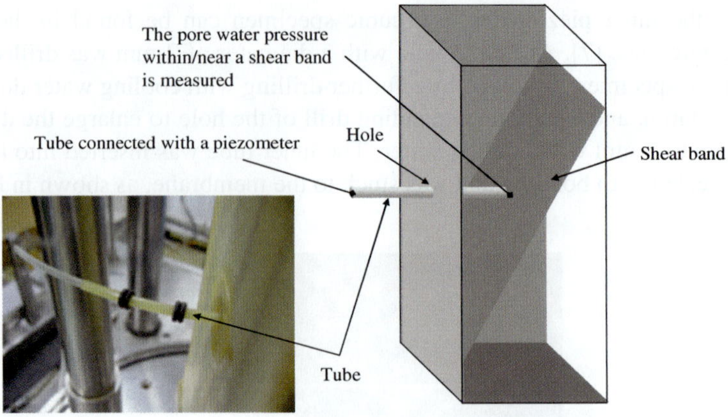

Fig. 5. Method of measuring pore water pressure inside specimen

3 Test Results

After being isotropically-consolidated for 24 hours under a prescribed confining pressure, the specimen was subjected to constant-strain-rate compression test. A constant back pressure of 0.5 MPa was maintained during all phases, i.e. from consolidation to shearing of all tests in this study.

3.1 Influence of a Hole for Measuring Pore Water Pressure Inside the Specimen

In this study, in order to measure the change in pore water pressure within/near the shear band, a small hole was drilled to the center of the specimen and an inner tube was inserted before testing. The influence of the hole on the mechanical strength properties of the specimen was investigated. Tests which used two types of specimen, one with and without a hole, were conducted under identical conditions. The former is called "with" and the latter is called "without". The test conditions were: confining pressure of 1.0 MPa, axial strain rate of 0.001 %/min and drained condition. Figure 6 shows comparison of the test results between "with" and "without".

The effect of the intermediate principal stress on the specimen's behavior is illustrated in Fig. 6(b). It was considered that the observed behavior depended on the form of the shear band, as shown in Fig. 7, because the difference of the intermediate principal stress appeared after the strength reached the peak. However, there was no significant difference in the mechanical strength properties of the specimen between both tests which involved the two types of specimens. Therefore, it was concluded that the method, used in this study, for measuring the pore water pressure inside the specimen had no effect on the mechanical behavior of the specimen.

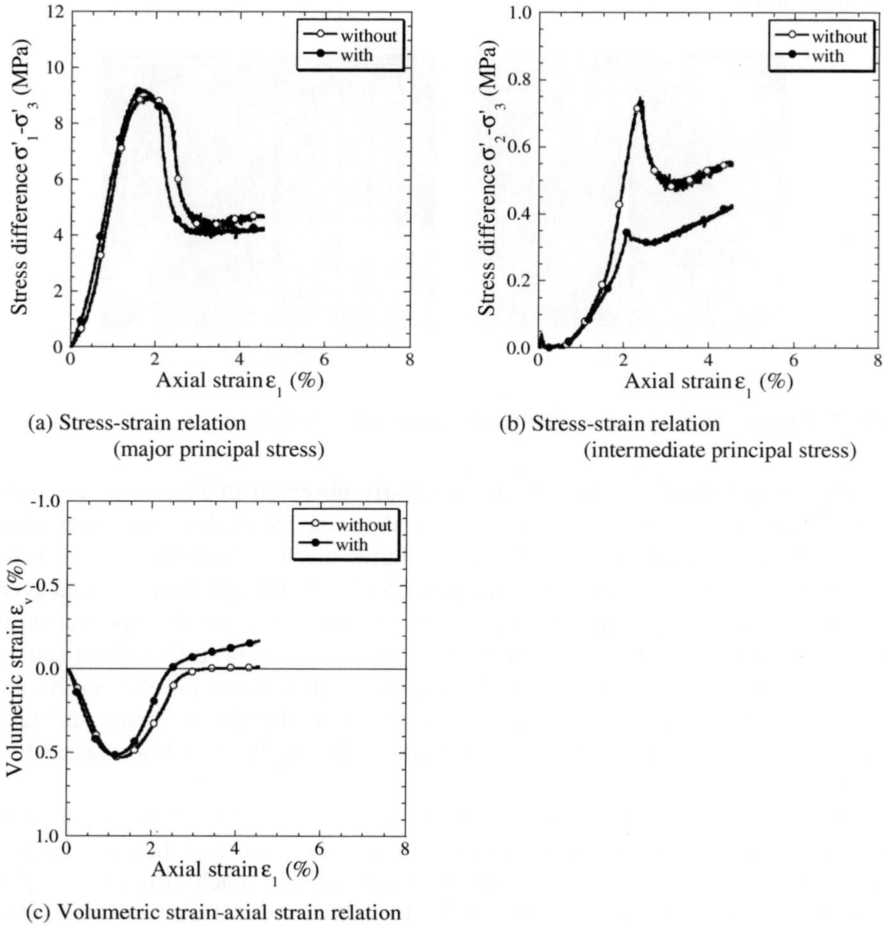

Fig. 6. Comparison of test result between with hole and without hole

3.2 Drained Plane-Strain Compression Tests

Plane-strain compression tests were carried out five times under the same test conditions. The test conditions were a confining pressure of 1.0 MPa, an axial strain rate of 0.001 %/min and drained conditions. Figure 8 shows the plane-strain compression test results.

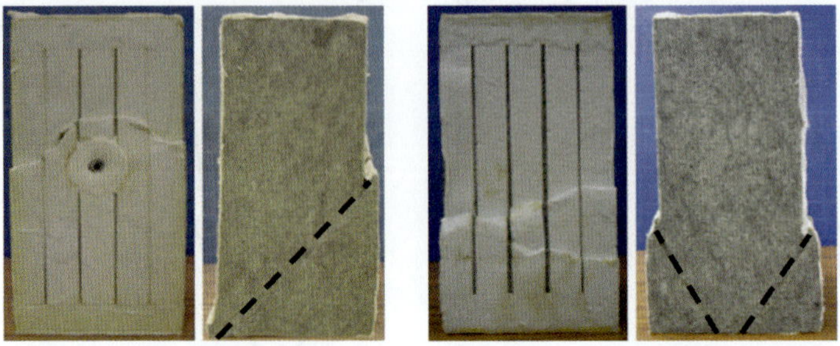

With Without

Fig. 7. Comparison of specimen after test between with hole and without hole

The strain softening behavior can be clearly observed in Fig. 8. It was confirmed that the residual stresses were almost the same for all three specimens (test 3, 4 and 5). The peak strength, however, shows much a difference between the five specimens. The excess pore water pressures inside the specimens existed even under drained condition and showed much difference for the five specimens. It was very clear that the plane-strain tests of soft rock in this study did not satisfy the condition of a drained test because the excess pore water pressure inside the specimen existed during shearing process, even in the slower strain rate test. Therefore, the plane-strain tests of soft rock in this study should be regarded as model tests.

Figure 9 shows the five specimens after the tests. A shear band and the point of the internal pore water pressure measurement are shown in each figure. As noted above, the negative excess pore water pressures showed much difference for the five specimens as shown in Fig. 8(d). From Figs. 8(d) and 9, it was considered that the results of the excess pore water pressures inside the specimens depended on the distance between the measure point and the shear band. For example, the excess pore water pressure of test 3, the measure point which was just within the shear band, was largest among the five specimens. The excess pore water pressure of test 5, where the measure point was outside the shear band, was the smallest one. However, the excess pore water pressures before the strength reached the peak, before shear band formation, were the same for four specimens.

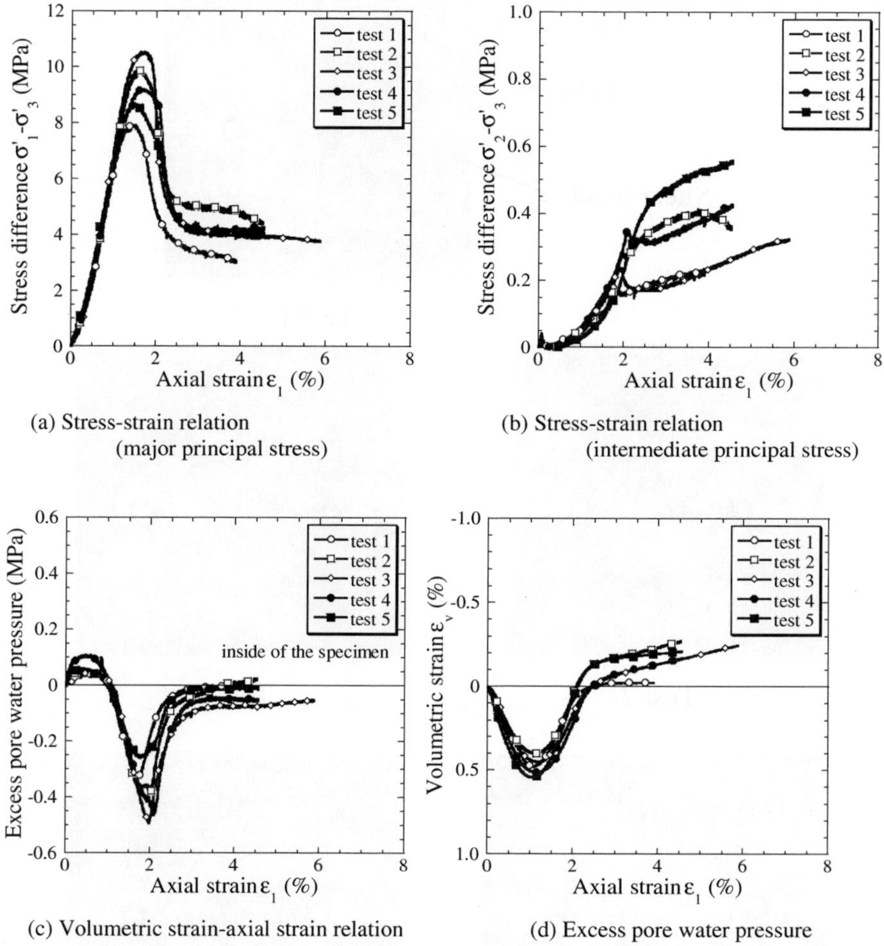

(a) Stress-strain relation (major principal stress)

(b) Stress-strain relation (intermediate principal stress)

(c) Volumetric strain-axial strain relation

(d) Excess pore water pressure

Fig. 8. Plane-strain compression test results (Confining stress=1.0MPa, axial strain rate=0.001%/min, drained condition)

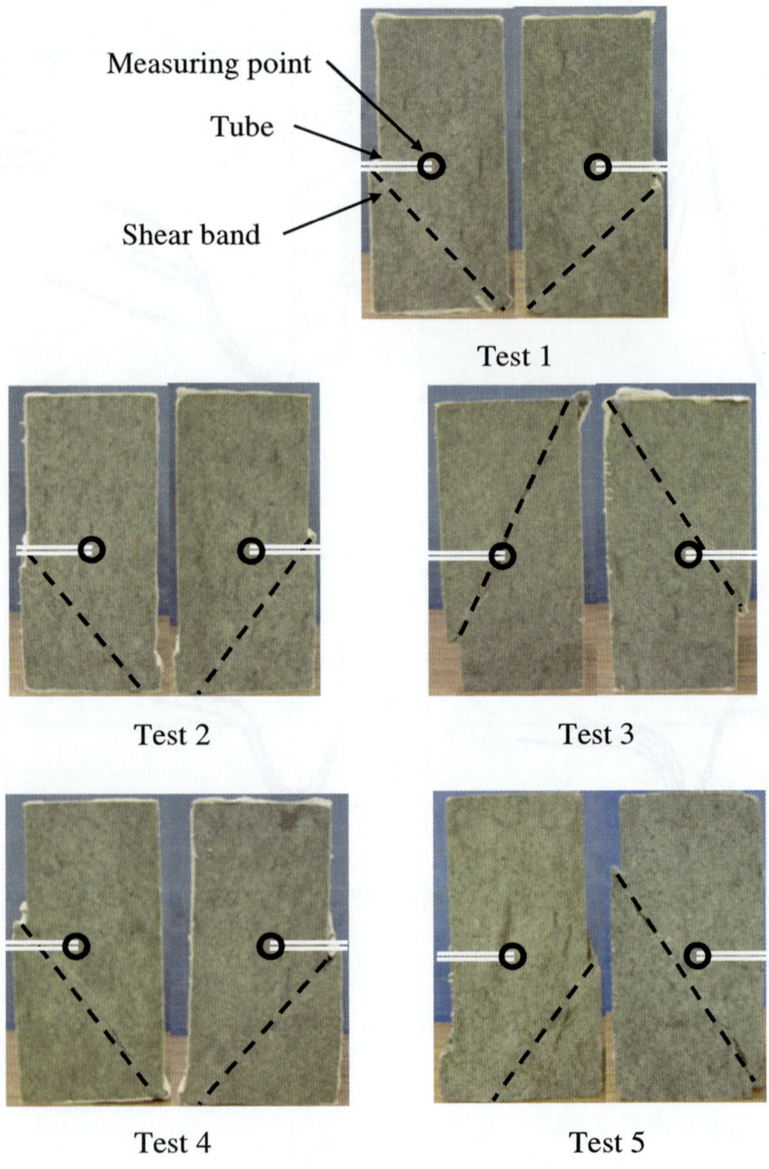

Fig. 9. Specimens after the test (Shear band and point of pore water pressure measurement)

3.2 *Plane-Strain Compression Tests under Different Test Conditions*

Plane-strain compression tests were carried out under different axial strain rates as follows: 0.001, 0.01, 0.1 and 1.0 %/min. Both drained and undrained tests were carried out and the confining stress was 1.0 MPa in all experiments. Figures 10 and 11 show the results of the drained and undrained plane-strain compression tests respectively. The result shown for the 0.001 %/min strain rate in Fig. 10 is identical to one obtained for test 4 in Fig. 8.

Strain softening behavior was clearly observed in all tests. The results showed that the faster the strain rate is, the larger the strength, see Fig. 10(a). This trend was of particular note in the undrained tests in Fig. 11.

From Figs. 10(b) and 11(b), it was found that the intermediate principal stress behaviors of the drained test under the axial strain rate of 0.01 %/min and those of the undrained test at 0.1 %/min differed from the results of the other tests. The cause for such a discrepancy can be explained by simple observation of the specimens at the end of each test (see Fig. 12). As a reference, it is noted that the surface *without* the filter paper was confined by the plain strain platens. Figure 12(a) shows the case of the undrained test where a shear plane forms normal to the confining platens as indicated by the trace of the failure plane on each one of them. In contrast, the shear plane formed at an angle with respect to the confining platens for the undrained case as clearly shown in Fig. 12(b) by the traces of the failure plane. Similarly, for the specimen of the undrained test under axial strain rate of 0.1 %/min, it was found that the shear plane was not normal to the confining platens. Therefore, from the results of the two cases in which the shear planes were not normal to the confining platens, the behavior of the intermediate principal stress differed from the results of the other tests and continued to increase. It was considered that the residual strength increased under the influence of an increasing intermediate principal stress in the two cases. As for the undrained test under axial strain rate of 0.01 %/min, the shear plane was normal to the confining platen, the intermediate principal stress became almost constant at residual state.

Turning to the series of drained tests (except for the test under axial strain rate of 1.0 %/min) it is clear that the faster the axial strain rate is, the larger is the positive excess pore water pressure, as shown in Figure 10(d). Also, it is found that the faster the strain rate, the smaller the volumetric strain with the specimen always contracting during the test with a strain rate larger than 0.001 %/min.

Pore water pressure changes at the bottom and inside of the specimen were measured in the undrained tests. The results are shown in Figs. 11(c) and (d). It is clearly found that the faster the strain rate is, the larger is the positive pore water pressure at the bottom the specimen, similar to the results obtained for the inside of the specimen in the drained tests. On the other hand, inside the specimen, it was confirmed that the faster the strain rate, the smaller the positive pore water pressure. The behaviors of the excess pore water pressure inside and at the bottom of the specimen were almost the same under axial strain rate of 0.001 %/min, but differed under axial strain rate of 1.0 %/min. From these results, the pore water pressure in the specimen was homogeneous in the undrained tests under slower strain rate. However, the pore water pressure in the specimen was not homogeneous in the undrained tests under faster strain rate and in the drained tests. It was very clear that the plane-strain tests of soft rock did not satisfy the condition of element test, not only in drained tests but also in undrained tests. Therefore, the plane-strain tests on soft rock should be regarded as model tests.

Figure 13 shows the comparison of results between drained and undrained tests. The left graphs refer to tests conducted at an axial strain rate of 0.001 %/min, while the right graphs show the results for the 1.0 %/min axial strain rate.

The results of tests under strain rate of 0.001 %/min show that the strength of the drained test was larger than that of the undrained test. This is probably due to the fact that the effective stress is decreased by the positive excess pore water pressure in the undrained test. Compared with the results of the plane-strain tests under axial strain rate of 0.001 %/min in this study, the positive excess pore water pressure existed in the drained test, but the positive excess pore water pressure of the undrained test was larger than that of the drained test.

As for the results of tests under strain rate of 1.0 %/min, the mechanical strength properties and the behavior of the pore water pressure inside the specimen were similar between the drained and undrained tests. This is not the case for tests carried out at 0.001 %/min. It was considered that there was no significant difference in the mechanical strength properties of the specimen between drained and undrained tests because the behavior of the pore water pressure near the shear band of drained test was similar to that of undrained test in the faster strain rate test.

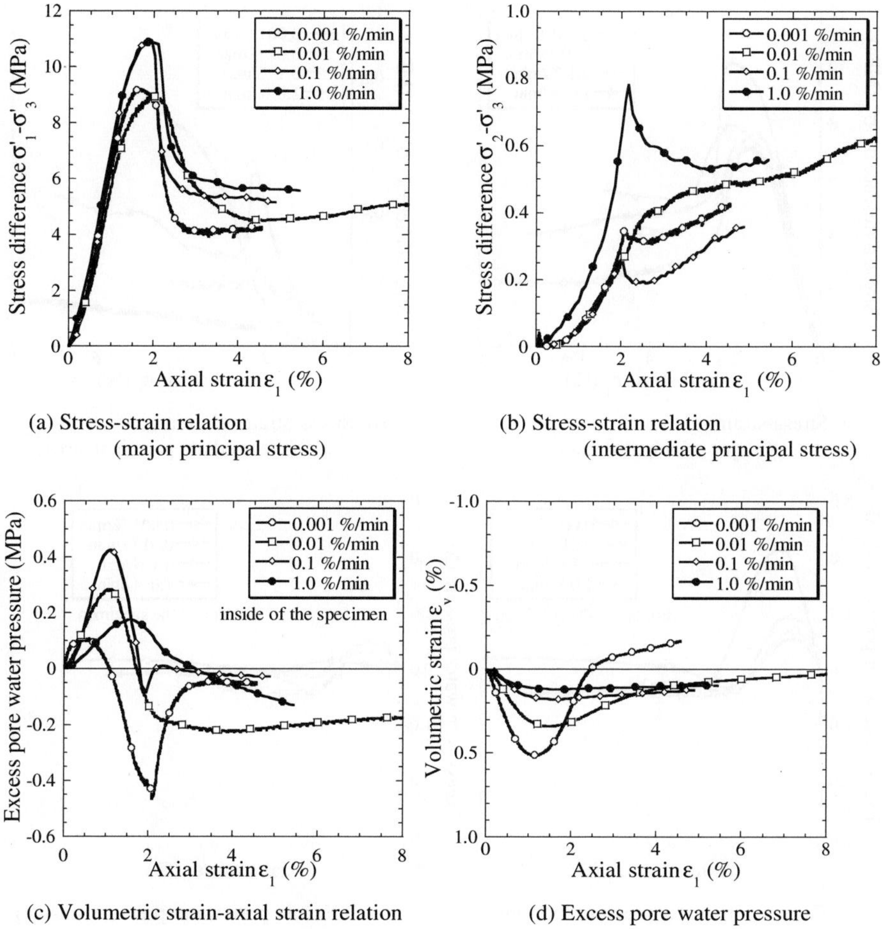

(a) Stress-strain relation (major principal stress)

(b) Stress-strain relation (intermediate principal stress)

(c) Volumetric strain-axial strain relation

(d) Excess pore water pressure

Fig. 10. Drained plane-strain compression test results (Confining stress=1.0MPa)

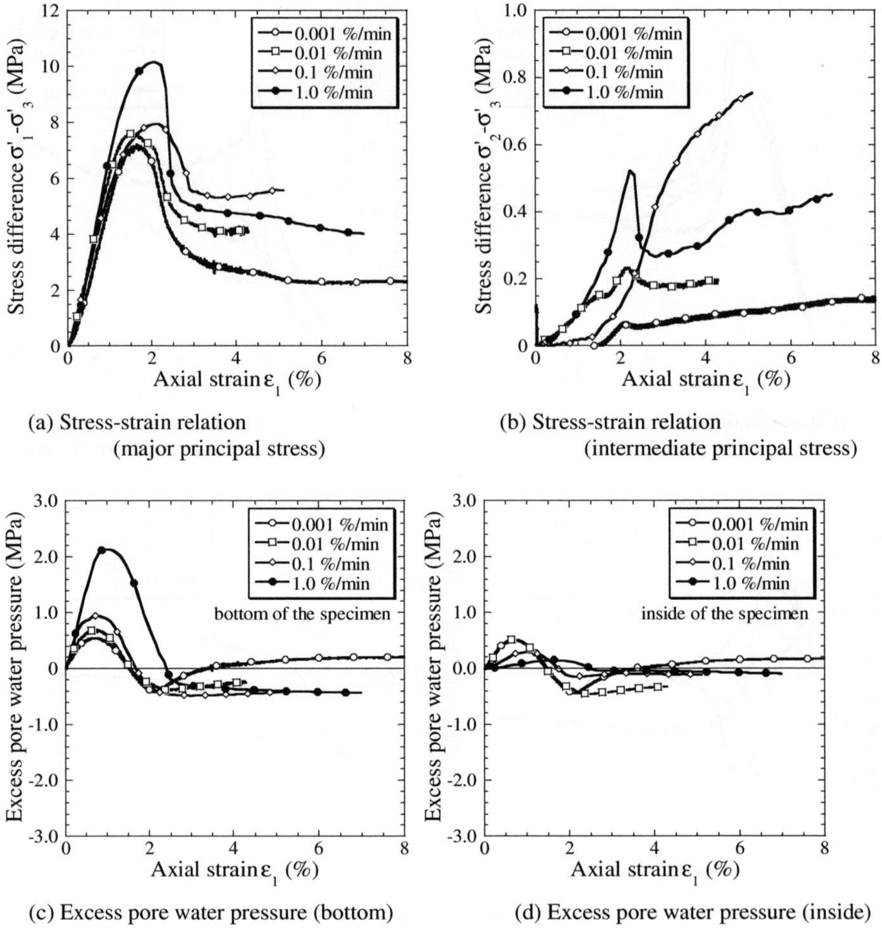

(a) Stress-strain relation (major principal stress)

(b) Stress-strain relation (intermediate principal stress)

(c) Excess pore water pressure (bottom)

(d) Excess pore water pressure (inside)

Fig. 11. Undrained plane-strain compression test results (Confining stress=1.0MPa)

(a) undrained test under axial strain rate of 0.01%/min

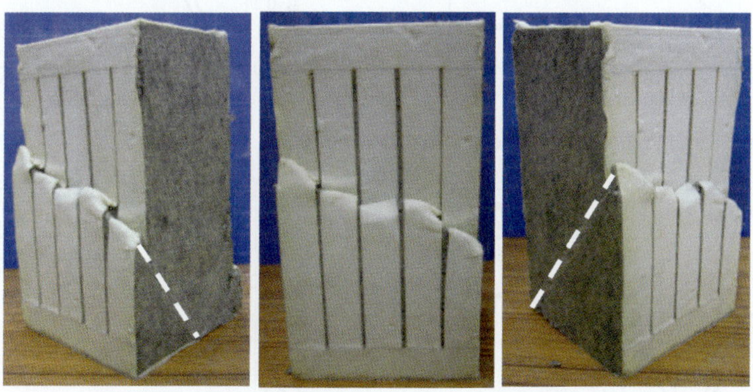

(b) drained test under axial strain rate of 0.01 %/min

Fig. 12. Specimen after test

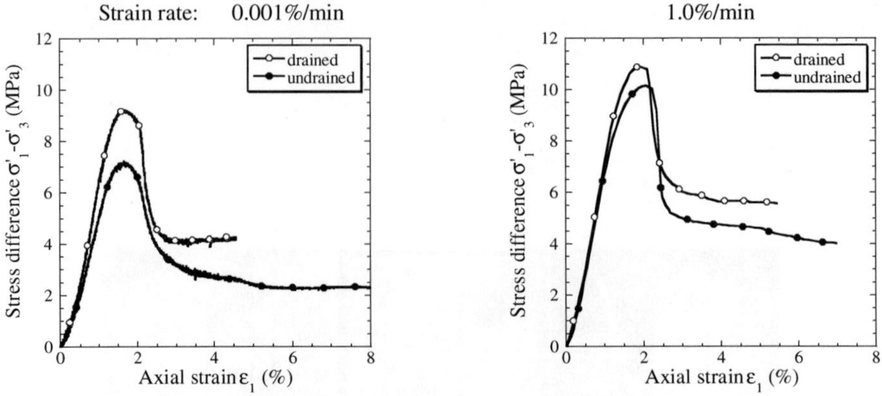

(a) Stress-strain relation (major principal stress)

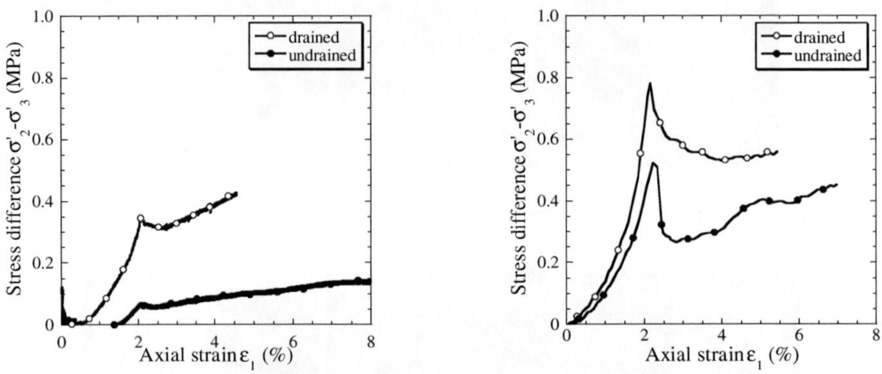

(b) Stress-strain relation (intermediate principal stress)

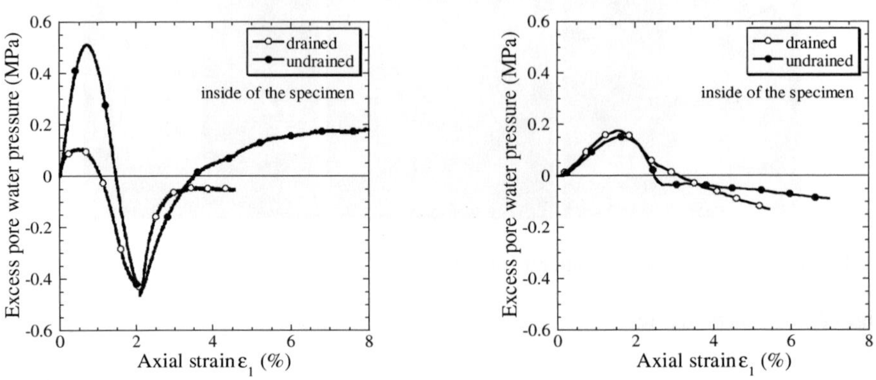

(c) Excess pore water pressure (inside the specimen)

Fig. 13. Comparison of result between drained test and undrained test (Confining stress=1.0MPa)

4 Conclusions

In this study, plane-strain compression tests of sedimentary soft rock were carried out in order to understand the mechanical behavior of soft rock based under the influence of the intermediate principal stress. In addition, the pore water pressure within/near shear band was measured during shearing.

The test results confirmed that the excess pore water pressure inside the specimen existed in the tests even under drained condition. Moreover, in the tests under undrained condition, the pore water pressure was not the same inside and at the bottom of the specimen. This clearly showed that the pore water pressure inside the specimen was not homogeneous during plane-strain test. Therefore, these tests should be regarded as model tests rather than element tests.

The plane-strain test results obtained in this study can serve as a basis for the understanding of soft sedimentary rock under different drainage and axial strain rate conditions. It is important to understand the issue of loss of homogeneity in the sample with regard to the deformation and pore water pressure fields and their interpretations to objectively develop constitutive models for soft rock.

References

[1] Zhang, F., Yashima, A., Nakai, T., Ye, G.L., Aung, H.: An elasto-viscoplastic model for soft sedimentary rock based on tij concept and subloading yield surface. Soils and Foundations 45(1), 65–73 (2005)

[2] Aung, H.: Modeling of time-dependent behavior of sedimentary soft rock and its applications to progressive failure of slope. Doctoral thesis, Gifu University (2006)

[3] Ye, G.L., Naito, K., Sawada, K., Zhang, F., Yashima, A.: Experimental study on soft sedimentary rock under plane-strain. In: Proc. Int. Conf. Contribution of Rock Mechanics to the New Century, vol. 2, pp. 865–870 (2004)

[4] Adachi, T., Ogawa, T.: Mechanical properties and failure criteria of soft rock. In: Proc. JSCE, vol. 295, pp. 51–62 (1980) (in Japanese)

[5] Adachi, T., Takase, A.: Prediction of long term strength of soft sedimentary rock. In: Proc. Int. Symp. on Weak Rock, Tokyo, pp. 21–24 (1981)

[6] Koike, M.: Elasto-viscoplastic constitutive model with strain softening for soft rock. Master Thesis, Kyoto University (1997) (in Japanese)

[7] Ye, G.L., Zhang, F., Naito, K., Aung, H., Yashima, A.: Test on soft sedimentary rock under different loading paths and its interpretation. Soils and Foundations 47(5), 897–909 (2007)

FE Investigations of Dynamic Shear Localization in Granular Bodies within Non-local Hypoplasticity Using ALE Formulation

M. Wójcik and J. Tejchman

Faculty of Civil and Environmental Engineering,
Gdańsk University of Technology, 80-952 Gdańsk, Poland
e-mail: mwojcik@pg.gda.pl, tejchmk@pg.gda.pl

Abstract. Dynamic shear localization in granular bodies during plane strain compression and plane strain silo flow was investigated. Finite element calculations were carried out with a hypoplastic constitutive model enhanced by a characteristic length of micro-structure by means of a non-local theory. The FE-dynamic analyses were performed with the help of an Arbitrary Lagrangian-Eulerian formulation using an explicit time integration approach. Emphasis was given to the formation of the pattern of shear zone in the interior of granular specimens.

Keywords: ALE formulation, finite element method, granular body, hypoplasticity, forces, non-local continuum, shear zone.

1 Introduction

Localization of deformation in the form of narrow zones of intense shearing is an inherent phenomenon observed during granular flow [1, 2, 3]. Localization under shear occurs either in the interior domain in the form of spontaneous shear zones [1] or at interfaces in the form of induced shear zones where structural members are interacting and stresses are transferred from one member to the other [4]. The localized shear zones inside of the granular material are closely related to its unstable behaviour and they can be considered as a symptom of the initiation of structural failure. Thus, it is of primary importance to take it into account while modeling the behaviour of granulates.

The mechanism of strain localization under quasi-static conditions has been comprehensively described both theoretically and experimentally. Extensive experimental studies conducted on shear localization in granular materials investigated various aspects of shear localization, such as shear resistance, localization criteria and analytical expressions for shear zone orientations, thickness of shear zones and distribution of void ratio. Theoretically, strain localization was treated analytically as a bifurcation problem [1, 5]. Numerically, shear localization was

R. Wan et al. (Eds.): Bifurcations, Instabilities and Degradations in Geomaterials, SSGG, pp. 229–250.
springerlink.com

studied using enhanced continuum models including a characteristic length of micro-structure within elasto-plasticity [6, 7, 8, 9] and hypoplasticity [10, 11, 12] based on the micro-polar, non-local, gradient and viscous approach. In addition, the so-called strong discontinuity approaches were used to capture shear localization which allow a finite element with a displacement discontinuity [13]. Shear localization has been also investigated using a micromechanically-based model [14] and a discrete element method in order to gain some insight into the microscopic mechanism [15].

However, the mechanism of shear zone patterning in granular bodies has not been comprehensively investigated in a dynamic regime, in particular within enhanced hypoplasticity.

The aim of our FE analyses is to describe the propagation of a pattern of shear zones in the interior of sand specimens during dynamic plane strain compression under constant lateral pressure and dynamic plane strain granular flow in a bin and in a hopper. In the calculations, a finite element method based on a hypoplastic constitutive law enhanced by a characteristic length of micro-structure means of a non-local theory to obtain mesh-insensitive results was used. The calculations were carried out using an Arbitrary Lagrangian-Eulerian (ALE, in short) formulation which enabled us to simulate the entire discharge process. Such FE analyses have not been performed yet.

2 Hypoplastic Model

Granular materials consist of grains in contact, and of voids. Their micromechanical behaviour is inherently discontinuous, heterogeneous and non-linear. Despite the discrete nature of granular materials, their mechanical behaviour can be reasonably described by continuum models, in particular elastoplastic and hypoplastic ones. Non-polar hypoplastic constitutive models formulated at Karlsruhe University [16, 17, 18] describe the evolution of the effective stress tensor depending on the current void ratio, stress state and rate of deformation by isotropic non-linear tensorial functions according to a representation theorem by Wang [19]. These constitutive models were formulated by a heuristic process considering the essential mechanical properties of granular materials undergoing homogeneous deformations.

Hypoplastic models are capable of describing a number of significant properties of granular materials: non-linear stress-strain relationship, dilatant and contractant behaviour, pressure dependence, density dependence and material softening. A further feature of hypoplastic models is the inclusion of critical states, i.e. states in which a grain aggregate can deform continuously be deformed at constant stress and constant volume. In contrast to elasto-plastic models, a decomposition of deformation components into elastic and plastic parts, the formulation of a yield surface, plastic potential, flow rule and hardening rule are not needed. Moreover, both the coaxiality (understood as a coincidence of the directions of the principal stresses and principal plastic strain increments) and stress-dilatancy rule are not assumed in advance [20]. The hallmark of these models is their simple formulation and procedure for determining material parameters with standard laboratory experiments. The material parameters are related to granulometric properties, viz. size distribution, shape, angularity and hardness of grains [21]. A further advantage lies in the fact that one single set of material parameters is valid for a wide range of pressures and densities. To increase the application range, a hypoplastic constitutive law has been extended for an elastic strain range [22], anisotropy [23, 24] and for viscosity [25, 26]. It can be used for soils with low friction angles [27] and clays [28]. In addition, it is also suitable to investigate size effects [29].

The summary of a hypoplastic constitutive law [16, 17] is given in the Appendix (Eqs.4-18). The changes of the values of e_i, e_d and e_c decrease with the pressure σ_{kk} according to the exponential functions (Eqs.10-12, Appendix). The constitutive relationship requires the following eight material parameters: e_{i0}, e_{d0}, e_{c0}, ϕ_c, h_s, β, n and α. The parameters h_s and n are estimated from a single oedometric compression test with an initially loose specimen (h_s reflects the slope of the curve in a semi-logarithmic representation, and n its curvature). The constants α and β are found from a triaxial or plane strain test with a dense specimen and trigger the magnitude and position of the peak friction angle. The critical friction angle ϕ_c is determined from the angle of repose or measured in a triaxial test with a loose specimen. The values of e_{i0}, e_{d0}, e_{c0} and d_{50} are obtained with conventional index tests ($e_{c0}\approx e_{max}$, $e_{d0}\approx e_{min}$, $e_{i0}\approx(1.1\text{-}1.5)e_{max}$). The parameter a_1^{-1} lies in the range of 3.0-4.3 for usual critical friction angles between 25° and 35° [10]. The material parameters were determined for a pressure range of 1 kPa $< p_s=-\sigma_{kk}/3 <$ 1000 kPa. Below the stress level of 1 kPa, additional capillary forces due to the air humidity and van der Waals forces may become important. Above the stress level of 1000 kPa, grain crushing is expected. The calibration procedure for the non-polar model and the material parameters for different sands were given by Bauer [17] and Herle and Gudehus [21]. The FE-analyses were carried out with the material constants for the so-called Karlsruhe sand [17], [21]: e_{i0}=1.3, e_{d0}=0.51, e_{c0}=0.82, ϕ_c=30°, h_s=190 MPa, β=1, n=0.5, α=0.3.

Hypoplastic constitutive models without a characteristic length can describe realistically the onset of shear localization, but not its further evolution. An enhancement of the underlying constitutive model via a characteristic length is necessary for problems involving shear localization to regularize the boundary value problem, to achieve objective and properly convergent numerical solutions (mesh-insensitive load-displacement diagram and mesh-insensitive deformation pattern) and to take into account microscopic inhomogeneities triggering shear localization (e.g. grain size). The solution is unique and stable and the type of the system incremental equations describing the process remains unchanged. A characteristic length can be introduced into hypoplasticity by means of micro-polar, non-local or second-gradient theories [11, 12]. In this paper, a non-local theory was adopted due to its simple implementation into existing commercial FE codes.

3 Non-local Approach

A non-local approach has been proposed for concrete [30] and for soils [31] to regularise a boundary value problem and to calculate strain localisation in the form of shear zones and cracks. It is based on spatial averaging of tensor or scalar state variables in a certain neighbourhood of a given point, i.e. material response at a point depends both on the state of its neighbourhood and on the state in the point itself. Thus, a characteristic length can be incorporated and softening can spread over material points. In contrast, in classical continuum mechanics, the principle of local action holds (i.e. the dependent variables in each material point depend only upon the values of the independent variables at the same point). The advantages of a non-local approach are: it is suitable for both shear and tension dominated applications and is easy to implement in existing commercial FE-codes. The disadvantages are: long computation time and the characteristic length is not directly related to the micro-structure of materials (as e.g. in micro-polar hypoplasticity).

Our hypoplastic FE calculations were carried out with a non-local modulus of the deformation rate d^*. The modulus of the deformation rate (Eq.5, Appendix) expressed by

$$d = \sqrt{d_{kl} d_{kl}} \tag{1}$$

was treated non-locally:

$$d^*(\mathbf{x}) = \frac{\int_V \omega(\|\mathbf{x}-\xi\|)\, d(\xi)\, \mathrm{d}\xi}{\int_V \omega(\|\mathbf{x}-\xi\|)\, \mathrm{d}\xi}, \tag{2}$$

where d^* - the non-local modulus of deformation, V – the volume of the body, $\boldsymbol{x}$ – the coordinates of the considered (actual) point, $\boldsymbol{\xi}$ – the coordinates of the surrounding points and $\boldsymbol{\omega}$ - the weighting function. The chosen formula (Eq.2) does not alter a uniform field which means that it satisfies the normalizing condition [32]. As a weighting function, a Gauss distribution function was used:

$$\omega(r) = \frac{1}{l_c \sqrt{\pi}} e^{-\left(\frac{r}{l_c}\right)^2}, \tag{3}$$

wherein the parameter l_c is a characteristic length of micro-structure and r is a distance between two points. The parameter l_c determines the size of the neighbourhood influencing the state at a given point. Generally, it is not directly related to dimensions of the material microstructure since it depends on the constitutive model and the weighting function [32]. It is usually determined with an inverse identification process of experimental data [33]. The averaging in Eq.3 is restricted to a small representative area around each material point (the influence of points at the distance of $r=3l_c$ is only of 0.01%). A non-local model was implemented in the commercial finite element code Abaqus Explicit [34].

4 Arbitrary Lagrangian–Eulerian Formulation

The calculations were carried out in the environment of Abaqus program [34] taking advantage of the so-called uncoupled ALE-method [35] which is a certain extension of the Lagrangian formulation by taking into account the so-called operator splitting. The uncoupled approach [35] has some advantages with respect to the coupled approach where nodal point and material values are calculated by solving a global assembled set of equations [36]. It is not necessary to describe the mesh velocity in a set of equations. In addition, a greater freedom is allowed to determine a new element mesh. Moreover, standard time integration schemes can be used.

In this approach, the state variables are calculated in two steps. First, an updated Lagrangian phase is performed which results in calculating the material displacements (convective effects are neglected). Secondly, the Eulerian phase combined with a smoothing phase is performed. In the smoothing phase, boundary nodes remain on the boundary by allowing only a tangential movement to the boundary and mesh distortion is controlled by moving inner nodes in an appropriate way. Thus, the boundary after the remeshing approximately coincides with boundary obtained with the Lagrangian calculation; the mesh topology remains similar and the number of the nodes and elements through is kept constant. In turn, in the Eulerian phase, the remap of the solution of the Lagrangian phase onto the new mesh is performed by taking into account all convective effects.

A non-linear dynamic analysis was performed with an explicit time integration method [34]. The time increment was of order 1×10^{-5}-1×10^{-7} to obtain a stable solution. Thus, several million time increments were performed. To suppress the hourglass modes, a Kelvin-type viscoelastic approach was used [34]. In order to limit numerical oscillations, linear bulk viscosity pressure was used [34] which was not included in the material point stresses. The frequency of adaptive meshing was every 10 increments.

5 Plane Strain Compression

The dynamic FE calculations of a plane strain compression test were performed with a sand specimen $b_o\times h_o$=40×140 mm^2 (b_o – initial width, h_o – initial height). The specimen dimensions were similar as in the quasi-static experiments by

Vardoulakis [1]. The specimen depth was l=1.0 m due to plane strain conditions. In all cases, quadrilateral elements with a reduced integration scheme were used to avoid volumetric locking due to dilatancy effects. To properly capture shear localization inside of the granular specimen, the size of the finite elements was always 2.5 mm. i.e. equal to $5\times d_{50}$ (d_{50}=0.5 mm) to obtain mesh-independent results.

Dynamic deformation in sand was imposed through a constant vertical velocity v prescribed at all nodes of the upper edge of the specimen. The boundary conditions implied no shear stress imposed at the smooth top and bottom of the specimen. Both vertical boundaries were free. To preserve the stability of the specimen against horizontal sliding, the node in the middle of the bottom was kept fixed. Constant confining pressure of σ_c=200 kPa was prescribed along boundaries. The specimen was initially dense with an initial void ratio of e_0=0.55.

The calculations were carried out with a small material imperfection and without imperfections. For the time integration of stresses in finite elements, a one-step Euler forward scheme was applied. The geometric non-linearity was taken into account [34].

Figs.1 and 2 demonstrate the effect of the vertical loading velocity (in the range of v=5-1000 mm/s) on the load-displacement diagram and shear localization for two different characteristic lengths (l_c=1 mm or l_c=2.5 mm) using a weak element. The peak shear resistance and material ductility decrease almost linearly with decreasing loading velocity v and ratio h_o/l_c (Fig.2). The inertial forces increase the material shear resistance by decreasing lateral strains due to inertial forces acting in an opposite direction and accelerate the occurrence of strain localization. The shear zone inclination and residual shear resistance do not depend on v. The shear zone thickness slightly increases with increasing v since the rate of softening decreases (Fig.2). In addition, the larger the velocity v (l_c=1.0-2.5 mm), the more visible is the second shear zone which is created at the beginning of loading.

In addition, Figs.3-5 show the effect of the loading velocity (v=5-1000 mm/s) on shear localization in the case of the uniform distribution of the initial void ratio (without a weak element, l_c=1.0 mm - 2.5 mm).

The results show that shear localization occurs without imposition of any material imperfections (it is caused by the propagation and reflection of waves). Depending upon v, a different pattern of shear zones can occur in the case of l_c=1 mm: one or several intersecting shear zones occur. An increase of the loading velocity v contributes to the number growth of shear zones in the specimen (with l_c=1.0 mm). In the case of l_c=2.5 mm, deformation has a form of a shear zone reflecting from the top or bottom edge independently of v. A large loading velocity significantly improves the material ductility. The relationship between the peak shear resistance and loading velocity in the range of v=5 mm/s – 1000 mm/s is almost linear and is similar independently of l_c (Fig.5).

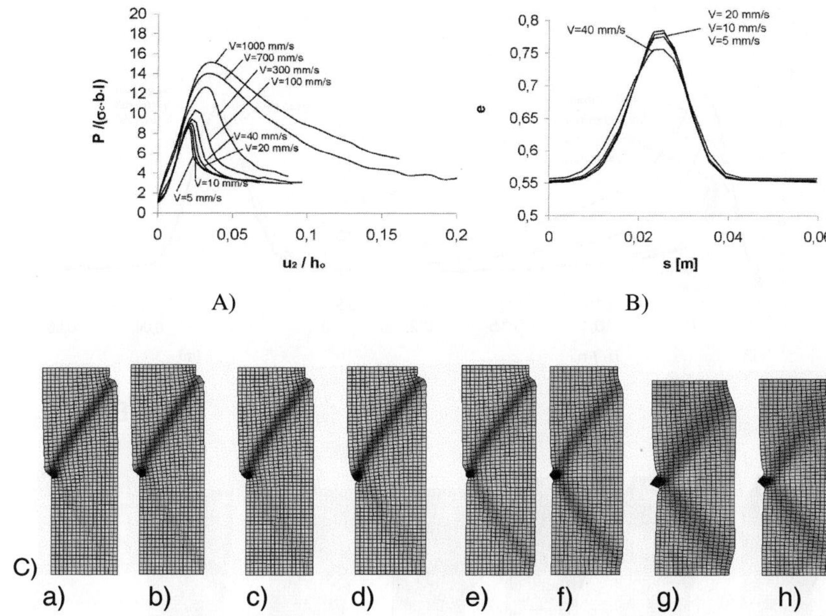

Fig. 1. Calculated normalized load-displacement curves (A), distribution of void ratio along the line perpendicular to the shear zone (B) and deformed meshes (C) for different deformation velocities v: a) v=5 mm/s, b) v=10 mm/s, c) v=20 mm/s, d) v=40 mm/s, e) v=100 mm/s, f) v=300 mm/s, g) v=700 mm/s, h) v=1000 mm/s (specimen 4×14 cm^2, 16×56 elements, initially dense sand e_o = 0.55, l_c=1 mm, ALE-formulation, weak element)

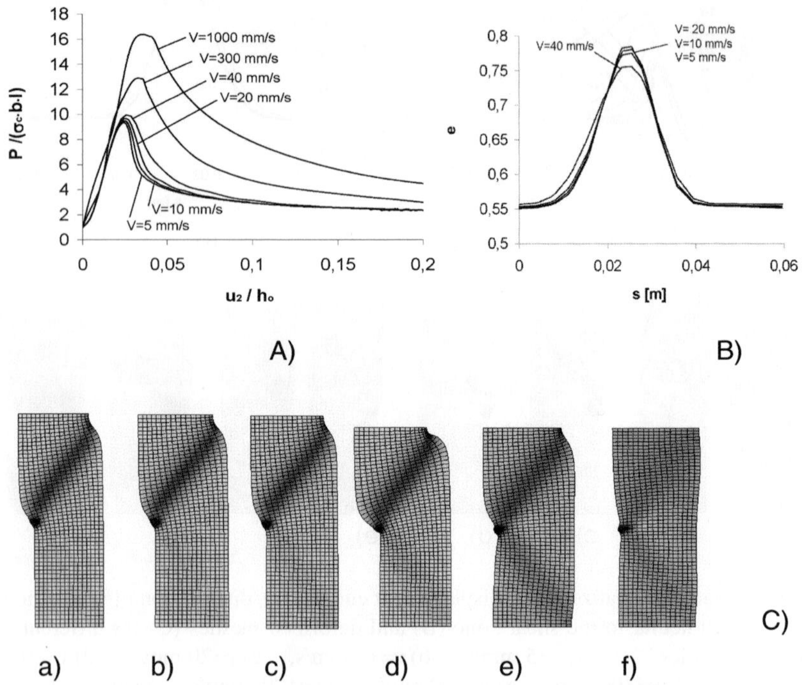

Fig. 2. Calculated normalized load-displacement curves (A), distribution of void ratio along the line perpendicular to the shear zone (B) and deformed meshes (C) for different deformation velocities v: a) v=5 mm/s, b) v=10 mm/s, c) v=20 mm/s, d) v=40 mm/s, e) v=300 mm/s, f) v=1000 mm/s (initially dense sand $e_o = 0.55$, l_c=2.5 mm, ALE-formulation, weak element)

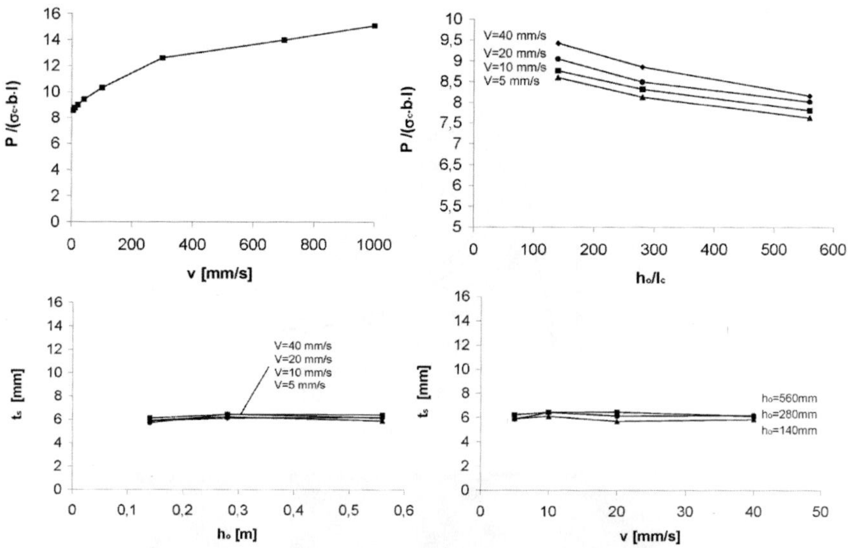

A)

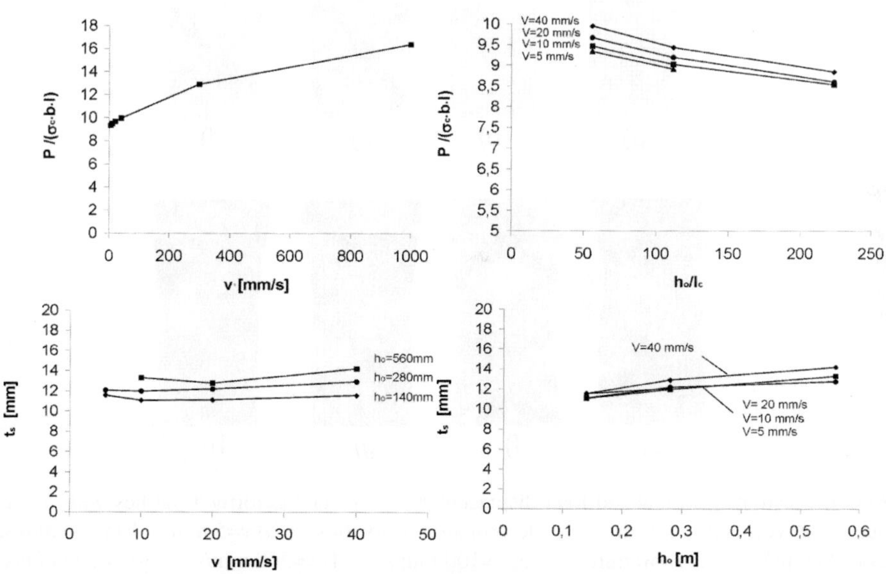

B)

Fig. 3. Calculated peak normalized vertical force $P/(\sigma_c bl)$ versus loading velocity v, peak normalized vertical force $P/(\sigma_c bl)$ versus ratio h_o/l_c, and shear zone thickness t_s versus loading velocity v and specimen height h_o (initially dense sand $e_o = 0.55$, weak element): A) l_c=1 mm, B) l_c=2.5 mm

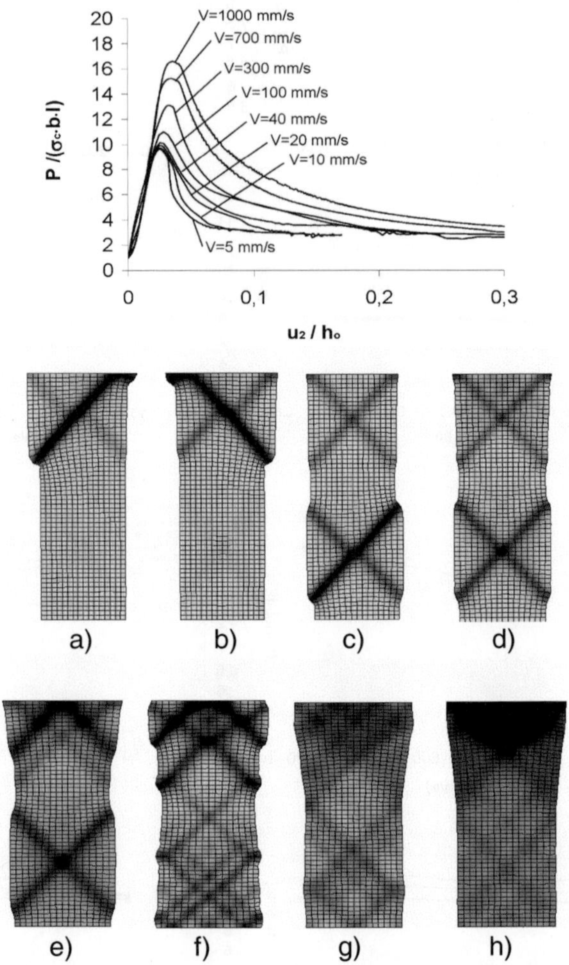

Fig. 4. Calculated normalized load-displacement curves and deformed meshes with distribution of void ratio for different deformation velocities v: a) v=5 mm/s, b) v=10 mm/s, c) v=20 mm/s, d) v=40 mm/s, e) v=100 mm/s, f) v=300 mm/s, g) v=700 mm/s, h) v=1000 mm/s (initially dense sand $e_o = 0.55$, l_c=1.0 mm, ALE-formulation, specimen without weak element)

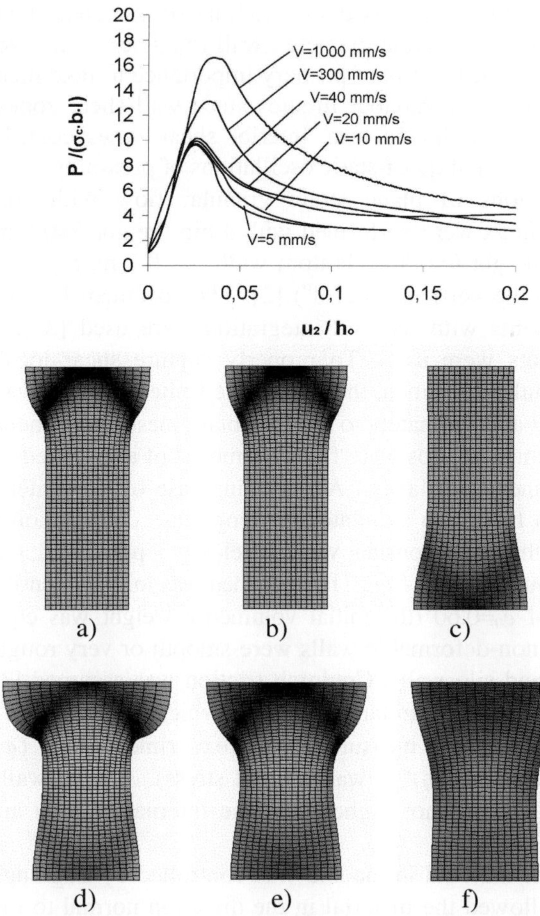

Fig. 5. Calculated normalized load-displacement curves and deformed meshes with distribution of void ratio in the specimen for different deformation velocities v: a) v=5 mm/s, b) v=10 mm/s, c) v=20 mm/s, d) v=40 mm/s, e) v=300 mm/s, f) v=1000 mm/s (initially dense sand e_o = 0.55, l_c=2.5 mm, ALE-formulation, specimen without weak element)

6 Granular Flow in Silos

Shear zones during confined granular flow in laboratory silos were observed among others with the aid of colored sand layers [37], x-rays [38] and PIV technique [29]. It was recognized that shear zones occur always during silo flow along walls and in the interior of initially dense granular solids. Their thickness depends on the initial density, mean grain diameter, wall roughness, silo size, grain roughness and emptying velocity. It is of primary importance to take them into account while modeling the flow behaviour in silos since wall shear zones influence directly loads acting on walls. In turn, interior shear zones contribute to a non-symmetric distribution and quasi-static oscillations of pressures.

The FE-calculations of plane strain granular flow with controlled outlet velocity of v=0.1 mm/s were performed with a bin (height h=0.5 m, width b=0.2 m) and a hopper (height h=0.5 m, bottom width b_1=0.1 m, top width b_2=0.30 m, wall inclination to the vertical α=11.4°) [2]. The specimen length was l=0.6 m. Quadrilateral elements with reduced integration were used [34]. Totally, 1250-1600 finite elements were used. To properly capture shear localization in the interior of the granular specimen, the size of the finite elements was always about 5-10 mm. i.e. equal approximately to $5\times l_c$ to obtain mesh-independent results.

First, the experimental silos were filled by means of a so-called ‘layer-by-layer' method using 8 sand equal layers. A linear increase of the material weight was prescribed to each layer. To simulate silo emptying, deformation in sand during flow was imposed through a constant vertical velocity v prescribed at all nodes of the bottom. The top boundary was free. The specimen was initially dense with a uniform initial void ratio of e_0=0.60 (the initial volumetric weight was equal to γ_d=16.75 kN/m^3). The rigid non-deformable walls were smooth or very rough. Between the granular material and silo walls, Coulomb friction was assumed [34] with a constant wall friction angle $\varphi_w=\arctan(\sigma_{12}^w/(\sigma_{22}^w)$ which was equal to the maximum resultant wall friction angle measured in the experiments [37] (φ_w=arctan(T/N)) (σ_{11}^w - wall shear stress, σ_{11}^w – wall normal stress). A local wall friction angle $\varphi_w=\arctan(\sigma_{12}^w/(\sigma_{22}^w)$ was not higher than the internal friction angle calculated from principal stresses.

The shape of the material in the silo was controlled by Lagrangian boundaries (the boundaries followed the material in the direction normal to the material surface). Along the outlet, the Eulerian condition with a constant vertical velocity was assumed allowing the material to flow throughout the mesh. The geometric non-linearity was taken into account [34].

Figs.6 and 7 presents the results of the resultant forces T, N and P during bottom displacement compared to experiments for initially dense sand with smooth and very rough walls in the bin and hopper. In the calculations, the constant wall friction angles from experiments were used. The evolution of the mobilized resultant wall friction angles, φ_w=arctan(T/N), for initially dense sand during emptying of the bin and hopper is demonstrated in Fig.8.

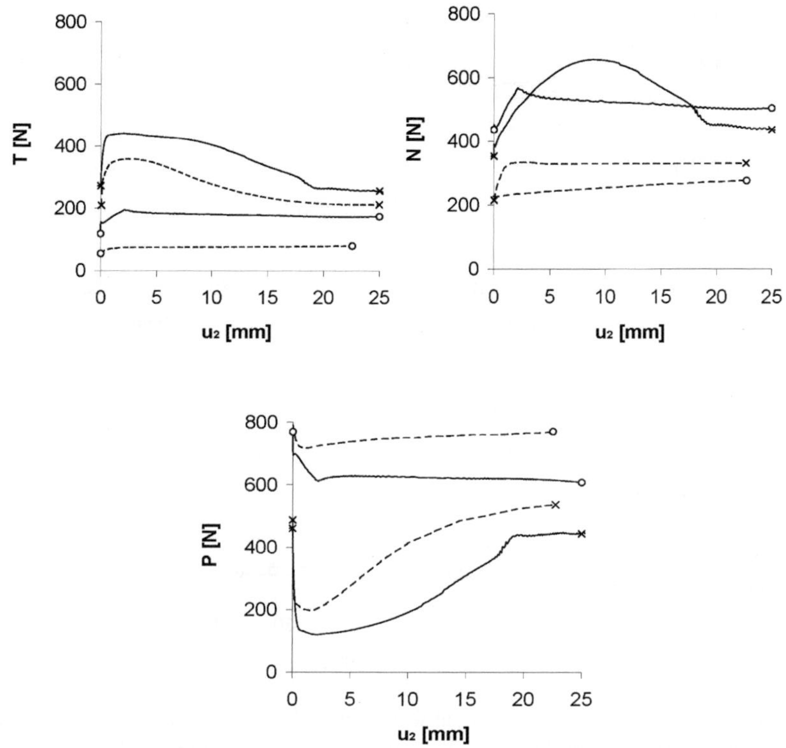

Fig. 6. Evolution of the resultant vertical wall friction force T, horizontal wall force N and vertical force on the bottom P for initially dense sand during bottom displacement u2 in experimental bin (—- FE calculations, ---- - experiments [2], o - smooth walls, × - very rough walls)

The theoretical maximum wall friction force *T* in a bin and hopper during filling and emptying increases with increasing wall roughness. The theoretical maximum vertical bottom force *P* behaves obviously in the opposite way. Both forces reach their residual state as in the experiments. The highest calculated normal wall force *N* occurs with smooth walls during filling. In the case of emptying, the theoretical maximum force *N* increases in a bin with increasing wall roughness and increases in a hopper with decreasing wall roughness. The evolutions of theoretical forces *T*, *P* and *N* are qualitatively similar as in the experiments. However, there exist some differences between experimental and theoretical forces. The differences are of order of: 10% (bin) and 30% (hopper) during filling and 10% (bin) and 5% (hopper) during flow in the case of the wall force *T*, and 5% (bin) and 30% (hopper) during filling and 10% (bin) and 10% (hopper) during flow in the case of the bottom force *P*. Significantly larger discrepancies occur in the case of the maximum force *N*. The theoretical values are higher even by 80% (bin) and 100% (hopper) during filling and by 80% (bin) and 10% (hopper) during flow.

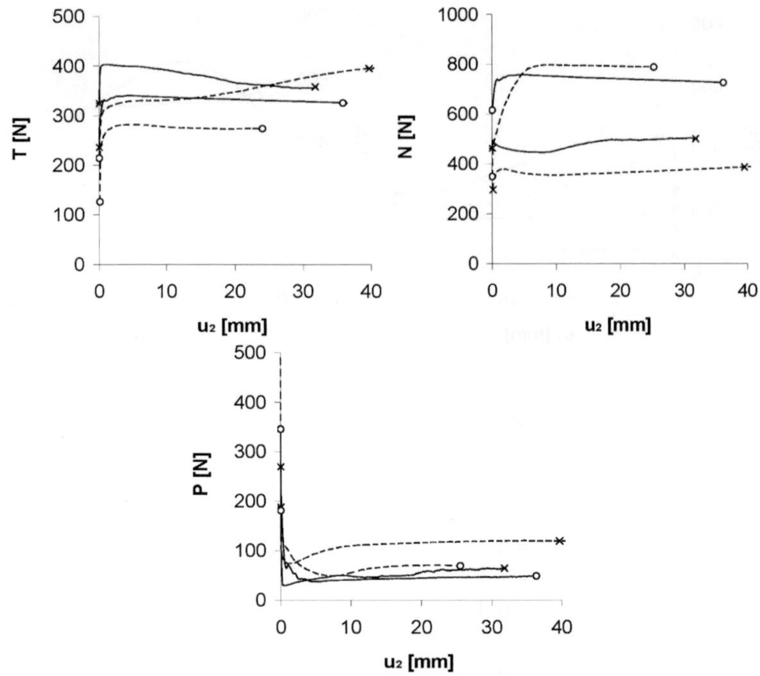

Fig. 7. Evolution of the resultant vertical wall friction force T, horizontal wall force N and vertical force on the bottom P for initially dense sand during bottom displacement u2 in experimental hopper (— - FE calculations, ---- - experiments [2], o - smooth walls, × - very rough walls)

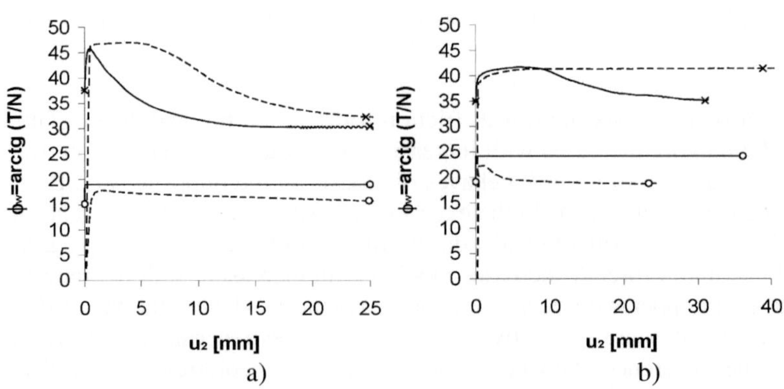

Fig. 8. Evolution of the mobilized resultant wall friction angle φw=arctan(T/N) for initially dense sand during bottom displacement u2: a) experimental bin, b) experimental hopper (— - FE calculations, ---- - experiments [2], o - smooth walls, × - very rough walls)

In Figs.9 and 10, the distribution of void ratio after bottom displacement u_2 is shown in a bin with smooth (Fig.9a) and very rough walls (Fig.9b), and in a hopper with smooth (Fig.10a) and very rough walls (Fig.10b). The higher is the void ratio, the lighter the areas are (the scale of grey intensity was enclosed).

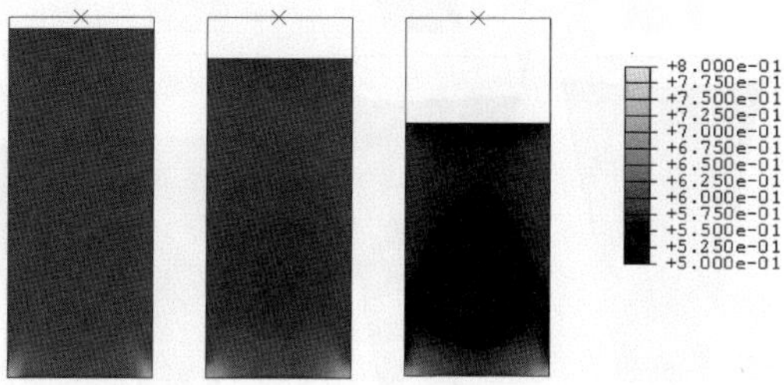

a)

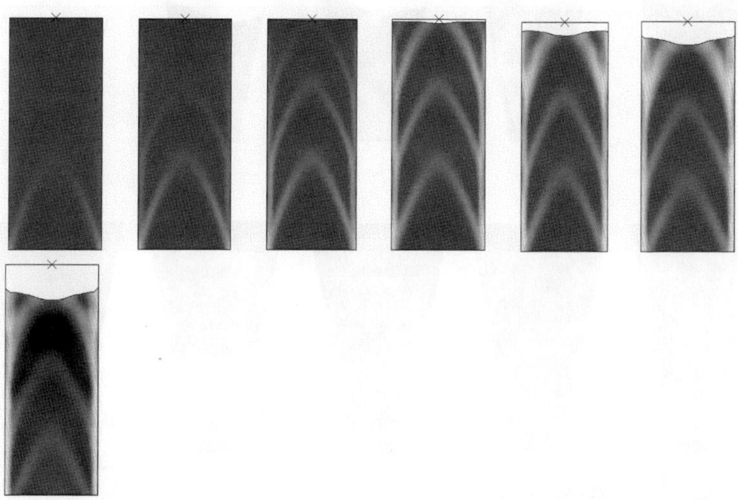

b)

Fig. 9. Evolution of void ratio in initially dense sand during silo flow (experimental bin): a) smooth walls, b) very rough walls

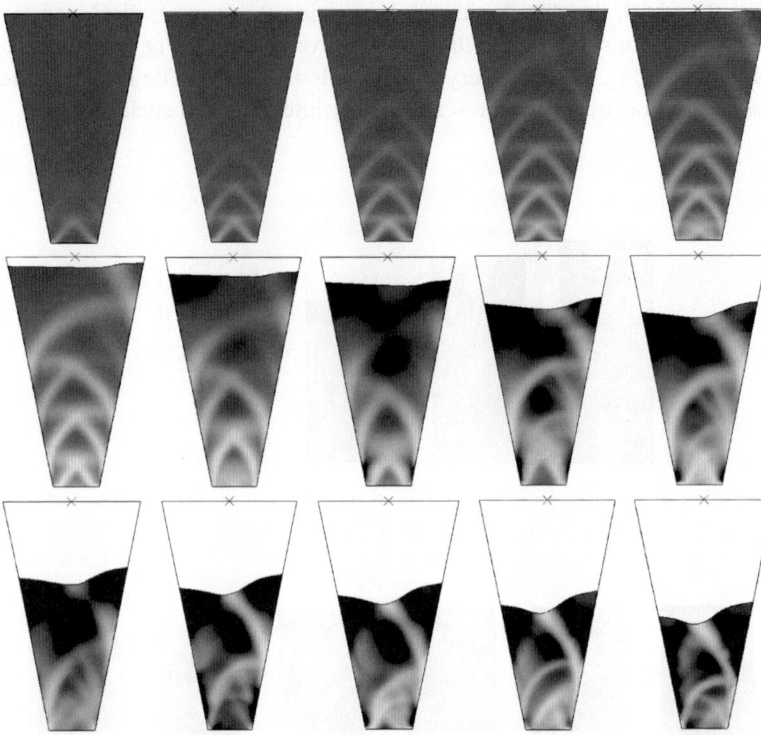

a)

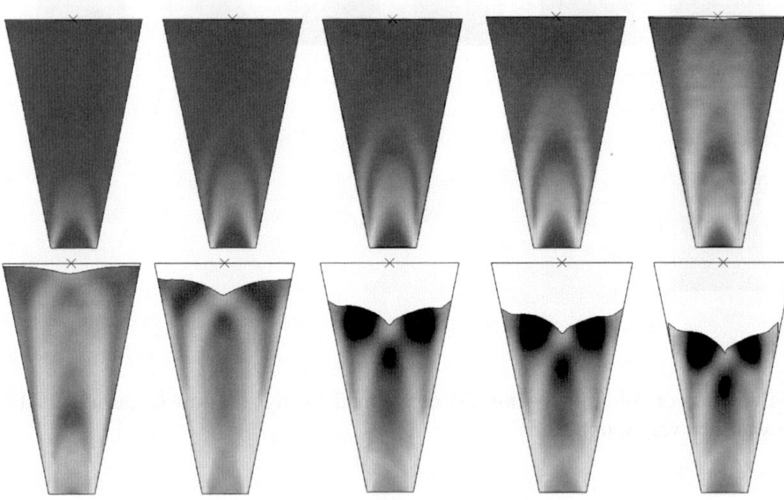

b)

Fig. 10. Evolution of void ratio in initially dense sand during silo flow (experimental hopper): a) smooth walls, b) very rough walls

The results show that during granular flow in silo with controlled outlet velocity several dilatant shear zones occur with high void ratio. They appear along walls and in the interior of the material. The interior curvilinear shear zones are created only in initially dense granular solid in a bin with very rough walls and in a hopper independently of the wall roughness. They are created always at the outlet and propagate upward until they reach the free boundary. In addition, the interior shear zones in a hopper with smooth walls cross each other around the silo symmetry, reach the walls and subsequently are reflected from them. In a hopper with very rough walls, the curvilinear dilatant zones propagate upwards only in the material core. Some of them cross each other. They do not reach the walls. The thickness of wall shear zones is about 3 mm ($6\times d_{50}$) (smooth walls), 12.5 mm ($25\times d_{50}$) (very rough walls). In turn, the thickness of interior shear zones is approximately 10-12.5 mm [$(20\text{-}25)\times d_{50}$]. The shape and thickness of shear zones in a hopper with smooth walls is almost identical as in experiments [37] with a mass flow hopper. In turn, the shape and thickness of shear zones in a hopper with very rough walls resembles those from experiments [37] with a funnel flow hopper. For smooth walls in a bin, the internal shear zones occur only close to the bottom-wall region due to insignificant deformation in sand (Fig.9a).

Figures 11 and 12 present the effect of a characteristic length of micro-structure on the evolution of interior shear zones during silo flow in a hopper with smooth and very rough walls, respectively. An increase of a characteristic length increases the thickness of shear zones up to their almost complete disappearance in the interior of sand.

In general, a satisfactory agreement between theoretical and experimental results was achieved in particular with respect to the shape and thickness of multiple shear zones in the interior of flowing sand and the evolution of the global vertical wall friction and vertical bottom force. Significant discrepancies between theoretical and experimental results mainly concerned the global wall normal force. They were probably caused due to the following reasons: a) the assumed evolution of the mobilized wall friction angle was different than the experimental one, b) simplified plane strain symmetric calculations were carried out (experimental flow was always non-symmetric), c) wall friction forces along perspex silo walls were not taken into account, d) the measured global wall friction angle was not equivalent to the local one assumed in FE calculations, and e) material hypoplastic constants were determined for moderate pressures (in experimental silos, very low pressures occurred). To obtain a better agreement, new calculations will be performed under 3D conditions by assuming a stochastic distribution of material properties (e.g. initial void ratio and mean grain diameter). The hypoplastic constants will be more precisely calibrated in the range of low pressures. To describe free silo outflow (due to gravity), viscosity will be considered. The main problem to consistently describe the material behaviour during silo flow concerns the boundary condition along the silo walls. An assumption of a constant wall friction angle is not true when the wall is rough or very rough. In turn, a direct transfer of the mobilized wall friction angle from wall shear tests is not possible since it depends on the boundary conditions of the entire system. Thus, a development of novel wall conditions depending on the wall roughness only is of a primary importance.

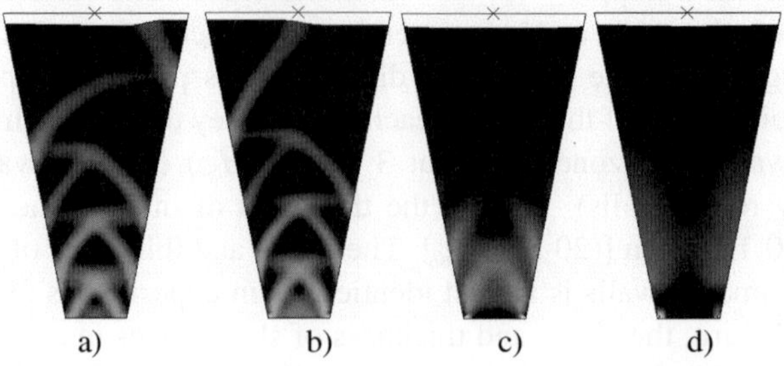

Fig. 11. Evolution of void ratio in dense sand during silo flow in hopper with smooth walls for a different characteristic length l_c: a) l_c=1.5 mm, b) l_c=3.5 mm, c) l_c=5 mm, d) l_c=10 mm, (φ_w=24°, e_o=0.6, u2=90 mm)

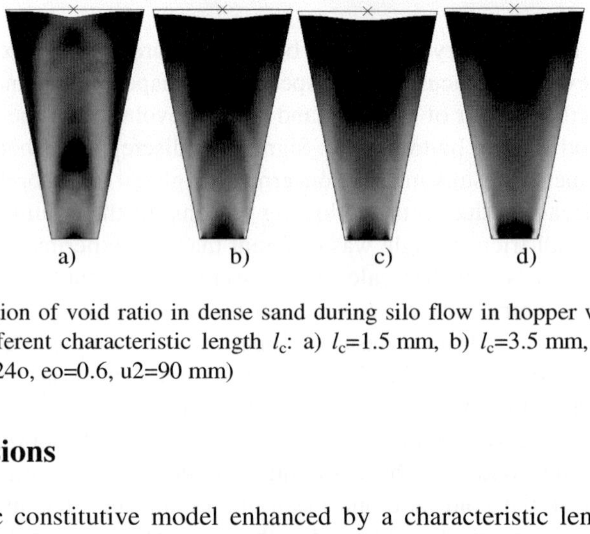

Fig. 12. Evolution of void ratio in dense sand during silo flow in hopper with very rough walls for a different characteristic length l_c: a) l_c=1.5 mm, b) l_c=3.5 mm, c) l_c=5 mm, d) l_c=10 mm, (φ=24o, eo=0.6, u2=90 mm)

7 Conclusions

A hypoplastic constitutive model enhanced by a characteristic length of microstructure was able to capture the most important properties of granular solids during plane strain compression and granular silo flow. It was very useful to investigate the changes of void ratio in the interior of flowing sand.

A significant advantage of an Arbitrary Lagrangian-Eulerian formulation was a moderate mesh deformation in the interior of the flowing solid.

A geometrical or material imperfection was not necessary to induce shear localization in a dynamic regime during plane strain compression due to propagations, interactions and reflections of dispersive waves. A number of shear zones increased with increasing loading velocity at a small ratio of h_o/l_c (h_o/l_c=140). In the case of an imperfection in the specimen, one shear zone occurred at a small loading velocity and two shear zones were created at a large loading velocity at the residual state. The peak shear resistance increased almost linearly with increasing loading velocity in the range of v=5-1000 mm/s. The shear zone thickness slightly increased with v. In turn, the residual shear resistance and shear zone inclination only insignificantly depended on v.

A pattern of periodic internal shear zones occurred in the interior of cohesionless sand during plane strain silo flow. Shear zones were created only in initially dense granular solids in an experimental bin with very rough walls and in an experimental hopper independently of the wall roughness. The shape of internal shear zones depended on the silo form, wall roughness, initial solid density and characteristic length of micro-structure. In the hopper with smooth walls, the shear zones reflected from walls. In turn, in the hopper with very rough walls, they occurred only in the material core. The interior shear zones contributed to oscillation and non-uniform distribution of pressure and density.

The calculated global wall normal force in experimental silos was very sensitive to changes of the wall friction angle. Thus, a realistic description of the interface behavior (evolution of the wall friction angle and thickness of the wall shear zone) is a key to realistically describe the distribution of wall stresses.

References

[1] Vardoulakis, I.: Shear band inclination and shear modulus in biaxial tests. Int. J. Num. Anal. Meth. Geomech. 4, 103–119 (1980)

[2] Desrues, J., Chambon, R., Mokni, M., Mazerolle, F.: Void ratio evolution inside shear bands in triaxial sand specimens studied by computed tomography. Géotechnique 46(3), 529–546 (1996)

[3] Yoshida, T., Tatsuoka, F., Siddiquee, M.: Shear banding in sands observed in plane strain compression. In: Chambon, R., Desrues, J., Vardoulakis, I. (eds.) Localisation and Bifurcation Theory for Soils and Rocks, Balkema, Rotterdam, pp. 165–181 (1994)

[4] Tejchman, J., Wu, W.: Experimental and numerical study of sand-steel interfaces. Int. Journal of Numerical and Anal. Methods in Geomechanics 19(8), 513–537 (1995)

[5] Rudnicki, J.W., Rice, J.R.: Conditions of the localization of deformation in pressure-sensitive dilatant materials. Journal of Mechanics and Physics of Solids 23, 371–394 (1975)

[6] Mühlhaus, H.-B.: Application of Cosserat theory in numerical solutions of limit load problems. Ing. Arch. 59, 124–137 (1989)

[7] Sluys, L.: Wave propagation, localization and dispersion in softening solids. PhD Thesis, Delft University of Technology (1992)

[8] Tejchman, J., Wu, W.: Numerical study on shear band patterning in a Cosserat continuum. Acta Mechanica 99, 61–74 (1993)

[9] Brinkgreve, R.: Geomaterial models and numerical analysis of softening. Dissertation, Delft University, 1-153 (1994)

[10] Tejchman, J., Gudehus, G.: Shearing of a narrow granular strip with polar quantities. I. J. Num. and Anal. Methods in Geomechanics 25, 1–28 (2001)

[11] Maier, T.: Comparison of non-local and polar modelling of softening in hypoplasticity. Int. Journal for Numerical and Analytical Methods in Geomechanics 28(3), 251–268 (2004)

[12] Tejchman, J.: Influence of a characteristic length on shear zone formation in hypoplasticity with different enhancements. Computers and Geotechnics 31(8), 595–611 (2004)

[13] Regueiro, R.A., Borja, R.I.: Plane strain finite element analysis of pressure sensitive plasticity with strong discontinuity. Int. J. Solids and Structures 38(21), 3647–3672 (2001)

[14] Gardiner, B.S., Tordesillas, A.: Micromechanics of shear bands. Int. J. Solids and Structures 41, 5885–5901 (2004)

[15] Pena, A.A., García-Rojo, R., Herrmann, H.J.: Influence of particle shape on sheared dense granular media. Granular Matter 3-4, 279–292 (2007)

[16] Gudehus, G.: A comprehensive constitutive equation for granular materials. Soils and Foundations 36(1), 1–12 (1996)

[17] Bauer, E.: Calibration of a comprehensive hypoplastic model for granular materials. Soils and Foundations 36(1), 13–26 (1996)

[18] von Wolffersdorff, P.A.: A hypoplastic relation for granular materials with a predefined limit state surface. Mechanics Cohesive-Frictional Materials 1, 251–271 (1996)

[19] Wang, C.C.: A new representation theorem for isotropic functions. J. Rat. Mech. Anal. 36, 166–223 (1970)

[20] Tejchman, J., Wu, W.: Non-coaxiality and stress-dilatancy rule in granular materials: FE investigation within micro-polar hypoplasticity. Int. J. Num. Anal. Meths. in Geomech. (2008), doi:10.10002/nag.715

[21] Herle, I., Gudehus, G.: Determination of parameters of a hypoplastic constitutive model from properties of grain assemblies. Mechanics of Cohesive-Frictional Materials 4(5), 461–486 (1999)

[22] Niemunis, A., Herle, I.: Hypoplastic model for cohesionless soils with elastic strain range. Mechanics of Cohesive-Frictional Materials 2, 279–299 (1997)

[23] Tejchman, J., Bauer, E., Wu, W.: Effect of texturial anisotropy on shear localization in sand during plane strain compression. Acta Mechanica 1-4, 23–51 (2007)

[24] Tejchman, J., Niemunis, A.: FE-studies on shear localization in an anisotropic micro-polar hypoplastic granular material. Granular Matter 8(3-4), 205–220 (2006)

[25] Niemunis, A.: Extended hypoplastic models for soils. Habilitation Monography, Gdansk University of Technology (2003)

[26] Gudehus, G.: Seismo-hypoplasticity with a granular temperature. Granular Matter 8, 93–102 (2006)

[27] Herle, I., Kolymbas, D.: Hypoplasticity for soils with low friction angles. Computers and Geotechnics 31, 365–373 (2004)

[28] Masin, D.: A hypoplastic constitutive model for clays. Int. J. Numer. and Anal. Meths. in Geomech. 29, 311–336 (2005)

[29] Tejchman, J., Górski, J.: Computations of size effects in granular bodies within micro-polar hypoplasticity during plane strain compression. Int. J. for Solids and Structures 45(6), 1546–1569 (2008)

[30] Bazant, Z.P., Lin, F., Pijaudier-Cabot, G.: Yield limit degradation: non-local continuum model with local strain. In: Owen (ed.) Proc. Int. Conf. Computational Plasticity, Barcelona, pp. 1757–1780 (1987)

[31] Brinkgreve, R.: Geomaterial models and numerical analysis of softening. Dissertation, Delft University, 1-153 (1994)

[32] Bazant, Z., Jirasek, M.: Nonlocal integral formulations of plasticity and damage: survey of progress. J. Engng. Mech. 128(11), 1119–1149 (2002)

[33] Mahnken, R., Kuhl, E.: Parameter identification of gradient enhanced damage models. Eur. J. Mech. A/Solids 18, 819–835 (1999)

[34] Hibbitt, Karlsson & Sorensen, Inc. Abaqus. User's manual, version 6.4 (2004)

[35] Stoker, C.: Developments of the Arbitrary Lagrangian-Eulerian Method in Non-linear Solid Mechanics. PhD Thesis, University of Twente (1999)

[36] Belytschko, T., Liu, W.K., Moran, B.: Nonlinear Finite Elements for Continua and Structures. Wiley, Chichester (2000)

[37] Tejchman, J., Gudehus, G.: Verspannung, Scherfugenbildung und Selbsterregung bei der Siloentleerung. In: Eibl, J., Gudehus, G. (eds.) Silobauwerke und ihre spezifischen Beanspruchungen, pp. 245–284. Deutsche Forschungsgemeinschaft, Wiley-VCH (2000)

[38] Drescher, A., Cousens, T., Bransby, T.P.L.: Kinematics of the mass flow of granular material through a plane hopper. Geotechnique 28(1), 27–42 (1978)

[39] Slominski, C., Niedostatkiewicz, M., Tejchman, J.: Application of particle image velocimetry (PIV) for deformation measurement during granular silo flow. Powder Technology 173(1), 1–18 (2007)

Appendix

The hypoplastic constitutive law can be summarized as follows [16, 17]:

$$\overset{o}{\sigma}_{ij} = F(e, \sigma_{kl}, d_{kl}), \tag{4}$$

$$\overset{o}{\sigma}_{ij} = f_s\,[\,L_{ij}(\hat{\sigma}_{kl}, d_{kl}) + f_d N_{ij}(\hat{\sigma}_{ij})\sqrt{d_{kl}d_{kl}}\,]\,, \tag{5}$$

$$L_{ij} = a_1^2 d_{ij}^c + \hat{\sigma}_{ij}\,\hat{\sigma}_{kl}\,d_{kl}\,, \qquad N_{ij} = a_1(\hat{\sigma}_{ij} + \hat{\sigma}_{ij}^*), \tag{6}$$

$$\hat{\sigma}_{ij} = \frac{\sigma_{ij}}{\sigma_{kk}}\,, \qquad \overset{o}{\sigma}_{ij} = \dot{\sigma}_{ij} - w_{ik}\,\sigma_{kj} + \sigma_{ik}\,w_{kj}\,, \tag{7}$$

$$d_{ij} = (v_{i,j} + v_{j,i})/2, \qquad w_{ij} = (v_{i,j} - v_{j,i})/2, \tag{8}$$

$$\dot{e} = (1+e) d_{kk}, \tag{9}$$

$$e_i = e_{i0} \exp[-(-\sigma_{kk} / h_s)^n], \tag{10}$$

$$e_d = e_{d0} \exp[-(-\sigma_{kk} / h_s)^n], \tag{11}$$

$$e_c = e_{c0} \exp[-(-\sigma_{kk} / h_s)^n], \tag{12}$$

$$f_s = \frac{h_s}{n h_i}\left(\frac{1+e_i}{e_i}\right)\left(\frac{e_i}{e}\right)^{\beta}\left(-\frac{\sigma_{kk}}{h_s}\right)^{1-n}, \tag{13}$$

$$h_i = \frac{1}{c_1^2} + \frac{1}{3} - \left(\frac{e_{i0}-e_{d0}}{e_{c0}-e_{d0}}\right)^{\alpha} \frac{1}{c_1\sqrt{3}}, \tag{14}$$

$$f_d = \left(\frac{e-e_d}{e_c-e_d}\right)^{\alpha}, \tag{15}$$

$$a_1^{-1} = c_1 + c_2 \sqrt{\hat{\sigma}^*_{kl}\,\hat{\sigma}^*_{lk}}\,[1+\cos(3\theta)], \tag{16}$$

$$\cos(3\theta) = -\frac{\sqrt{6}}{[\hat{\sigma}^*_{pq}\,\hat{\sigma}^*_{pq}]^{1.5}}(\hat{\sigma}^*_{kl}\,\hat{\sigma}^*_{lm}\,\hat{\sigma}^*_{mk}), \tag{17}$$

$$c_1 = \sqrt{\frac{3}{8}}\,\frac{(3-\sin\phi_c)}{\sin\phi_c}, \qquad c_2 = \frac{3}{8}\,\frac{(3+\sin\phi_c)}{\sin\phi_c}, \tag{18}$$

wherein: F - isotropic tensor-valued function of its arguments, σ_{ij} – Cauchy stress tensor, σ^*_{ij} - deviatoric part of σ_{ij}, $\overset{o}{\sigma}_{ij}$ - Jaumann stress rate tensor (objective stress rate tensor), e - current void ratio, d_{kl} - rate of deformation tensor (stretching tensor), w_{ij} – spin tensor, v - velocity, d^c_{ij} - polar rate of deformation tensor, f_s - stiffness factor, h_s - granulate hardness, σ_{kk} – mean stress, f_d - density factor, e_c - critical void ratio (e_{c0} - value of e_c for σ_{kk}=0), e_d - void ratio at maximum densification (e_{d0} - value of e_d for σ_{kk}=0), e_i - maximum void ratio (e_{i0} - value of e_i for σ_{kk}=0), α - pycnotropy coefficient, n - compression coefficient, β - stiffness coefficient, a_1 - parameter representing the deviatoric part of the normalized stress in critical states [17], ϕ_c - critical angle of internal friction during stationary flow, θ - Lode angle (angle on the deviatoric plane $\sigma_1+\sigma_2+\sigma_3=0$ between the stress vector and the axis σ_3; σ_i - principle stress vector).

Concurrent Multiscale Computational Modeling for Dense Dry Granular Materials Interfacing Deformable Solid Bodies

Richard A. Regueiro and Beichuan Yan

University of Colorado at Boulder, Boulder, CO, USA
e-mail: regueiro@colorado.edu, beichuan.yan@colorado.edu

Abstract. A method for concurrent multiscale computational modeling of interfacial mechanics between granular materials and deformable solid bodies is presented. It involves two main features: (1) coupling discrete element and higher order continuum finite element regions via an overlapping region; and (2) implementation of a finite strain micromorphic pressure sensitive plasticity model as the higher order continuum model in the overlap region. The third main feature, adaptivity, is not currently addressed, but is considered for future work. Single phase (solid grains) and dense conditions are limitations of the current modeling. Extensions to multiple phases (solid grains, pore liquid and gas) are part of future work. Applications include fundamental grain-scale modeling of interfacial mechanics between granular soil and tire, tool, or penetrometer, while properly representing far field boundary conditions for quasi-static and dynamic simulation.

1 Introduction

Granular materials are commonly found in nature and industrial processes, and are composites of three phases: solids, liquids, and gases. We limit the modeling currently to single phase (solid grains) and dense materials (average coordination number ≈ 5). Examples include metallic powders (for powder metallurgy), pharmaceutical pills, agricultural grains (in silo flows), dry soils (sand, silt, gravel), and lunar and martian regolith (soil found on the surface of the Moon and Mars), for instance. We are interested primarily in modeling the grain to macro-continuum scale response in the large shear deformation interface region between a granular material and deformable solid body. Such interface can be between a granular soil (e.g., sand, Fig.1(a)) and a tire(Fig.2(a)), tool (e.g., bucket, Fig.2(b)), or cone penetrometer (Fig.1(b)).

Granular materials remain an unmastered class of materials with regard to modeling their spectrum of mechanical behavior in a physically-based manner across several orders of magnitude in length-scale. They may transition in an instant from deforming like a solid to flowing like a fluid or gas and vice versa. Examples of such physical transition are the flow of quartz grains around and at the tip of a driven cone penetrometer penetrating sand, the shoveling of sand or gravel by a tractor bucket, and the flow of agricultural grains from the bulk top region through the bottom chute

R. Wan et al. (Eds.): Bifurcations, Instabilities and Degradations in Geomaterials, SSGG, pp. 251–273.
springerlink.com

(a) image courtesy of Khalid Alshibli, Louisiana State University

(b) [1]

Fig. 1. (a) Sand grains at 150×. (b) Cone penetrometers.

in a silo, for instance. These examples each involve material regions where relative neighbor particle motion is 'large' (flowing like a fluid or gas) and regions where relative neighbor particle motion is 'small' (deforming like a solid).

It is too computationally intensive to account for the grain-scale properties and intergranular constitutive behavior within a physics-based simulation (e.g., discrete element (DE) model) to understand fundamentally the mechanics in a large shear deformation interface region between deformable solid bodies and granular materials. Grain-scale properties include grain size, shape, sphericity, morphology, stiffness, strength, and surface friction, while intergranular constitutive behavior accounts for contact behavior and grain fracture/crushing, for instance. High fidelity particle DE computations that account for these features are expensive, requiring their application be restricted to regions of large shearing at the interface between granular media and a solid body. Boundary effects on the outer simulation boundaries of an assembly of particles interacting with the solid body will render the computational results questionable, because fictitious forces and wave reflections will occur at these outer boundaries of the box of particles, thus influencing in a numerically-artificial manner the actual interface-region mechanics (see section 1.1). To resolve the issue properly, it is necessary to introduce multiscale methods that correctly combine (1) efficient finite element (FE) and/or meshfree based continuum methods used in regions where phenomenological constitutive relationships are accurate, with (2) DE models used in regions where granular physics must be represented accurately (e.g., in the granular soil-tool, soil-tire, or soil-track interface region). The use of multiscale methods offers immediate payoff because the fewer discrete particles needed to simulate the interaction, the faster physics-based simulations can be conducted. As a result, more "what-if" scenarios can be simulated and more uncertainties in grain-scale material parameters can be investigated by simulation,

(a) Mars exploration rover: tire interaction with Martian soil (photo source NASA)

(b) loader bucket scooping gravel (www.dymaxinc.com)

Fig. 2. (a) Soil-tire, and (b) soil-tool interface problems

providing "error bars" on the physics-based simulation results. To make the multiscale approach feasible for granular media, an open research question must be addressed: how to maintain a fundamental granular physics representation in the large shear deformation interface region as the solid body shears through the granular material. At the heart of the question is how to achieve adaptability and coupling of the computational scheme to convert from continuum to particle representation around the solid body, as it shears through the granular material, and perform particle to continuum conversion in spatial regions where particles are less sheared or have stopped flowing, and thus a continuum representation is appropriate.

Therefore, the focus of the current research is to bridge grain-scale properties and mechanics to the macro-scale continuum behavior in a large shear deformation interface region between a deformable solid body (e.g., metal scoop, rubber tire, or metal track) and a dense cohesionless granular material (e.g., dry sand or gravel). A multiscale approach is presented to provide fundamental physics-based simulation consisting of (i) FE or rigid body mechanics for the solid body (scoop/tire/track) and DE for the granular material in the large shear deformation interface region (cf. Fig.3), and (ii) FE-DE for the representation of the granular material in the transition/coupling region. The transition (overlap region in the Fig.3) provides proper boundary conditions (BCs) on the physics-based computational discretization (i.e., proper BCs on the DE simulation region).

Ultimately, a fundamental understanding of granular physics interacting with a solid body can lead to improved design of devices for granular soil-machine tool and soil-tire interaction, and the interpretation of granular soil-penetrometer shear resistance interaction.

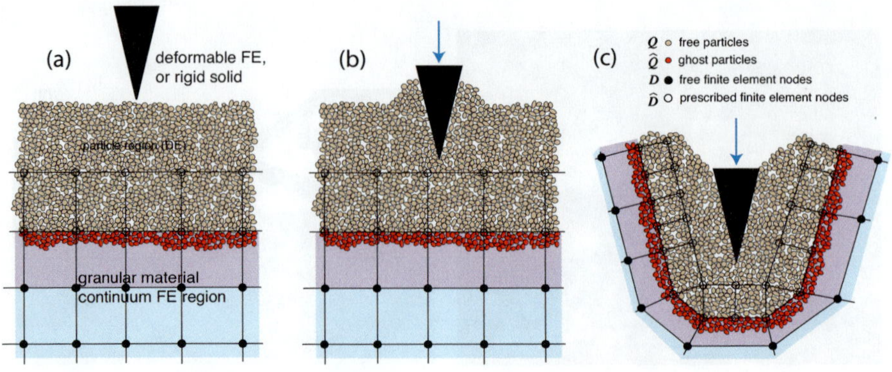

Fig. 3. Illustration of adaptivity and coupling. In (a), a deformable or rigid solid body approaches the granular material, and in (b) it begins to shear/penetrate the granular material in the DE particle region. In (c), the solid body has sheared the particle region enough that the FE mesh is re-meshed adaptively and the particle region is extended. Adaptivity is addressed in future work.

1.1 Motivation: Artificial Boundary Effects

A penetration test is simulated quasi-statically to demonstrate artificial boundary effects on a DE simulation. The penetrator is modeled using a larger ellipsoidal particle, and the boundaries are composed of fixed spherical particles, shown in Fig.4. Three different-sized containers are used, number of equal-sized particles being 2760, 4260 and 6088, respectively, with ellipsoidal particle radii $2.5 \times 2.0 \times 1.5$mm. Parameters for the DE simulation are shown in Table 1.

Table 1. Parameters of particles and numerical computation

Young's modulus E (Pa)	2.9×10^{10}
Poisson's ratio ν	0.25
specific gravity G_s	2.65
interparticle coef. of friction μ	0.5
interparticle contact damping ratio ξ	5%
particle radii (m)	$0.0015 \sim 0.0025$
background damping ratio	dynamic relaxation
time step $\triangle t$ (sec)	5.0×10^{-6}

The vertical force-displacement curves are plotted in Fig.4(c) for the penetrator particle. It can be found that the penetrator force increases as penetration increases. For a smaller container, the force has a larger value because of the boundary effect, as expected. The question then becomes how to make the shearing DE domain around a deformable solid body as small as possible without introducing artificial boundary effects. This is the overall goal of the research.

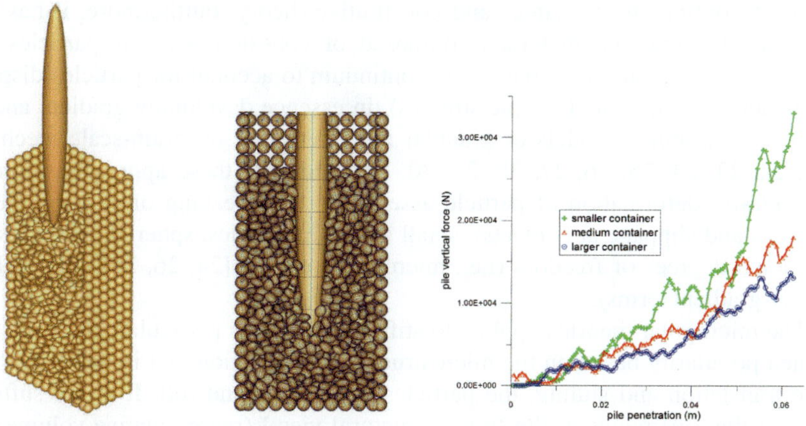

(a) 3D view (b) 2D view (c) force-displacement curves

Fig. 4. Cross-sectional view of penetration, and force-displacement curves

An outline of the remainder of the chapter is as follows: section 2 provides a literature review; 3 a summary of balance equations for a particle and micropolar continuum representation of a granular material and their coupling [2]; 4 a method for coupling DE to FE facets [3] and numerical example; 5 a summary; and 6 mention of ongoing and future work.

2 Literature Review

The literature review briefly covers work done on micromechanical modeling for granular materials, and computational methods for coupling particle and continuum representations of granular materials.

2.1 Micromechanical Continuum Models for Dense Dry Granular Materials

Apparently Reynolds [4] was the first to study granular materials at the grain scale, and coined the term "dilatancy" in the process. Others followed [5, 6] with attempts to relate continuum concepts like stress and strain to grain-scale behavior. Conferences were held to focus on micromechanical modeling of granular materials (this is not a complete list) [7, 8, 9, 10, 11]. The development of continuum relations like stress-strain equations based on micromechanical models of granular materials has spanned nearly five decades and continues today [12, 13, 14, 15, 16, 17, 18, 19, 20]. These micromechanically-based models attempt to bridge the grain to continuum scale mechanics of granular materials within the framework of continuum mechanics and constitutive theory. Furthermore, it has been proposed for granular materials composed of cohesionless, stiff particles (like spherical glass beads) to enhance the continuum to account for particles displacements and rotations (and couple stresses), in essence developing gradient and micropolar continuum models of granular material based on grain-scale mechanics [21, 22, 23, 24, 25, 26, 27, 28, 29, 30, 31]. Many of these approaches consider only elastic deformation of particle assemblies (no breaking of existing particle contacts and slippage at contacts), small strain kinematics, spherical particles, and rotational degrees of freedom (i.e., micropolar; except [24, 26, 28] who included higher-gradient terms).

The micropolar theories applied to stiff, cohesionless particulate materials have gained popularity based on the microstructural observation that in addition to particle translation and sliding, the particles may rotate and roll. It is not sufficient to limit the kinematics of the 'microstructural view' (representative volume) of a single particle or cluster of particles to rigid rotation. A representative microscopic volume of granular material—whether the particles are nearly rigid or deformable—will exhibit not only micro-rotation but also micro-shear and micro-stretch (micro-dilatation and micro-compaction). Such additional degrees of freedom within the mathematical framework for micromorphic continuum theories [32] give more realistic bridging kinematics between deformable and rigid particle mechanics and its continuum representation than a micropolar theory would provide.

2.2 Computational Particle/Continuum Coupling

As continuum micromechanical models were being developed, many recognized the role computers could play in simulating the discrete grain-scale response of granular materials. Such an approach has been called a Distinct Element Method or Discrete Element Method (DEM) [33, 34, 35, 36, 37, 38, 39, 40] (not a complete list). Certain DEM approaches model directly the physical grain size of the material, while others approximate the continuum as an assembly of particles approximating the continuum response discretely, wherein the particles have arbitrary size and thus provide an arbitrary internal length scale. Few approaches have coupled DEM and FEM for modeling deformation and flow of dense dry granular materials accounting for the physical particle size, i.e. truly micromechanically coupled models [41, 42, 43, 44]. These methods approach the coupling issue, however, as a contact/interface problem between discrete particles and finite element facets and not as overlapping regions of the same material, which an approach coupling particle and continuum representations of the same material should do. Examples of such approaches have been demonstrated for coupled atomistic-continuum regions [45, 46, 47]. Section 3.3 shows the extension of the approach by Klein and Zimmerman [47] to coupled overlapping particle and continuum regions, wherein significant differences have mainly to do with the DE representation of particles (with rotation and frictional sliding, as opposed to molecular dynamics for atoms) and inelastic micromorphic constitutive model for the continuum (and its associated FE implementation). The approach by Xiao and Belytschko [46] is also being considered, which could be somewhat simpler to implement.

Unit cell methods like that by Feyel and Chaboche [48] provide a method to up-scale underlying micromechanical simulations (such as DE) to a macro-scale simulation (such as FE). Belytschko et al. [49] extended the method to modeling fracture. They recognized the complexities and limitations of unit cell methods as they are currently formulated, implemented, and applied. Feyel [50] stated that, in addition to the periodicity assumption for the micro-structure (impossible to model localized deformation), the mechanical response near boundaries was not modeled properly. As a result, these methods are not well suited for modeling the interfacial mechanics of soil-tire, tool, or penetrometer interface conditions. The overlaying FE mesh would quickly become too distorted and require continuous remeshing, aside from the fact that the grain-scale DE mechanics would be influenced by the overlaying continuum mechanical response (through their coupling). The methods are useful, however, in up-scaling fracture or shear banding in a material, but not for interfacial mechanics, as far as we can tell.

3 Particle and Continuum Representations and Their Coupling

The balance of linear and angular momentum equations are presented for particle and continuum representations of a dense dry granular material. A strategy for coupling these equations within an overlap region (Fig.6) is summarized in section 3.3.

3.1 Particle Mechanics and Discrete Element Method

The balance of linear and angular momentum for a system of stiff elastic particles in contact may be written as [33]

$$
\begin{aligned}
&\boldsymbol{M}^Q\ddot{\boldsymbol{Q}}+\boldsymbol{C}^Q\dot{\boldsymbol{Q}}+\boldsymbol{F}^{INT,Q}(\boldsymbol{Q})=\boldsymbol{F}^{EXT,Q} \\
&\boldsymbol{M}^Q=\mathop{\mathbf{A}}_{\delta=1}^{N}\boldsymbol{m}_\delta^Q\,;\quad \boldsymbol{m}_\delta^Q=\begin{bmatrix} m_\delta & \mathbf{0} \\ \mathbf{0} & \boldsymbol{I}_\delta \end{bmatrix} \\
&\boldsymbol{F}^{INT,Q}=\mathop{\mathbf{A}}_{\delta=1}^{N}\boldsymbol{f}_\delta^{INT,Q}\,;\quad \boldsymbol{f}_\delta^{INT,Q}=\sum_{\varepsilon=1}^{n_c}\begin{bmatrix} \boldsymbol{f}^{\varepsilon,\delta} \\ \boldsymbol{r}^{\varepsilon,\delta}\times\boldsymbol{f}^{\varepsilon,\delta} \end{bmatrix} \\
&\boldsymbol{F}^{EXT,Q}=\mathop{\mathbf{A}}_{\delta=1}^{N}\boldsymbol{f}_\delta^{EXT,Q}\,;\quad \boldsymbol{f}_\delta^{EXT,Q}=\begin{bmatrix} \boldsymbol{f}^{EXT,\delta} \\ \boldsymbol{\ell}^{EXT,\delta} \end{bmatrix}
\end{aligned}
\tag{1}
$$

where $\boldsymbol{M}^Q$ is the mass and rotary inertia matrix for a system of N particles, $\boldsymbol{m}_\delta^Q$ is the mass and rotary inertia matrix for particle δ, $\boldsymbol{m}_\delta$ is the mass matrix for particle δ, $\boldsymbol{I}_\delta$ is the rotary inertia matrix for particle δ, $\mathbf{A}_{\delta=1}^{N}$ is an assembly operator to obtain the system matrices from the individual particle matrices and contact vectors, $\boldsymbol{C}^Q = a\boldsymbol{M}^Q$ the mass and rotary inertia proportional damping matrix with proportionality constant a (used in a dynamic relaxation solution method for quasi-static problems, but otherwise set to zero), $\boldsymbol{F}^{INT,Q}$ the internal force and moment vector associated with n_c particle contacts which is a nonlinear function of particle displacements and rotations when particles slide with friction, $\boldsymbol{f}_\delta^{INT,Q}$ the resultant internal force and moment vector for particle δ, $\boldsymbol{f}^{\varepsilon,\delta}$ the internal force vector for particle δ at contact ε, $\boldsymbol{r}^{\varepsilon,\delta} \times \boldsymbol{f}^{\varepsilon,\delta}$ the internal moment vector at the centroid of particle δ due to force at contact ε with moment arm $\boldsymbol{r}^{\varepsilon,\delta}$, $\boldsymbol{F}^{EXT,Q}$ the assembled external force and moment vector, $\boldsymbol{f}_\delta^{EXT,Q}$ the external body force and moment vector for particle δ, $\boldsymbol{f}^{EXT,\delta}$ the external body force vector at the centroid of particle δ, and $\boldsymbol{\ell}^{EXT,\delta}$ the external body moment vector at the centroid of particle δ. $\boldsymbol{Q}$ is the generalized degree of freedom (dof) vector for particle displacements and rotations

$$\boldsymbol{Q} = [\boldsymbol{q}_\delta, \boldsymbol{q}_\varepsilon, \ldots, \boldsymbol{q}_\eta, \boldsymbol{\theta}_\delta, \boldsymbol{\theta}_\varepsilon, \ldots, \boldsymbol{\theta}_\eta]^T \ , \quad \delta, \varepsilon, \ldots, \eta \in \mathscr{A} \tag{2}$$

where $\boldsymbol{q}_\delta$ is the displacement vector of particle δ, $\boldsymbol{\theta}_\delta$ its rotation vector, and $\mathscr{A}$ is the set of free particles. In general, a superscript Q denotes a variable associated with particle motion, whereas a superscript D will denote a variable associated with continuum deformation. Further details of assembling the matrices and vectors in (1) from the individual particle and particle contact contributions are not given here, as they are well established in the literature.

With regard to putting the particle mechanics and DE implementation into a form amenable to energy partitioning in the coupled particle-continuum overlap region, we consider an energy formulation of the balance equations using Lagrange's equation of motion. It may be stated as

$$\frac{d}{dt}\left(\frac{\partial T^Q}{\partial \dot{\boldsymbol{Q}}}\right) - \frac{\partial T^Q}{\partial \boldsymbol{Q}} + \frac{\partial F^Q}{\partial \dot{\boldsymbol{Q}}} + \frac{\partial U^Q}{\partial \boldsymbol{Q}} = \boldsymbol{F}^{EXT,Q} \tag{3}$$

where T^Q is the kinetic energy, F^Q the dissipation function, and U^Q the potential energy, such that

$$T^Q = \frac{1}{2}\dot{\boldsymbol{Q}}\boldsymbol{M}^Q\dot{\boldsymbol{Q}} \ , \ F^Q = aT^Q \ , \ U^Q(\boldsymbol{Q}) = \int_{\boldsymbol{0}}^{\boldsymbol{Q}} \boldsymbol{F}^{INT,Q}(\boldsymbol{S})d\boldsymbol{S} \tag{4}$$

The dissipation function F^Q is written as a linear function of the kinetic energy T^Q, which falls within the class of damping called Rayleigh damping (pg. 130 [51]). Carrying out the derivation in (3), and using the Second Fundamental Theorem of Calculus for $\partial U^Q/\partial \boldsymbol{Q}$, leads to (1).

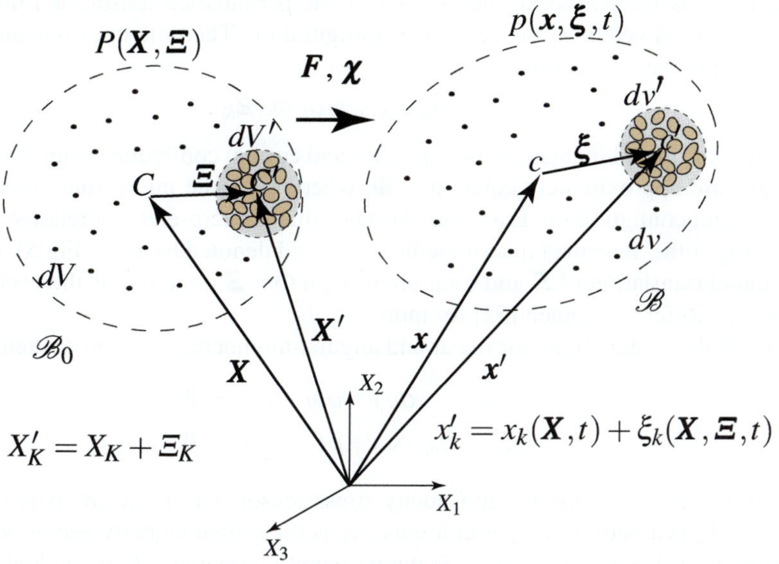

Fig. 5. Material points $P(\boldsymbol{X},\boldsymbol{\Xi})$ and $p(\boldsymbol{x},\boldsymbol{\xi},t)$ in reference and current configurations $\mathscr{B}_0$ and $\mathscr{B}$, respectively, centroids of macro-element C and c and micro-element C' and c', relative micro-element position vectors $\boldsymbol{\Xi}$ and $\boldsymbol{\xi}$, differential macro-element volumes dV and dv and micro-element volumes dV' and dv'. Because of linear kinematics assumption, $\mathscr{B}_0 \approx \mathscr{B}$, etc.

3.2 Micropolar Continuum and Finite Element Method

Following the formulation of Eringen [52], we present the balance of linear and angular momentum equations and finite element formulation for a small strain micropolar continuum (i.e., stiff particles with small frictional sliding in overlap region). For clarity of presentation, index tensor notation is used, and Cartesian coordinates are assumed.

The kinematics are reviewed in Fig.5. A micro-element differential volume dv' (and dV' in reference configuration[1]) is located by a relative position vector ξ_k from the centroid c of the macro-element material point with position x_k in the current configuration (and relative position vector Ξ_K from the centroid C of the macro-element material point with position X_K in the reference configuration). A micro-deformation tensor χ_{kK} relates the reference to current relative position vectors as $\xi_k = \chi_{kK}(\boldsymbol{X},t)\Xi_K$ (summation of repeated indices implied). For small strain micropolar kinematics, the micro-deformation tensor takes the form

$$\chi_{kK} = \delta_{kK} + \varepsilon_{kMK}\Phi_M \tag{5}$$

[1] Because of the assumption of linear kinematics, small rotations and strains, the reference and current configurations are nearly the same.

where δ_{kK} is the Kronecker delta, ε_{kMK} is the permutation tensor, and Φ_M is the micro-rotation vector in the reference configuration. Then the micro-element relative position vector becomes

$$\xi_k = \delta_{kK}\Xi_K + \varepsilon_{kMK}\Phi_M\Xi_K \tag{6}$$

Because of linear kinematics, the reference and current configuration micro-rotation vectors are approximately equal $\boldsymbol{\varphi} \approx \boldsymbol{\Phi}$, where φ_k is the micro-rotation vector in the current configuration. Equation (6) states that a micro-element relative position vector $\boldsymbol{\xi}$ at the deformed macro-element centroid denoted by $\boldsymbol{x}$ (cf. Fig.5), involves a parallel translation of $\boldsymbol{\Xi}$ and rotation through $\boldsymbol{\Phi} \times \boldsymbol{\Xi}$ (where $\times$ is the vector cross product). Refer to Eringen [52] for more details.

The balance equations for linear and angular momentum may be written as

$$\sigma_{lk,l} + \rho b_k - \rho \dot{v}_k = 0 \tag{7}$$

$$m_{lk,l} + \varepsilon_{kmn}\sigma_{mn} + \rho \ell_k - \rho \dot{\beta}_k = 0 \tag{8}$$

where σ_{lk} is the unsymmetric Cauchy stress tensor over body $\mathscr{B}$, ρ is the mass density, b_k is a body force per unit mass, v_k is the spatial velocity vector, m_{lk} is the unsymmetric couple stress, ε_{kmn} is the permutation operator, ℓ_k is the body couple per unit mass, β_k is the intrinsic spin per unit mass, indices $k,l,\cdots = 1,2,3$, and $(\bullet)_{,l} = \partial(\bullet)/\partial x_l$ denotes partial differentiation with respect to the spatial coordinate x_l. The micro-gyration vector ν_l for linear kinematics is written as

$$\nu_l = \dot{\varphi}_l \, , \; \dot{\nu}_l = \ddot{\varphi}_l \tag{9}$$

Introducing w_k and η_k as weighting functions for the macro-displacement vector u_k and micro-rotation vector φ_k, respectively, we apply the Method of Weighted Residuals to formulate the partial differential equations in (7) and (8) into weak form [53]. The weak, or variational, equations then result as

$$\int_{\mathscr{B}} \rho w_k \dot{v}_k dv + \int_{\mathscr{B}} w_{k,l}\sigma_{lk} dv = \int_{\mathscr{B}} \rho w_k b_k dv + \int_{\Gamma_t} w_k t_k da \tag{10}$$

$$\int_{\mathscr{B}} \rho \eta_k \dot{\beta}_k dv + \int_{\mathscr{B}} \eta_{k,l} m_{lk} dv - \int_{\mathscr{B}} \eta_k \varepsilon_{kmn}\sigma_{mn} dv = \int_{\mathscr{B}} \rho \eta_k \ell_k dv + \int_{\Gamma_r} \eta_k r_k da \tag{11}$$

where $\mathscr{B}$ is the volume of the continuum body, $t_k = \sigma_{lk} n_l$ is the applied traction on the portion of the boundary Γ_t with outward normal vector n_l, and $r_k = m_{lk} n_l$ is the applied surface couple on the portion of the boundary Γ_r.

The weak equations (10) and (11) may be approximated in Galerkin form [53], whereby the discretization parameter h implies a discrete approximation, in this case finite element discretization. Introducing shape functions N_a^u and N_b^φ for the macro-displacement u_k^h and micro-rotation φ_k^h vectors, respectively, and assuming the micro-inertia is approximately constant for small strains and rotations (microinertia j_{lk} is nearly constant, and $\dot{\beta}_k^h \approx j_{lk}\ddot{\varphi}_l^h$), we may write the interpolations and derivatives as

$$u_k^h = \sum_{a=1}^{n_{en}^u} N_a^u d_{k(a)} \ , \ \dot{v}_k^h = \sum_{a=1}^{n_{en}^u} N_a^u \ddot{d}_{k(a)} \tag{12}$$

$$w_k^h = \sum_{a=1}^{n_{en}^u} N_a^u c_{k(a)} \ , \ w_{k,l}^h = \sum_{a=1}^{n_{en}^u} (N_a^u)_{,l} c_{k(a)} \tag{13}$$

$$\varphi_l^h = \sum_{b=1}^{n_{en}^\varphi} N_b^\varphi \phi_{l(b)} \ , \ \ddot{\varphi}_l^h = \sum_{b=1}^{n_{en}^\varphi} N_b^\varphi \ddot{\phi}_{l(b)} \tag{14}$$

$$\eta_k^h = \sum_{b=1}^{n_{en}^\varphi} N_b^\varphi e_{k(b)} \ , \ \eta_{k,l}^h = \sum_{b=1}^{n_{en}^\varphi} (N_b^\varphi)_{,l} e_{k(b)} \tag{15}$$

where $d_{k(a)}$ is the displacement vector at node a, $\phi_{l(b)}$ is the rotation vector at node b, $c_{k(a)}$ is the displacement weighting function vector at node a, $e_{k(b)}$ is the rotation weighting function vector at node b, n_{en}^u is the number of element nodes associated with interpolating the continuum macro-displacement vector, and n_{en}^φ is the number of element nodes associated with interpolating the continuum micro-rotation vector. It is assumed that the shape functions and integrals are expressed in natural coordinates for an isoparametric formulation, but such details are omitted and can be found in the textbook by Hughes [53]. Substituting these approximations into the Galerkin form, accounting for essential boundary conditions, and recognizing that the nodal weighting function values are arbitrary (except where essential boundary conditions are applied, and nodal weighting function values are zero), we arrive at a coupled matrix form of the linear and angular momentum balance equations as

$$\boldsymbol{M}^u \ddot{\boldsymbol{d}} + \boldsymbol{F}^{INT,u}(\boldsymbol{d},\boldsymbol{\phi}) = \boldsymbol{F}_b + \boldsymbol{F}_t \tag{16}$$

$$\boldsymbol{M}^\varphi \ddot{\boldsymbol{\phi}} + \boldsymbol{F}^{INT,\varphi}(\boldsymbol{d},\boldsymbol{\phi}) = \boldsymbol{F}_\ell + \boldsymbol{F}_r \tag{17}$$

where matrices and vectors are assembled from their element contributions using a finite element assembly operator [53] as

$$\boldsymbol{M}^u = \mathop{\mathbf{A}}_{e=1}^{n_{\mathrm{el}}} \boldsymbol{m}^{e,u} \ , \ \boldsymbol{m}^{e,u} = \int_{\mathscr{B}^e} \rho (\boldsymbol{N}^{e,u})^T \boldsymbol{N}^{e,u} dv \tag{18}$$

$$\boldsymbol{M}^\varphi = \mathop{\mathbf{A}}_{e=1}^{n_{\mathrm{el}}} \boldsymbol{m}^{e,\varphi} \ , \ \boldsymbol{m}^{e,\varphi} = \int_{\mathscr{B}^e} \rho (\boldsymbol{N}^{e,\varphi})^T \boldsymbol{j} \boldsymbol{N}^{e,\varphi} dv \tag{19}$$

$$\boldsymbol{F}^{INT,u}(\boldsymbol{d},\boldsymbol{\phi}) = \mathop{\mathbf{A}}_{e=1}^{n_{\mathrm{el}}} \boldsymbol{f}^{e,INT,u} \ , \ \boldsymbol{f}^{e,INT,u} = \int_{\mathscr{B}^e} (\boldsymbol{B}^{e,u})^T \boldsymbol{\sigma}(\boldsymbol{d}^e,\boldsymbol{\phi}^e) dv \tag{20}$$

$$\boldsymbol{F}^{INT,\varphi}(\boldsymbol{d},\boldsymbol{\phi}) = \mathop{\mathbf{A}}_{e=1}^{n_{\mathrm{el}}} \boldsymbol{f}^{e,INT,\varphi} \tag{21}$$

$$\boldsymbol{f}^{e,INT,\varphi} = \int_{\mathscr{B}^e} (\boldsymbol{B}^{e,\varphi})^T \boldsymbol{m}(\boldsymbol{d}^e,\boldsymbol{\phi}^e) dv - \int_{\mathscr{B}^e} (\boldsymbol{N}^{e,\varphi})^T \boldsymbol{\sigma}^{\varepsilon}(\boldsymbol{d}^e,\boldsymbol{\phi}^e) dv$$

$$\boldsymbol{F}_b = \mathop{\mathbf{A}}_{e=1}^{n_{\text{el}}} \boldsymbol{f}_b^{e,EXT,u} \; , \; \boldsymbol{f}_b^{e,EXT,u} = \int_{\mathscr{B}^e} \rho (\boldsymbol{N}^{e,u})^T \boldsymbol{b} dv \tag{22}$$

$$\boldsymbol{F}_\ell = \mathop{\mathbf{A}}_{e=1}^{n_{\text{el}}} \boldsymbol{f}_\ell^{e,EXT,\varphi} \; , \; \boldsymbol{f}_\ell^{e,EXT,\varphi} = \int_{\mathscr{B}^e} \rho (\boldsymbol{N}^{e,\varphi})^T \boldsymbol{\ell} dv \tag{23}$$

$$\boldsymbol{F}_t = \mathop{\mathbf{A}}_{e=1}^{n_{\text{el}}} \int_{\Gamma_t^e} (\boldsymbol{N}^{e,u})^T \boldsymbol{t} da \; , \; \boldsymbol{F}_r = \mathop{\mathbf{A}}_{e=1}^{n_{\text{el}}} \int_{\Gamma_r^e} (\boldsymbol{N}^{e,\varphi})^T \boldsymbol{r} da \tag{24}$$

where $\mathbf{A}_{e=1}^{n_{\text{el}}}$ is the element assembly operator, n_{el} is the number of elements, $\boldsymbol{N}_e^u$, $\boldsymbol{N}_e^\varphi$, $\boldsymbol{j}$, $\boldsymbol{B}_e^u$, $\boldsymbol{\sigma}$, $\boldsymbol{d}^e$, $\boldsymbol{\phi}^e$, $\boldsymbol{B}_e^\varphi$, $\boldsymbol{m}$, $\boldsymbol{\sigma}^\varepsilon$, $\boldsymbol{b}$, $\boldsymbol{\ell}$, $\boldsymbol{t}$, and $\boldsymbol{r}$ are the element matrix and vector forms of N_a^u, N_b^φ, j_{lk}, $(N_a^u)_{,l}$, σ_{lk}, $d_{k(a)}$, $\phi_{l(b)}$, $(N_b^\varphi)_{,l}$, m_{lk}, $\varepsilon_{kmn}\sigma_{mn}$, b_k, ℓ_k, t_k, and r_k, respectively.

Introducing a generalized nodal degree of freedom vector $\boldsymbol{D}$, the coupled micropolar linear and angular momentum balance equations are written as

$$\boldsymbol{M}^D \ddot{\boldsymbol{D}} + \boldsymbol{F}^{INT,D}(\boldsymbol{D}) = \boldsymbol{F}^{EXT,D} \tag{25}$$

$$\boldsymbol{M}^D = \begin{bmatrix} \boldsymbol{M}^u & \boldsymbol{0} \\ \boldsymbol{0} & \boldsymbol{M}^\varphi \end{bmatrix} \quad \boldsymbol{D} = \begin{bmatrix} \boldsymbol{d} \\ \boldsymbol{\phi} \end{bmatrix}$$
$$\boldsymbol{F}^{INT,D} = \begin{bmatrix} \boldsymbol{F}^{INT,u} \\ \boldsymbol{F}^{INT,\varphi} \end{bmatrix} \quad \boldsymbol{F}^{EXT,D} = \begin{bmatrix} \boldsymbol{F}_b + \boldsymbol{F}_t + \boldsymbol{F}_g^u \\ \boldsymbol{F}_\ell + \boldsymbol{F}_r + \boldsymbol{F}_g^\varphi \end{bmatrix} \tag{26}$$

With regard to putting the continuum micropolar mechanics and finite element implementation into a form amenable to energy partitioning in the coupled particle-continuum overlap region, we consider an energy formulation of the balance equations using Lagrange's equation of motion. It may be stated as

$$\frac{d}{dt}\left(\frac{\partial T^D}{\partial \dot{\boldsymbol{D}}}\right) - \frac{\partial T^D}{\partial \boldsymbol{D}} + \frac{\partial F^D}{\partial \dot{\boldsymbol{D}}} + \frac{\partial U^D}{\partial \boldsymbol{D}} = \boldsymbol{F}^{EXT,D} \tag{27}$$

where T^D is the kinetic energy, F^D the dissipation function, and U^D the potential energy, such that

$$T^D = \frac{1}{2}\dot{\boldsymbol{D}}\boldsymbol{M}^D\dot{\boldsymbol{D}} \; , \; F^D = 0 \; , \; U^D(\boldsymbol{D}) = \int_{\boldsymbol{0}}^{\boldsymbol{D}} \boldsymbol{F}^{INT,D}(\boldsymbol{S}) d\boldsymbol{S} \tag{28}$$

Carrying out the derivation in (27) leads to (25), assuming constant inertia $\boldsymbol{M}^D$.

3.3 Coupling Method

An aspect of the computational concurrent multiscale modeling approach is to couple regions of material represented by particle DE to regions of material represented by continuum FE. Another aspect is to bridge the particle mechanics to a continuum representation using finite strain micromorphic plasticity (see [54, 55]), whereas the small strain micropolar continuum is a simple approximation of stiff particles with small frictional sliding in the overlap region. The coupling implementation will allow arbitrarily overlapping particle and continuum regions in a single "handshaking" or overlap region such that fictitious forces and wave reflections are minimized in the overlap region. In theory, for nearly homogeneous deformation, if the particle and continuum regions share the same region (i.e., are completely overlapped), the results should be the same as if the overlap region is a subset of the overall problem domain (cf. Fig.6). This will serve as a future benchmark problem for the numerical implementation. The coupling implementation extends to particle mechanics and micropolar continuum the "bridging scale decomposition" proposed by Wagner and Liu [45] and modifications thereof by Klein and Zimmerman [47] (see references therein for further background on these atomistic continuum methods).

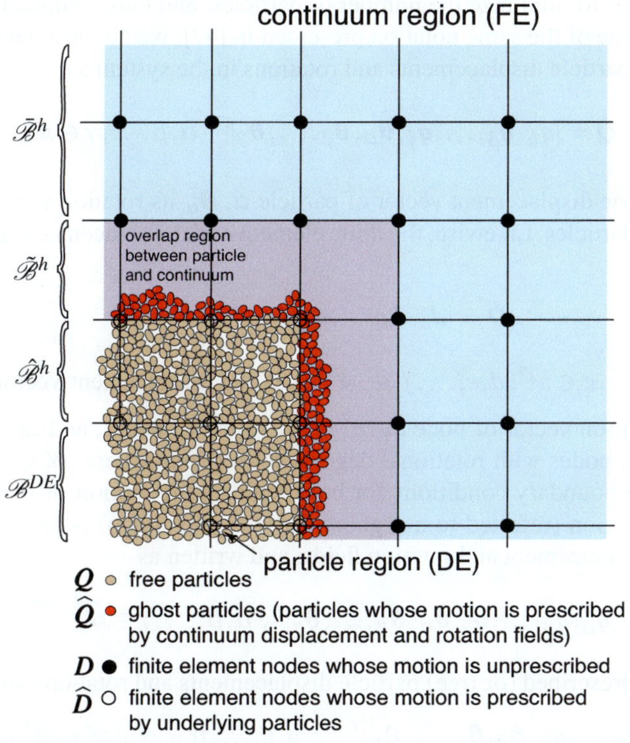

Fig. 6. Two-dimensional illustration of the coupling between particle and continuum regions. The purple background denotes the FE overlap region $\tilde{\mathscr{B}}^h$ with underlying ghost particles, aqua blue the FE continuum region $\bar{\mathscr{B}}^h$ with no underlying particles, and white background (with brown particles) the free particle region $\hat{\mathscr{B}}^h \cup \mathscr{B}^{DE}$. In summary, the finite element domain $\mathscr{B}^h$ is the union of pure continuum FE domain $\bar{\mathscr{B}}^h$, overlapping FE domain with underlying ghost particles $\tilde{\mathscr{B}}^h$, and overlapping FE domain with underlying free particles $\hat{\mathscr{B}}^h$, such that $\mathscr{B}^h = \bar{\mathscr{B}}^h \cup \tilde{\mathscr{B}}^h \cup \hat{\mathscr{B}}^h$. The pure particle domain with no overlapping FE domain is indicated by $\mathscr{B}^{DE}$.

3.3.1 Kinematics

Here, a summary of the kinematics of the coupled regions is given, following the illustration shown in Fig.6. It is assumed that the finite element mesh covers the domain of the problem in which the material is behaving more solid-like, whereas in regions of large relative particle motion (fluid-like), a particle mechanics representation is used (DE). In Fig.6, discrete domains are defined, such as the pure particle domain (no overlapping FE mesh) as $\mathscr{B}^{DE}$, the FE domain $\mathscr{B}^h = \hat{\mathscr{B}}^h \cup \tilde{\mathscr{B}}^h \cup \bar{\mathscr{B}}^h$, where $\hat{\mathscr{B}}^h$ is the overlapping FE domain where nodal dofs are completely prescribed by the underlying particle DE, $\tilde{\mathscr{B}}^h$ the overlapping FE domain where particle DE motions and nodal dofs are prescribed and free nodal dofs exist, and $\bar{\mathscr{B}}^h$ the pure continuum FE domain with no underlying particles. The goal is to have the overlap region $\hat{\mathscr{B}}^h \cup \tilde{\mathscr{B}}^h$ as close to the region of interest (e.g., penetrometer skin, bucket, or tire tread) as to minimize the number of particles, and thus computational effort. Following some of the same notation presented in [47], we define a generalized dof vector $\breve{\boldsymbol{Q}}$ for particle displacements and rotations in the system as

$$\breve{\boldsymbol{Q}} = [\boldsymbol{q}_\alpha, \boldsymbol{q}_\beta, \ldots, \boldsymbol{q}_\gamma, \boldsymbol{\theta}_\alpha, \boldsymbol{\theta}_\beta, \ldots, \boldsymbol{\theta}_\gamma]^T \,,\; \alpha, \beta, \ldots, \gamma \in \breve{\mathscr{A}} \tag{29}$$

where $\boldsymbol{q}_\alpha$ is the displacement vector of particle α, $\boldsymbol{\theta}_\alpha$ its rotation vector, and $\breve{\mathscr{A}}$ is the set of all particles. Likewise, the finite element nodal displacements and rotations are written as

$$\breve{\boldsymbol{D}} = [\boldsymbol{d}_a, \boldsymbol{d}_b, \ldots, \boldsymbol{d}_c, \boldsymbol{\phi}_d, \boldsymbol{\phi}_e, \ldots, \boldsymbol{\phi}_f]^T \tag{30}$$

where $a, b, \ldots, c \in \breve{\mathscr{N}}$, $d, e, \ldots, f \in \breve{\mathscr{M}}$, $\boldsymbol{d}_a$ is the displacement vector of node a, $\boldsymbol{\phi}_d$ is the rotation vector of node d, $\breve{\mathscr{N}}$ is the set of all nodes, and $\breve{\mathscr{M}}$ is the set of finite element nodes with rotational degrees of freedom, where $\breve{\mathscr{M}} \subset \breve{\mathscr{N}}$. In order to satisfy the boundary conditions for both regions, the motion of the particles in the overlap region (referred to as "ghost particles," cf. Fig.6) is prescribed by the continuum displacement and rotation fields, and written as

$$\widehat{\boldsymbol{Q}} = [\boldsymbol{q}_\alpha, \boldsymbol{q}_\beta, \ldots, \boldsymbol{q}_\gamma, \boldsymbol{\theta}_\alpha, \boldsymbol{\theta}_\beta, \ldots, \boldsymbol{\theta}_\gamma]^T \,,\; \alpha, \beta, \ldots, \gamma \in \widehat{\mathscr{A}} \,,\; \widehat{\mathscr{A}} \in \tilde{\mathscr{B}}^h \tag{31}$$

while the unprescribed (or free) particle displacements and rotations are

$$\boldsymbol{Q} = [\boldsymbol{q}_\delta, \boldsymbol{q}_\varepsilon, \ldots, \boldsymbol{q}_\eta, \boldsymbol{\theta}_\delta, \boldsymbol{\theta}_\varepsilon, \ldots, \boldsymbol{\theta}_\eta]^T \,, \quad \delta, \varepsilon, \ldots, \eta \in \mathscr{A} \,,\; \mathscr{A} \in \tilde{\mathscr{B}}^h \cup \mathscr{B}^{DE} \tag{32}$$

where $\widehat{\mathscr{A}} \cup \mathscr{A} = \breve{\mathscr{A}}$ and $\widehat{\mathscr{A}} \cap \mathscr{A} = \emptyset$. Likewise, the displacements and rotations of nodes overlaying the particle region are prescribed by the particle motion and written as

$$\widehat{\boldsymbol{D}} = [\boldsymbol{d}_a, \boldsymbol{d}_b, \ldots, \boldsymbol{d}_c, \boldsymbol{\phi}_d, \boldsymbol{\phi}_e, \ldots, \boldsymbol{\phi}_f]^T \tag{33}$$

where $a,b,\ldots,c \in \widehat{\mathscr{N}}, d,e,\ldots,f \in \widehat{\mathscr{M}}, \widehat{\mathscr{N}}, \widehat{\mathscr{M}} \in \tilde{\mathscr{B}}^h \cup \hat{\mathscr{B}}^h$, while the unprescribed (or free) nodal displacements and rotations are

$$\boldsymbol{D} = [\boldsymbol{d}_m, \boldsymbol{d}_n, \ldots, \boldsymbol{d}_s, \boldsymbol{\phi}_t, \boldsymbol{\phi}_u, \ldots, \boldsymbol{\phi}_v]^T \tag{34}$$

where $m,n,\ldots,s \in \mathscr{N}, t,u,\ldots,v \in \mathscr{M}, \mathscr{N}, \mathscr{M} \in \tilde{\mathscr{B}}^h \cup \bar{\mathscr{B}}^h$, $\widehat{\mathscr{N}} \cup \mathscr{N} = \breve{\mathscr{N}}$, $\widehat{\mathscr{N}} \cap \mathscr{N} = \emptyset$, $\widehat{\mathscr{M}} \cup \mathscr{M} = \breve{\mathscr{M}}$, and $\widehat{\mathscr{M}} \cap \mathscr{M} = \emptyset$. Referring to Fig.6, the prescribed particle motions $\widehat{\boldsymbol{Q}}$ can be viewed as constrained boundary particles on the free particle region, and likewise the prescribed finite element nodal displacements and rotations $\widehat{\boldsymbol{D}}$ can be viewed as constrained boundary nodes on the finite element mesh in the overlap region.

In general, the displacement vector of a particle α can be represented by the finite element interpolation of the continuum macro-displacement field $\boldsymbol{u}^h$ evaluated at the particle centroid $\boldsymbol{x}_\alpha$, such that

$$\boldsymbol{u}^h(\boldsymbol{x}_\alpha, t) = \sum_{a \in \breve{\mathscr{N}}} N_a^u(\boldsymbol{x}_\alpha) \boldsymbol{d}_a(t) \quad \alpha \in \breve{\mathscr{A}} \tag{35}$$

where N_a^u are the shape functions associated with the continuum displacement field $\boldsymbol{u}^h$. Recall that N_a^u have compact support and thus are only evaluated for particles with centroids that lie within an element containing node a in its domain. In DE, particle dofs (translations and rotations) are tracked at the particle centroids, as are resultant forces and moments (from forces acting at contacts). For example, we can write the prescribed displacement of ghost particle α as

$$\boldsymbol{q}_\alpha(t) = \boldsymbol{u}^h(\boldsymbol{x}_\alpha, t) = \sum_{a \in \breve{\mathscr{N}}} N_a^u(\boldsymbol{x}_\alpha) \boldsymbol{d}_a(t) \quad \alpha \in \widehat{\mathscr{A}} \tag{36}$$

Likewise, particle rotation vectors can be represented by the finite element interpolation of the continuum micro-rotation field $\boldsymbol{\varphi}^h$ evaluated at the particle centroid $\boldsymbol{x}_\alpha$, such that

$$\boldsymbol{\varphi}^h(\boldsymbol{x}_\alpha, t) = \sum_{b \in \breve{\mathscr{M}}} N_b^\varphi(\boldsymbol{x}_\alpha) \boldsymbol{\phi}_b(t) \quad \alpha \in \breve{\mathscr{A}} \tag{37}$$

where N_b^φ are the shape functions associated with the micro-rotation field $\boldsymbol{\varphi}^h$. For example, we can write the prescribed rotation of ghost particle α as

$$\boldsymbol{\theta}_\alpha(t) = \boldsymbol{\varphi}^h(\boldsymbol{x}_\alpha, t) = \sum_{b \in \breve{\mathscr{M}}} N_b^\varphi(\boldsymbol{x}_\alpha) \boldsymbol{\phi}_b \quad \alpha \in \widehat{\mathscr{A}} \tag{38}$$

For all ghost particles (cf. Fig.6), the interpolations can be written as

$$\widehat{\boldsymbol{Q}} = \boldsymbol{N}_{\widehat{Q}D} \boldsymbol{D} + \boldsymbol{N}_{\widehat{Q}\widehat{D}} \widehat{\boldsymbol{D}} \tag{39}$$

where $\boldsymbol{N}_{\widehat{Q}D}$ and $\boldsymbol{N}_{\widehat{Q}\widehat{D}}$ are shape function matrices containing individual nodal shape functions N_a^u and N_b^φ, but for now these matrices will be left general to increase our flexibility in choosing interpolation/projection functions (such as those used in meshfree methods). Overall, the particle displacements and rotations may be written as

$$\begin{bmatrix} \boldsymbol{Q} \\ \widehat{\boldsymbol{Q}} \end{bmatrix} = \begin{bmatrix} \boldsymbol{N}_{QD} & \boldsymbol{N}_{Q\widehat{D}} \\ \boldsymbol{N}_{\widehat{Q}D} & \boldsymbol{N}_{\widehat{Q}\widehat{D}} \end{bmatrix} \cdot \begin{bmatrix} \boldsymbol{D} \\ \widehat{\boldsymbol{D}} \end{bmatrix} + \begin{bmatrix} \boldsymbol{Q}' \\ \boldsymbol{0} \end{bmatrix} \tag{40}$$

where $\boldsymbol{Q}'$ is introduced [47] as the error (or "fine-scale" [45]) in the interpolation of the free particle displacements and rotations $\boldsymbol{Q}$, whose function space is not rich enough to represent the true free particle motion. The shape function matrices $\boldsymbol{N}$ are in general not square because the number of free particles are not the same as free nodes and prescribed nodes, and number of ghost particles not the same as prescribed and free nodes. A scalar measure of error in particle displacements and rotations is defined as [47] $e = \boldsymbol{Q}' \cdot \boldsymbol{Q}'$, which may be minimized with respect to prescribed continuum nodal dofs $\widehat{\boldsymbol{D}}$ to solve for $\widehat{\boldsymbol{D}}$ in terms of free particle and continuum nodal dofs as

$$\widehat{\boldsymbol{D}} = \boldsymbol{M}^{-1}_{\widehat{D}\widehat{D}} \boldsymbol{N}^T_{Q\widehat{D}} (\boldsymbol{Q} - \boldsymbol{N}_{QD}\boldsymbol{D}) \,, \quad \boldsymbol{M}_{\widehat{D}\widehat{D}} = \boldsymbol{N}^T_{Q\widehat{D}} \boldsymbol{N}_{Q\widehat{D}} \tag{41}$$

This is known as the "discretized L_2 projection" [47] of the free particle motion $\boldsymbol{Q}$ and free nodal dofs $\boldsymbol{D}$ onto the prescribed nodals dofs $\widehat{\boldsymbol{D}}$. Upon substituting (41) into (39), we may write the prescribed particle dofs $\widehat{\boldsymbol{Q}}$ in terms of free particle $\boldsymbol{Q}$ and continuum nodal $\boldsymbol{D}$ dofs. In summary, these relations are written as

$$\widehat{\boldsymbol{Q}} = \boldsymbol{B}_{\widehat{Q}Q}\boldsymbol{Q} + \boldsymbol{B}_{\widehat{Q}D}\boldsymbol{D} \,, \; \widehat{\boldsymbol{D}} = \boldsymbol{B}_{\widehat{D}Q}\boldsymbol{Q} + \boldsymbol{B}_{\widehat{D}D}\boldsymbol{D} \tag{42}$$

where

$$\begin{aligned} \boldsymbol{B}_{\widehat{Q}Q} &= \boldsymbol{N}_{\widehat{Q}\widehat{D}}\boldsymbol{B}_{\widehat{D}Q} \,, \; \boldsymbol{B}_{\widehat{Q}D} = \boldsymbol{N}_{\widehat{Q}D} + \boldsymbol{N}_{\widehat{Q}\widehat{D}}\boldsymbol{B}_{\widehat{D}D} \\ \boldsymbol{B}_{\widehat{D}Q} &= \boldsymbol{M}^{-1}_{\widehat{D}\widehat{D}}\boldsymbol{N}^T_{Q\widehat{D}} \,, \; \boldsymbol{B}_{\widehat{D}D} = -\boldsymbol{M}^{-1}_{\widehat{D}\widehat{D}}\boldsymbol{N}^T_{Q\widehat{D}}\boldsymbol{N}_{QD} \end{aligned} \tag{43}$$

As shown in Fig.6, for a finite element implementation of this dof coupling, we expect that free particle dofs $\boldsymbol{Q}$ will not fall within the support of free continuum nodal dofs $\boldsymbol{D}$, such that it can be assumed that $\boldsymbol{N}_{QD} = \boldsymbol{0}$. The assumption $\boldsymbol{N}_{QD} \neq \boldsymbol{0}$ would be valid for a meshfree projection of the particle motions to the FE nodal dofs, as in [47], where we could imagine that the domain of influence of the meshfree projection could encompass a free particle centroid; the degree of encompassment would be controlled by the chosen support size of the meshfree kernel function. The choice of meshfree projection in [47] was not necessarily to allow $\boldsymbol{Q}$ be projected to $\boldsymbol{D}$ (and vice versa), but to remove the computationally costly calculation of the inverse $\boldsymbol{M}^{-1}_{\widehat{D}\widehat{D}}$ in (42). Since we will also be using the Tahoe code `sourceforge.net/projects/tahoe` for the coupled multiscale particle-continuum implementation, where the meshfree projection has been implemented for atomistic-continuum coupling [47], we will also consider the meshfree projection in future implementations.

3.3.2 Kinetic and Potential Energy Partitioning and Coupling

We assume the total kinetic and potential energy and dissipation of the coupled particle-continuum system may be written as the sum of the energies

$$\begin{aligned} T(\dot{\boldsymbol{Q}},\dot{\boldsymbol{D}}) &= T^Q(\dot{\boldsymbol{Q}},\dot{\widehat{\boldsymbol{Q}}}(\dot{\boldsymbol{Q}},\dot{\boldsymbol{D}})) + T^D(\dot{\boldsymbol{D}},\dot{\widehat{\boldsymbol{D}}}(\dot{\boldsymbol{Q}})) \\ U(\boldsymbol{Q},\boldsymbol{D}) &= U^Q(\boldsymbol{Q},\widehat{\boldsymbol{Q}}(\boldsymbol{Q},\boldsymbol{D})) + U^D(\boldsymbol{D},\widehat{\boldsymbol{D}}(\boldsymbol{Q})) \\ F(\dot{\boldsymbol{Q}},\dot{\boldsymbol{D}}) &= F^Q(\dot{\boldsymbol{Q}},\dot{\widehat{\boldsymbol{Q}}}(\dot{\boldsymbol{Q}},\dot{\boldsymbol{D}})) \end{aligned} \tag{44}$$

where we have indicated the functional dependence of the prescribed particle motion and nodal dofs solely upon the free particle motion and nodal dofs $\boldsymbol{Q}$ and $\boldsymbol{D}$, respectively. Note that the dissipation function $F = F^Q$ only applies for the particle system, and only for static problems (dynamic relaxation DE simulation). For purely dynamical problems, $F^Q = 0$, and there is only dissipation in the particle system if particles are allowed to slide frictionally, and the continuum has plasticity or other inelastic constitutive response. Lagrange's equations may then be stated as

$$\begin{aligned} \frac{d}{dt}\left(\frac{\partial T}{\partial \dot{\boldsymbol{Q}}}\right) - \frac{\partial T}{\partial \boldsymbol{Q}} + \frac{\partial F}{\partial \dot{\boldsymbol{Q}}} + \frac{\partial U}{\partial \boldsymbol{Q}} &= \boldsymbol{F}^{EXT,Q} \\ \frac{d}{dt}\left(\frac{\partial T}{\partial \dot{\boldsymbol{D}}}\right) - \frac{\partial T}{\partial \boldsymbol{D}} + \frac{\partial F}{\partial \dot{\boldsymbol{D}}} + \frac{\partial U}{\partial \boldsymbol{D}} &= \boldsymbol{F}^{EXT,D} \end{aligned} \tag{45}$$

which lead to a coupled system of governing equations (linear and angular momentum) for the coupled particle-continuum mechanics. If the potential energy U is nonlinear with regard to particle frictional sliding and micropolar (or micromorphic) plasticity, then (45) may be integrated in time and linearized for solution by the Newton-Raphson method. The benefit of this multiscale method, as pointed out by Wagner and Liu [45], is that time steps are different for the DE and FE solutions. A multiscale time stepping scheme will follow an approach similar to [45].

4 DE-FE Facet Coupling

This section describes a preliminary method for coupling DE to FE codes, in this case through a single layer of ghost particles tied to FE facets. This is a code communication exercise, to ensure that ELLIP3D [3] can communicate with Tahoe, the DE and FE codes used in the coupling.

4.1 DE-FE Facet Coupling Method

A simple granular-continuum coupling scheme is used initially, illustrated in Fig.7. The FE mesh does not cover the entire domain. Instead, the FE and DE regions only overlap through a single layer of particles. This layer of particles is embedded on the surface of the FE domain with centroids constrained to FE facets and deform with FE mesh. We call these particles "ghost" particles, as done in atomistic-continuum coupling methods. Theoretically, the ghost particles can comprise multiple layers and extrude into/overlap with the FE mesh, but this is left for future work [2]. No energy partitioning is currently considered. Only force and kinematics are communicated between the FE and DE regions through the single layer of ghost particles constrained to follow the motion of the FE facets to which they are tied.

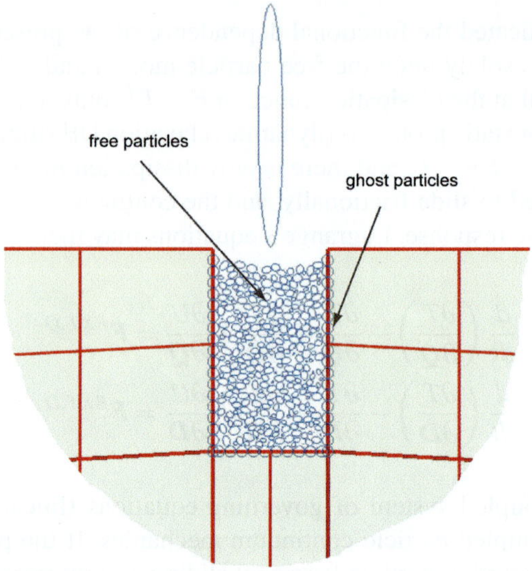

Fig. 7. Schematic illustration of granular-continuum coupling

Depending on the FE type, the ghost particles may or may not maintain rotational degree of freedom. Ideally, when a micropolar or micromorphic continuum model is used within the FE region, the ghost particles will have rotational degrees of freedom. If conventional FEM is adopted (like in this section), the ghost particles have constrained rotational degrees of freedom. Free particles in the DE domain carry both translational and rotational degrees of freedom.

The computational framework involves a two-way exchange of information: free particles in the DE simulation contribute to the boundary force in the FE domain through ghost particles, the FE domain provides information needed to compute the boundary condition on the free particles through ghost particles as well. The granular and continuum scales run simultaneously and exchange relevant information dynamically.

The ghost particles can be placed in such a manner that their centroids are exactly located on the surface FE facets. As ghost particles are discrete in space, the forces are discrete in space as well. Each force acts like a point load on the FE mesh, not necessarily acting at a finite element node. When a point force $\boldsymbol{P}$ acts in the interior (including boundary) of the element domain, the relation between the distributed force $\boldsymbol{b}(\boldsymbol{x})$ at point $\boldsymbol{x}$ and the point force can be denoted mathematically as

$$\boldsymbol{b}(\boldsymbol{x}) = \boldsymbol{P}\delta(\boldsymbol{x}-\boldsymbol{a}) \tag{46}$$

where $\delta(\boldsymbol{x}-\boldsymbol{a})$ is the Dirac delta function and $\boldsymbol{x}=\boldsymbol{a}$ the location of force action $\boldsymbol{P}$. The Dirac delta function has the property that for any vector function $\boldsymbol{g}(\boldsymbol{x})$

$$\int_{\Omega} \boldsymbol{g}(\boldsymbol{x})\delta(\boldsymbol{x}-\boldsymbol{a})dx = \begin{cases} \boldsymbol{g}(\boldsymbol{a}), & \boldsymbol{a}\in\Omega \\ \boldsymbol{0}, & \text{otherwise} \end{cases} \tag{47}$$

Thus the external nodal forces on an element e arising from a point force $\boldsymbol{P}$ at $\boldsymbol{a}$ can be obtained by

$$\boldsymbol{f}^e = \int_{\Omega^e} \boldsymbol{N}^{eT}(\boldsymbol{x})\boldsymbol{b}(\boldsymbol{x})dv = \int_{\Omega^e} \boldsymbol{N}^{eT}(\boldsymbol{x})\boldsymbol{P}\delta(\boldsymbol{x}-\boldsymbol{a})dv = \begin{cases} \boldsymbol{N}^{eT}(\boldsymbol{a})\boldsymbol{P}, & \boldsymbol{a}\in\Omega^e \\ \boldsymbol{0}, & \text{otherwise} \end{cases} \tag{48}$$

where $\boldsymbol{N}^e$ is the matrix of finite element shape functions for element e. Extending it to all finite elements over the entire domain we have

$$\boldsymbol{f} = \boldsymbol{N}^T\boldsymbol{P} \tag{49}$$

When the FE mesh deforms, the ghost particles move as well, maintaining their centroids on the surface of the FE mesh. Their centroid locations need to be mapped from global coordinates to local element natural coordinates using a Newton-Raphson iterative method. Once the natural coordinates are determined, the locations of ghost particles can be evaluated using the following relationship through shape functions $\boldsymbol{N}_{\widehat{Q}D}$ during the subsequent simulation: $\widehat{\boldsymbol{Q}} = \boldsymbol{N}_{\widehat{Q}D}\boldsymbol{D}$, where $\widehat{(\bullet)}$ denotes prescribed particle dofs. The DE code ELLIP3D is wrapped and integrated into FE code Tahoe using object-oriented programming methodology for the algorithm implementation.

4.2 DE-FE Facet Coupling Example

We revisit the penetration motivation example at the beginning of the chapter to demonstrate the effect of having a layer of ghost particles tied to FE facets.

4.2.1 Penetration with Coupled FE Facets

The particles from the penetration example with smaller "container" are combined with a finite element domain, shown in Fig.8.

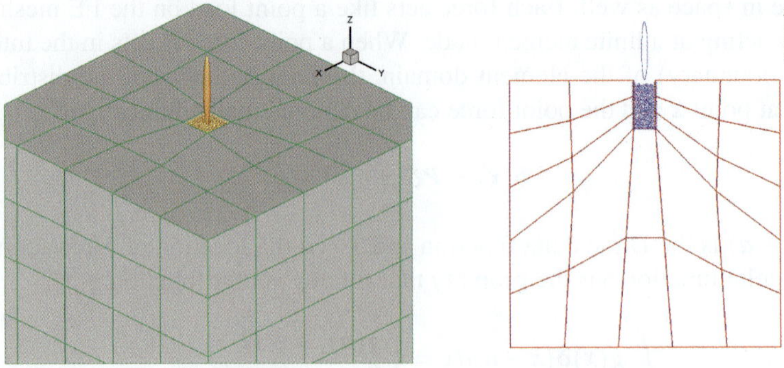

(a) opaque view (b) cross-sectional view

Fig. 8. 3D view of the DE and FE domains

As the penetrator particle is driven into the free particles, the ghost particles are squeezed outwards toward the FE domain. Figure 9(a) depicts the penetrator-induced displacement field of all ghost particles (rotations fixed because FE continuum is non-polar). It is noteworthy that the "container" formed by ghost particles swells at lower part, similar to the influence region for geotechnical pile excitation problems.

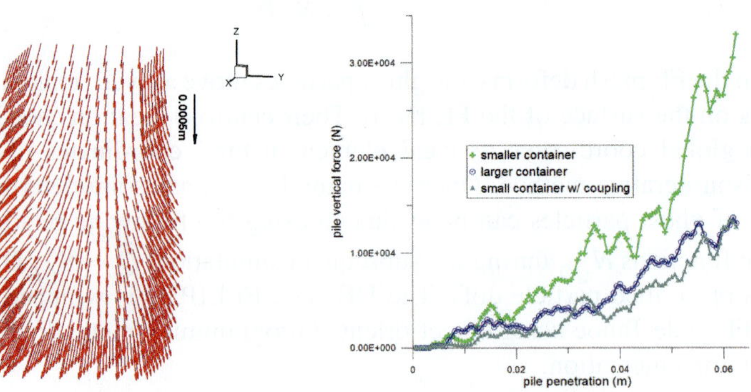

(a) (b)

Fig. 9. (a) Penetrator-induced displacement field of ghost particles using finer mesh. (b) Comparison of force-displacement curves.

To examine the effect of DE-FE coupling on force-displacement curves of penetration, the small container curve and large container curve in Fig.4(c) are plotted again, together with the curve obtained from small container with DE-FE coupling, shown in Fig.9(b). It is observed that the penetrator force of the small container with DE-FE coupling can be tuned to match the larger container with no coupling, by adjusting the elastic compliance of the FE continuum surrounding the container. The boundary effect difference shown in Fig.4(c) can be partially or completely eliminated by applying a more robust DE-FE coupling technique in future work, similar to the atomistic-continuum coupling methods, but accounting for differences with granular materials (see Section 3.3). Such work is ongoing.

5 Summary

The chapter presented a concurrent multiscale computational method for modeling at the grain-scale the interfacial mechanics between dense dry granular materials and deformable solid bodies. Section 3 presented the formulation for coupling particle and micropolar continuum mechanics regions of a granular material, following the lattice-structure-based approaches described in [45, 47], but extending to rotational dofs, and consideration of free particle domain $\mathscr{B}^{DE}$ with no overlain FE mesh. For the case of large particle motion and frictional sliding in the overlap region $\hat{\mathscr{B}}^h \cup \tilde{\mathscr{B}}^h$, a finite deformation micromorphic plasticity model is needed to couple to the particle mechanics and is presented in [54, 55]. Section 4 presented a preliminary DE-FE coupling via single ghost layer of particles tied to FE facets, which demonstrates a code communication between the DE and FE codes being used in the research.

6 Ongoing and Future Work

Various aspects of the research on ongoing, while others are considered for future work. Ongoing research includes: (1) implementing the micromorphic elastoplasticity model into Tahoe; (2) coupling the micromorphic FE to the DE code through an overlapping region; and (3) testing the computational implementations for a penetration example and other granular soil-solid body interface problems.

Future research entails: (4) extend micromorphic pressure sensitivity plasticity to more advanced constitutive models, such as critical state plasticity and including particle breakage; (5) address adaptivity of the multiscale scheme to be able to convert continuum to particle as a solid body shears through a granular material, or particle to continuum in particle regions that behave more like a continuum; and (6) extend to multiphase mechanics (solid grains, pore liquid and gas).

Acknowledgement. The authors acknowledge the support of NSF grant CMMI-0700648.

References

1. Cone penetrometers, `http://geosystems.ce.gatech.edu/Faculty/Mayne/Research/devices/cpt.htm`
2. Regueiro, R.: Proc. IMECE 2007, Seattle, WA, USA, 42717, pp. 1–6 (2007)
3. Yan, B., Regueiro, R., Sture, S.: Three dimensional discrete element modeling of granular materials and its coupling with finite element facets. Eng. Comput. 27(4), 519–550 (2010)

4. Reynolds, O.: Philosophical Mag. J. Science 20(127), 469 (1885)
5. Newland, P., Allely, B.: Geotechnique 7, 17 (1957)
6. Rowe, P.: Proc. R. Soc. Lond. A, Math. Phys. Eng. Sci. A269, 500 (1962)
7. Cowin, S., Satake, M.: Continuum Mechanical and Statistical Approaches in the Mechanics of Granular Materials. Gakujutsu Bunken Fukyu-Kai, Tokyo (1978)
8. Jenkins, J., Satake, M.: Mechanics of Granular Materials. Elsevier Science, Amsterdam (1983)
9. Satake, M., Jenkins, J.T.: Micromechanics of Granular Materials. Elsevier Science, Amsterdam (1988)
10. Chang, C.S., et al.: Mechanics of Deformation and Flow of Particulate Materials. American Society of Civil Engineers, Nashville (1997)
11. Kolymbas, D.: Constitutive Modelling of Granular Materials. Springer, Heidelberg (2000)
12. Duffy, J., Deresiewicz, H.: J. App. Mech. 24(4), 585 (1957)
13. Deresiewicz, H.: Adv. App. Mech. 5, 233 (1958)
14. Christoffersen, J., Mehrabadi, M., Nemat-Nasser, S.: J. App. Mech. 48, 339 (1981)
15. Rothenburg, L., Selvadurai, A.: In: Selvadurai, A. (ed.) Mechanics of Structured Media, pp. 469–486. Elsevier Scientific, Amsterdam (1981)
16. Goddard, J., Bashir, Y.: In: DeKee, D., Kaloni, P. (eds.) Recent Developments in Structured Continua, pp. 23–35. Longman Scientific and Technical, J. Wiley (1990)
17. Chang, C., Chang, Y., Kabir, M.: ASCE J. Geotech. Eng. Div. 118(12), 1959 (1992)
18. Gardiner, B., Tordesillas, A.: Int. J. Solids Struct. 41(21), 5885 (2004)
19. Luding, S.: Int. J. Solids Struct. 41(21), 5821 (2004)
20. Peters, J.: J. Eng. Math. 52(1-3), 231 (2005)
21. Kanatani, K.I.: Int. J. Engr. Sci. 17(4), 419 (1979)
22. Chang, C., Liao, C.: Int. J. Solids Struct. 26, 437 (1990)
23. Bardet, J., Proubet, J.: J. Eng. Mech. 118(2), 397 (1992)
24. Muhlhaus, H.B., Oka, F.: Int. J. Solids Struct. 33(19), 2841 (1996)
25. Ehlers, W., Diebels, S., Volk, W.: J. Phy. 8(8), 127 (1998)
26. Suiker, A., De Borst, R., Chang, C.: Acta Mech. 149(1-4), 161 (2001)
27. Tordesillas, A., Walsh, D.: Powder Technol. 124(1-2), 106 (2002)
28. Suiker, A., Chang, C.: J. Eng. Mech. 130(3), 283 (2004)
29. Walsh, S., Tordesillas, A.: Acta Mech. 167(3-4), 145 (2004)
30. Pasternak, E., Muhlhaus, H.B.: J. Eng. Math. 52(1), 199 (2005)
31. Gardiner, B., Tordesillas, A.: Powder Technol. 161(2), 110 (2006)
32. Eringen, A.: Microcontinuum Field Theories I: Foundations and Solids. Springer, Heidelberg (1999)
33. Cundall, P., Strack, O.: Geotechnique 29, 47 (1979)
34. Williams, J.: Eng. Comput. 5(3), 198 (1988)
35. Bardet, J., Proubet, J.: Comput. Struct. 39(3/4), 221 (1991)
36. Bashir, Y., Goddard, J.: J. Rheol. 35, 849 (1991)
37. Anandarajah, A.: ASCE J. Geotech. Eng. Div. 120(9), 1593 (1994)
38. Borja, R., Wren, J.: Comp. Meth. App. Mech. Engr. 127(1-4), 13 (1995)
39. Wren, J.R., Borja, R.I.: Comp. Meth. App. Mech. Engr. 141(3-4), 221 (1997)
40. Luding, S., Latzel, M., Volk, W., Diebels, S., Herrmann, H.: Comp. Meth. App. Mech. Engr. 191(1-2), 21 (2001)
41. Han, K., Peric, D., Crook, A., Owen, D.: Eng. Comput. 17(2), 593 (2000)
42. Kremmer, M., Favier, J.: Int. J. Numer. Methods Eng. 51(12), 1407 (2001)
43. Komodromos, P., Williams, J.: Eng. Comput. 21(2-4), 431 (2004)
44. Nakashima, H., Oida, A.: J. Terramech. 41(2-3), 127 (2004)

45. Wagner, G., Liu, W.: J. Comput. Phys. 190(1), 249 (2003)
46. Xiao, S., Belytschko, T.: Comp. Meth. App. Mech. Engr. 193(17-20), 1645 (2004)
47. Klein, P., Zimmerman, J.: J. Comput. Phys. 213(1), 86 (2006)
48. Feyel, F., Chaboche, J.L.: Comp. Meth. App. Mech. Engr. 183, 309 (2000)
49. Belytschko, T., Loehnert, S., Song, J.-H.: Int. J. Numer. Methods Eng. 73, 869 (2008)
50. Feyel, F.: Comp. Meth. App. Mech. Engr. 192, 3233 (2003)
51. Rayleigh, J.: The Theory of Sound, 1st edn., vol. 1. Dover Pub. Inc., New York (1945)
52. Eringen, A.: Theory of Micropolar Elasticity. Academic Press, New York (1968); Fracture, An Advanced Treatise, 1st edn., ch. 7, vol. 2, pp. 622–729
53. Hughes, T.J.R.: The Finite Element Method. Prentice-Hall, New Jersey (1987)
54. Regueiro, R.: Int. J. Solids Struct. 47, 786 (2010)
55. Regueiro, R.: J. Eng. Mech. 135, 178 (2009)

Performance of the SPH Method for Deformation Analyses of Geomaterials

H. Nonoyama, A. Yashima, K. Sawada, and S. Moriguchi

Gifu University, Gifu, Japan

Abstract. Various types of behaviors of different soils have been predicted by using the finite element method (FEM) with comprehensive constitutive models developed in geomechanics. There are, however, still some problems for the large deformation analyses within the framework of FEM. Numerical instabilities arise due to the distortion of the FE mesh. In this work, deformation analyses of geomaterials using Smoothing Particle Hydrodynamics (SPH) method are carried out. The SPH method belongs to the class of particle methods. In this paper, the analytical accuracy and the stability of SPH method are investigated for deformation analyses of geomaterials which are assumed to be solid or fluid.

1 Introduction

The modeling of large deformations in geostructures within the framework of FEM remains to be a major challenge, although there have been numerous comprehensive constitutive models developed in geomechanics. A common pathology is the occurrence of numerical instabilities due to the distortion of the FE mesh.

On the other hand, some numerical methods have been proposed to solve large deformation problems without FE mesh. The Eulerian method is one of the solutions for large deformation problems because it is not necessary to take the deformation of mesh into consideration. Simulations of large deformation problems of geomaterials, for example, the lateral flow of liquefied ground [1, 2] and the large deformation of slope failure [3] have been carried out using numerical schemes based on the Eulerian method. Numerical results obtained in the previous studies were in good agreement with theoretical solutions and experimental results. In the previous studies, the deformation behavior of geomaterials was expressed under the assumption that the geomaterials are a single-phase Bingham fluid with shear strength of soils.

Smoothed Particle Hydrodynamics (SPH) [4, 5], a kind of particle method, is also an effective method to solve large deformation problems because it does not require a structured mesh system. Recently, SPH method has been widely used in a variety of fields such as fluid dynamics [6] or solid mechanics [7]. The method has also been applied to geotechnical engineering [8]. The objectives of this work are to establish a computer program code for analyzing the deformation of geomaterials using SPH method. In order to use the nonstructural calculation points in the SPH method, it is possible to express the complex configuration without

R. Wan et al. (Eds.): Bifurcations, Instabilities and Degradations in Geomaterials, SSGG, pp. 275–290.
springerlink.com

particular treatment. Geomaterials can be assumed to be solid or fluid and thus soil-water coupled analysis can be envisaged. Moreover, it is possible to easily calculate the strain accumulation in the Lagrangian framework. In this paper, deformation analyses of geomaterials using SPH method are carried out. Based on comparisons between the simulated results using SPH method, theoretical solutions and numerical results of FE analysis, the analytical accuracy and stability of the SPH method are investigated for deformation analyses of geomaterials which are assumed to be solid or fluid.

2 Numerical Method

2.1 Basic Theory of SPH

The foundation of the SPH method is an interpolation theory with approximations being divided into two key steps. The first step is a kernel approximation of field functions which use neighbor particles β located at point x^β within the influence domain of a smoothing function W for a reference particle α located at point x^α. The second step is a particle approximation.

In the first step of interpolation, we define a smoothed physical quantity $\langle f(x^\alpha)\rangle$ for a physical quantity $f(x^\alpha)$ at reference particle α as below:

$$\left\langle f\left(x^{\alpha}\right)\right\rangle = \int_{\Omega} f(x^{\beta}) W(r,h) dx^{\beta} \tag{1}$$

where $r = |x^\alpha - x^\beta|$, h is a radius of the influence domain and Ω is the volume of the integral that contains x^α and x^β. Thus, the spatial derivative of Eq. (1) can be written as

$$\frac{\partial\left\langle f\left(x^{\alpha}\right)\right\rangle}{\partial x_i} = \frac{\partial}{\partial x_i}\int_{\Omega} f\left(x^{\beta}\right) W\left(r,h\right) dx^{\beta} = \int_{\Omega} f\left(x^{\beta}\right)\frac{\partial W\left(r,h\right)}{\partial x_i} dx^{\beta} \tag{2}$$

The second step considers a discrete distribution of particles for which Eqs. (1) and (2) are approximated by replacing the integral with the summation operation, i.e.

$$\left\langle f\left(x^{\alpha}\right)\right\rangle = \sum_{\beta}^{N}\frac{m^{\beta}}{\rho^{\beta}} f(x^{\beta}) W^{\alpha\beta} \tag{3}$$

$$\frac{\partial\left\langle f\left(x^{\alpha}\right)\right\rangle}{\partial x_i} = \sum_{\beta}^{N}\frac{m^{\beta}}{\rho^{\beta}} f\left(x^{\beta}\right)\frac{\partial W^{\alpha\beta}}{\partial x_i} \tag{4}$$

where m^β is the mass of a neighbor particle, ρ^β its density, and N is the number of particles in the influence domain.

Many kinds of smoothing functions have been proposed in the literature. In this work, the cubic B-spline function [9] is employed.

2.2 SPH Based on Solid Mechanics

2.2.1 Governing Equations

In general, the equation of continuity and the equation of motion are described as follows:

$$\frac{\partial \rho}{\partial t} + \frac{\partial(\rho u_i)}{\partial x_i} = 0 \tag{5}$$

$$\frac{Du_i}{Dt} = f_i + \frac{1}{\rho}\frac{\partial \sigma_{ij}}{\partial x_j} \tag{6}$$

where ρ is the density, u is the velocity, σ is the stress and f is the external force. The indices i and j denote the coordinate directions.

When applying the SPH interpolation theory to the gradients in Eq. (5), the SPH equation of continuity at particle α is expressed as

$$\frac{d\rho}{dt} = \sum_{\beta}^{N} m^{\beta}\left(u_i^{\alpha} - u_i^{\beta}\right)\frac{\partial W^{\alpha\beta}}{\partial x_i} \tag{7}$$

In place of Eq. (7), the density can also be directly determined from Eq. (3), i.e.

$$\rho^{\alpha} = \sum_{\beta}^{N} m^{\beta} W^{\alpha\beta} \tag{8}$$

When Eq. (8) is used, the density tends to be underestimated since there are a few numbers of fixed particles in the vicinity of the free surface. This problem can be avoided by using Eq. (7), the differential form. Alternatively, the next equation can be used, in which the smoothing functions are summed for normalization [10] and which allows for a solution even if the form is that of Eq. (8). Thus,

$$\rho^{\alpha} = \sum_{\beta}^{N} m^{\beta} W^{\alpha\beta} \Big/ \sum_{\beta}^{N}\left(\frac{m^{\beta}}{\rho^{\beta}}\right) W^{\alpha\beta} \tag{9}$$

Applying the SPH interpolation theory to the gradients into the general equation of motion, we get for particle α:

$$\frac{du_i}{dt} = \sum_{\beta}^{N} m^{\beta}\left(\frac{\sigma_{ij}^{\beta}}{\left(\rho^{\beta}\right)^2} + \frac{\sigma_{ij}^{\alpha}}{\left(\rho^{\alpha}\right)^2} + \delta_{ij}\Pi_{ij}\right)\frac{\partial W^{\alpha\beta}}{\partial x_j} + f_i \tag{10}$$

where δ is Kronecker's delta and Π is the artificial viscosity in order to decrease any numerical instability. In this study, an artificial viscosity proposed by the previous study [11] is used. The artificial viscosity is described as below:

$$\Pi_{ij} = \frac{1}{\bar{\rho}_{ij}}\left(-\alpha^{vis}\bar{c}\phi^{\alpha\beta} + \beta^{vis}\left(\phi^{\alpha\beta}\right)^2\right) \tag{11}$$

where α^{vis} and β^{vis} are the artificial viscosity parameters, $\bar{c}$ is the average of the sound speed of each particles, $\bar{\rho}$ is the average of the density of each particle, $u^{\alpha\beta}$ is the relative velocity and $\phi^{\alpha\beta}$ is described as follows:

$$\phi^{\alpha\beta} = \frac{hu_i^{\alpha\beta} \cdot x_i^{\alpha\beta}}{\left|x_i^{\alpha\beta}\right|^2 + 0.01h^2} \tag{12}$$

2.2.2 Constitutive Model

In this study, deformation analyses of elastic and elasto-plastic materials are carried out under the plane strain condition. For an elastic model, Hooke's law is used as a constitutive model such as

$$\sigma_{ij} = D_{ijkl}^{e}\,\varepsilon_{kl}^{e} \tag{13}$$

where σ_{ij} is the stress tensor, D_{ijkl}^{e} is the elastic modulus matrix, ε_{kl}^{e} is the elastic strain tensor. For an elasto-plastic model, the resulting constitutive model is typically given by

$$d\sigma_{ij} = \left(D_{ijkl}^{e} - \frac{D_{ijmn}^{e}\dfrac{\partial f}{\partial \sigma_{mn}}\dfrac{\partial f}{\partial \sigma_{pq}}D_{pqkl}^{e}}{\dfrac{\partial f}{\partial \sigma_{mn}}D_{mnpq}^{e}\dfrac{\partial f}{\partial \sigma_{pq}} - \dfrac{\partial f}{\partial L}\dfrac{\partial L}{\partial \varepsilon_{mn}^{p}}\dfrac{\partial f}{\partial \sigma_{mn}}} \right) d\varepsilon_{kl} \tag{14}$$

where L is the hardening parameter and f is the yield function. In this study, the yield function is based on the Cam-clay model [12] as follows:

$$f = \frac{\lambda - \kappa}{1 + e_0}\left(\ln\frac{p_c}{p_0} + \frac{1}{M}\frac{q}{p} \right) - \varepsilon_v^p = 0 \tag{15}$$

where λ is the compression index, κ is the swelling index, e_0 is the initial void ratio, p_c is the consolidation yield stress, p_0 is the initial mean stress, M is the stress ratio at failure, q is the deviatoric stress, p is the mean principal stress and ε_v^p is the plastic volumetric strain.

Additionally, the influence of rotational movement of the rigid body is taken into consideration by using the Jaumann stress rate:

$$\dot{\sigma}_{ij} = \hat{\sigma}_{ij} - \sigma_{ik}\omega_{kj} + \omega_{ik}\sigma_{kj} \tag{16}$$

where $\dot{\sigma}_{ij}$ is the Cauchy stress rate tensor, $\hat{\sigma}_{ij}$ is the Jaumann stress rate and ω_{ij} is the spin tensor.

3.1 SPH Based on Fluid Dynamics

3.1.2 Modeling of Geomaterials

In this study, the geomaterial is modeled as a single-phase Bingham fluid with shear strength of soils proposed by previous studies [3], i.e.

$$\tau = \eta_0\dot{\gamma} + c + p\tan\phi \tag{17}$$

Where τ is the shear stress, η_0 is the viscosity after yield, $\dot{\gamma}$ is the shear strain rate, c is the cohesion of soil, p is the hydraulic pressure and ϕ is the internal friction angle of soil. Because Eq. (17) cannot be directly calculated, we use an equivalent viscosity η' obtained for a Newtonian fluid as follows:

$$\eta' = \frac{\tau}{\dot{\gamma}} = \eta_0 + \frac{c + p\tan\phi}{\dot{\gamma}} \tag{18}$$

Figure 1 shows the equivalent viscosity of the Bingham fluid model. The value of the equivalent viscosity is dependently on the shear strain rate. The constitutive model used in this study can be obtained by introducing the equivalent viscosity into the constitutive model of the Newtonian fluid.

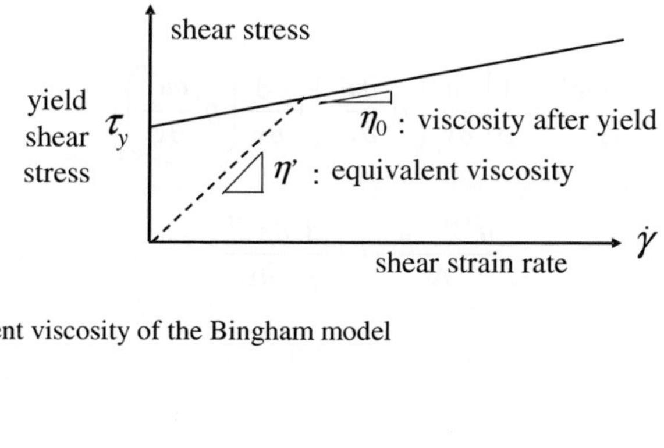

Fig. 1. Equivalent viscosity of the Bingham model

3.1.3 Governing Equations

In general, the equation of continuity and the equation of motion are given as

$$\frac{D\rho}{Dt} = 0 \tag{19}$$

$$\frac{Du_i}{Dt} = -\frac{1}{\rho}\frac{\partial p}{\partial x_i} + \frac{\eta}{\rho}\frac{\partial}{\partial x_j}\left(\frac{\partial u_i}{\partial x_j}\right) + f_i \tag{20}$$

where ρ is the density of fluid, u is the velocity, p is the hydraulic pressure, η is viscosity coefficient and f is external force.

In this study, we used SMAC-SPH method based on fluid dynamics proposed by previous studies [13]. The SMAC algorithm [14] is used to treat geomaterials as incompressible materials. Furthermore, in order to prevent numerical instabilities that may arise due to the large value of the Bingham viscosity for the quasi-rigid materials, an implicit calculation procedure is applied to the viscosity term of the equation of motion. Using the equivalent viscosity η' in place of the viscosity coefficient η in the equation of motion considering spatial gradient of the viscosity term, we get

$$\frac{Du_i}{Dt} = -\frac{1}{\rho}\frac{\partial p}{\partial x_i} + \frac{1}{\rho}\frac{\partial}{\partial x_j}\left[\eta'\left(\frac{\partial u_i}{\partial x_j} + \frac{\partial u_j}{\partial x_i}\right)\right] + f_i \tag{21}$$

For the case of a Newtonian fluid, the viscosity coefficient is treated as a constant value and its spatial derivatives are not considered. However, as we can see from Eq. (21), the equivalent viscosity has a distribution in space. Therefore, the effect of the spatial derivatives is taken into account in Eq. (21). Moreover, Eq. (21) can be discretized as follows:

$$\frac{u_i^{**} - u_i^k}{\Delta t} = f_i^k \tag{22}$$

$$\frac{u_i^* - u_i^{**}}{\Delta t} = \frac{1}{\rho}\left[\frac{\partial}{\partial x_j}\left(\eta'\frac{\partial u_i^*}{\partial x_j}\right) + \frac{\partial}{\partial x_j}\left(\eta'\frac{\partial u_j^*}{\partial x_i}\right)\right] \tag{23}$$

$$\frac{u_i^{k+1} - u_i^*}{\Delta t} = -\frac{1}{\rho}\frac{\partial p^{k+1}}{\partial x_i} \tag{24}$$

where Δt is the time increment, subscript k and k+1 indicate the quantities at each calculation time step and * indicates the temporal quality. Equations (22), (23) and (24) are the external force term, the viscous term and the pressure term of the equation of motion, respectively.

The following is an algorithm used in this study. A temporal value of the velocity u_i^{**} is obtained explicitly using the gravity f_i^k and the velocity u_i^k at the previous time from the Eq. (22):

$$u_i^{**} = u_i^k + \Delta t f_i^k \tag{25}$$

A temporal value of the velocity u_i^* is obtained using the temporal value u_i^{**} and the spatial derivative of u_i^* as

$$u_i^* = u_i^{**} + \frac{\Delta t}{\rho}\left[\frac{\partial}{\partial x_j}\left(\eta' \frac{\partial u_i^*}{\partial x_j}\right) + \frac{\partial}{\partial x_j}\left(\eta' \frac{\partial u_j^*}{\partial x_i}\right)\right] \tag{26}$$

According to previous studies [15], the right hand side of the temporal value u_i^* of the spatial derivative in Eq. (26) can be discretized as follows:

$$\frac{\partial}{\partial x_j}\left(\eta' \frac{\partial u_i^*}{\partial x_j}\right) = \sum_{\beta}^{N} \frac{4m^\beta}{\rho^\beta} \frac{\eta'^\alpha \eta'^\beta}{\eta'^\alpha + \eta'^\beta} \frac{\left(u_i^{\alpha,*} - u_i^{\beta,*}\right)\left(x_i^\alpha - x_i^\beta\right) \cdot \dfrac{\partial W^{\alpha\beta}}{\partial x_i}}{\left|x_i^{\alpha\beta}\right|^2} \tag{27}$$

The temporal value of the position x_i^* is obtained by the temporal value u_i^* obtained from Eq. (26), i.e.

$$x_i^* = x_i^k + \Delta t u_i^* \tag{28}$$

The continuity equation requires that the density of fluid be constant. This is equivalent to the particle number density being constant, n^0. When the temporal value of a particle number density n^* is not n^0, it is corrected to n^0 as

$$n^* + n' = n^0 \tag{29}$$

where n' is a correction value of the particle number density. A correction value of the velocity u_i' occurs in association with the pressure gradient term as follows:

$$u_i' = \Delta t \frac{1}{\rho} \frac{\partial p^{k+1}}{\partial x_i} \tag{30}$$

There is a relationship between the correction value of the velocity u_i' and the correction value of the particle number density n' from the equation of continuity, i.e.

$$\frac{1}{n^0}\frac{n'}{\Delta t}+\frac{\partial u_i'}{\partial x_i}=0 \tag{31}$$

The Poisson equation is obtained from Eqs. (29), (30) and (31):

$$\frac{\partial}{\partial x_i}\left(\frac{1}{\rho}\frac{\partial p^{k+1}}{\partial x_i}\right)=-\frac{1}{\Delta t^2}\frac{n^*-n^0}{n^0} \tag{32}$$

According to previous studies [15], the left hand side of Eq. (32) can be discretized as follows:

$$\frac{\partial}{\partial x_i}\left(\frac{1}{\rho}\frac{\partial p^{k+1}}{\partial x_i}\right)=\sum_{\beta}^{N}\frac{4m^\beta}{\rho^\beta}\left(\frac{1}{\rho^\alpha+\rho^\beta}\right)\frac{\left(p^\alpha-p^\beta\right)\left(x_i^\alpha-x_i^\beta\right)\cdot\dfrac{\partial W^{\alpha\beta}}{\partial x_i}}{\left|x_i^{\alpha\beta}\right|^2} \tag{33}$$

By solving Eq. (32), a pressure p^{k+1} at present time k+1 is obtained. In order to prevent numerical instabilities due to a negative pressure, the negative value of the pressure p^{k+1} set to zero. Using the pressure p^{k+1} from Eq. (30), the correction value of the velocity u_i' is obtained. Moreover, using the correction value of velocity u_i', a velocity u_i^{k+1} and a position x_i^{k+1} at present time k+1 are obtained as

$$u_i^{k+1}=u_i^*+u_i' \tag{34}$$

$$x_i^{k+1}=x_i^*+\Delta t u_i' \tag{35}$$

4 Solid Analysis

4.1 Simple Shear Test in Elastic Model

In order to validate the program for a solid, a simple shear test of elastic material with Jaumann stress rate is carried out. Figure 2 shows the numerical model used in this simulation. Here, the specimen is represented by a square object (10cm×10cm) with a virtual area set around it. The number of particles including the particles comprising the virtual area is 900. The initial interparticle distance is 1.0 cm and the radius of influence domain is 3.0 cm. The density is 1.5 g/cm^3. The time increment used in the calculations is 0.001s. The artificial viscosity parameter is 10.0.

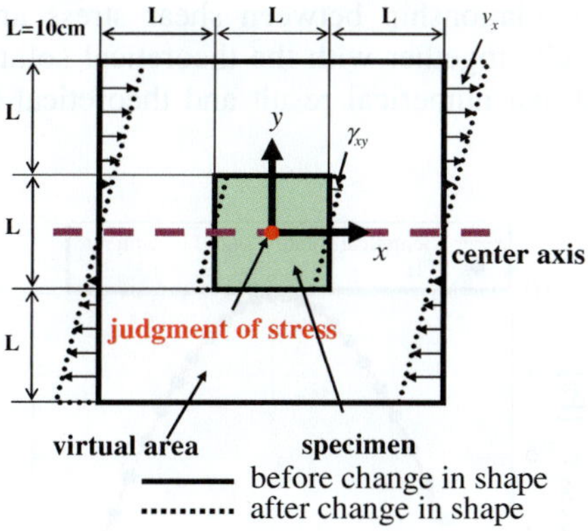

Fig. 2. Numerical model for simple shear test

Table 1 summarizes the numerical parameters used for this analysis. In the simulation, the virtual area moves with constant displacement to describe simple shear condition. The velocities of virtual area v_x are calculated according to

$$v_x = 0.10y \quad \text{cm/s} \tag{36}$$

where y is y-coordinate of each particle. A weightless condition is used for all analyses conducted in this work.

In the simple shear condition, a theoretical solution for elastic material can be described using a relationship between shear stress τ_{xy} and shear strain γ_{xy} , as follows:

$$\tau_{xy} = G\gamma_{xy} = \frac{E}{2(1+\nu)}\gamma_{xy} \tag{37}$$

where G is the shear modulus, E is the Young's modulus and ν is the Poisson's ratio. By using the concept of the Cauchy stress, the theoretical solution is given by:

$$\tau_{xy} = G\sin\gamma_{xy} = \frac{E}{2(1+\nu)}\sin\gamma_{xy} \tag{38}$$

which is used as the theoretical solution in this simulation.

Table 1. Material parameters

shear modulus G [Pa]	10.0
Poisson's ratio ν	0.30

Figure 3 shows the relationship between shear stress and shear strain. The obtained numerical results together with the theoretical solution are compared in Fig. 3. It is confirmed that numerical result and theoretical solution are in good agreement.

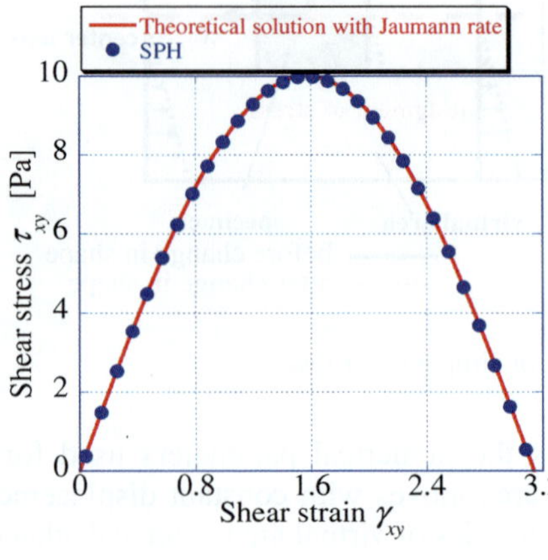

Fig. 3. Relationship between shear stress and shear strain at the center of specimen

4.2 Simple Shear Test in Elasto-Plastic Model

A simple shear test on an elasto-plastic material in undrained condition is simulated. Here, the Cam-clay model [12] is adopted as constitutive model for the specimen. The responses of constitutive model, such as stress, strain and stress path, are compared with the results from a FE analysis. The same numerical conditions such as the number of particles, interparticle distance, radius of influence and density used in the simulation of elastic material explained in the previous section are kept here. However, the time step has been changed to 0.0001s, while the artificial viscosity parameter is 25.0. Figure 4 shows the initial configuration used in the FE analysis. The shape of element is triangular element. The number of element is 2 and the number of nodal point is 4. The mesh size (Δx=Δy) is 10cm. The time increment used is 0.001s.

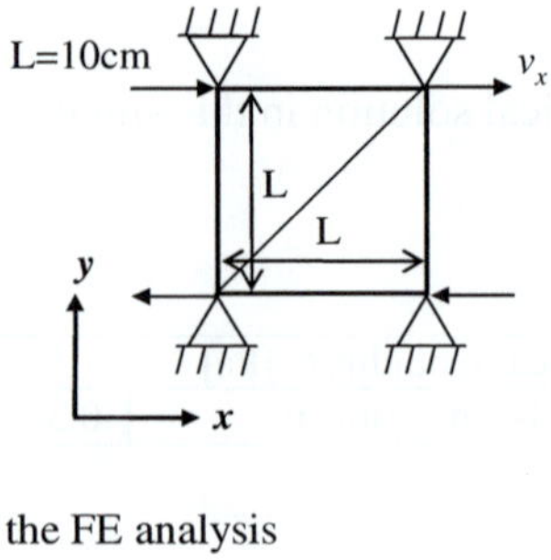

Fig. 4. Initial configuration of the FE analysis

Table 2 summarizes the material parameters used for both FE and SPH analyses. Two different values of the initial mean stress are used in this simulation.

Table 2. Material parameters

Case	1	2
compression index λ	0.355	
swelling index κ	0.0477	
initial void ratio e_0	2.0	
stress ratio at failure M	1.45	
consolidation yield stress p_c[kPa]	98.0	
initial mean stress p_0[kPa]	68.6	98.0
Poisson's ratio ν	0.33	

Figure 5 shows stress paths of Cam-clay model at the center of specimen. Figure 6 gives the relationship between shear stress and shear strain. Numerical results from both SPH and FE analyses are described together in Figs. 5 and 6. The critical state line (C.S.L.) obtained from the Cam-clay model is also plotted in Fig. 5. It is found that results from the SPH analysis are in good agreement with those obtained from the FE analysis. Therefore, we conclude that the SPH model can express quite well the behavior of soil.

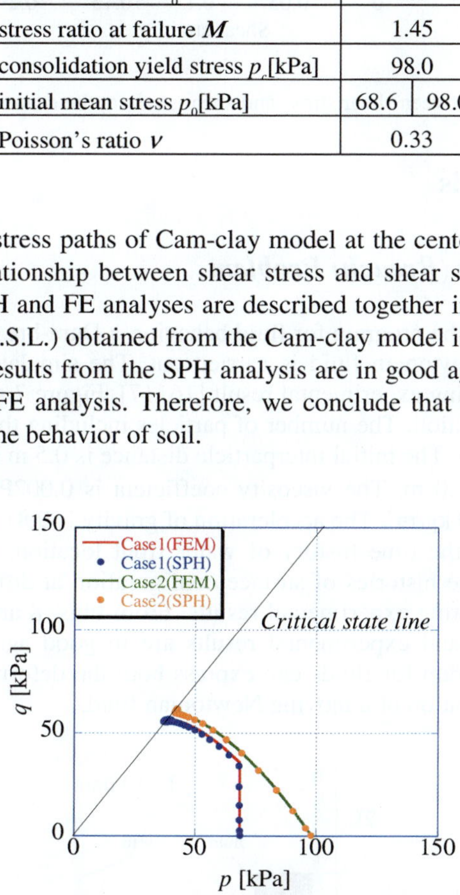

Fig. 5. Stress paths of Cam-clay model at the center of specimen

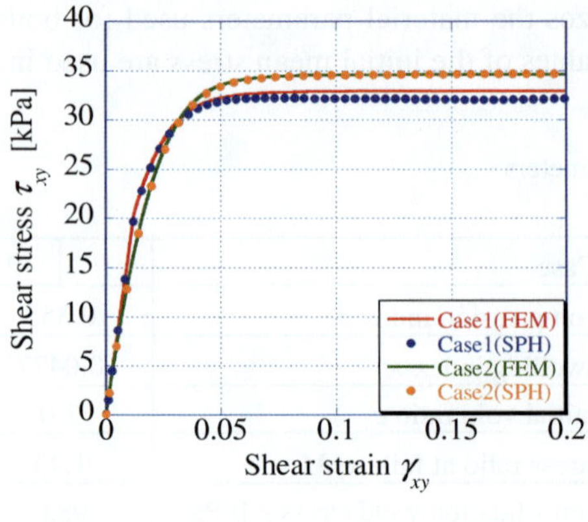

Fig. 6. Relationship between shear stress and shear strain at the center of specimen

5 Fluid Analysis

5.1 *Dam-Break (Breach) Problem*

In order to validate the program for fluid behavior, a Dam-break (breaching) problem involving a Newtonian fluid is carried out. The simulated results are compared with the existing experimental result [16, 17]. Figure 7 shows the numerical model in this simulation. The number of particles including the particles comprising the wall is 1,586. The initial interparticle distance is 0.5 m and the radius of influence domain is 1.0 m. The viscosity coefficient is 0.002Pa·s. The density of geomaterial is 1,000 kg/m^3. The acceleration of gravity is 9.81 m/s^2.

Figure 8 shows the time history of water front location. On the other hand, Fig. 9 gives the time histories of surface configuration at different time steps together with the existing experimental results. From Figs. 8 and 9, it is confirmed that the numerical and experimental results are in good agreement. Thus, it is found that the program for fluids can express both the deformation behavior and the surface configuration of a moving Newtonian fluid.

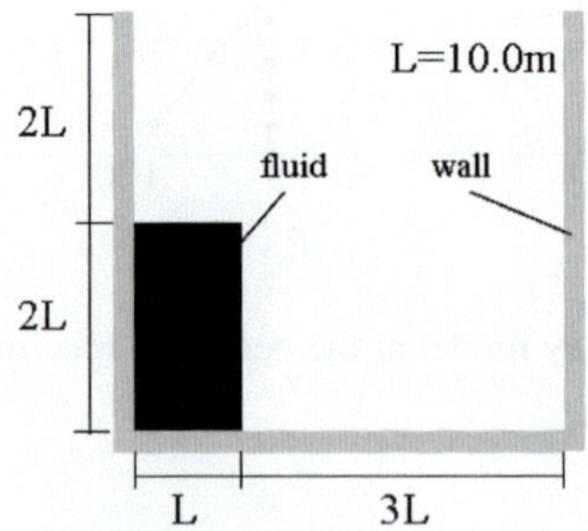

Fig. 7. Numerical model for Dam-break problem

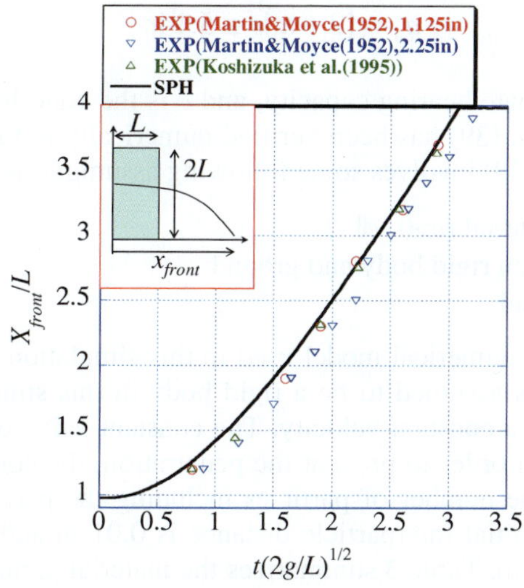

Fig. 8. Time history of water front location

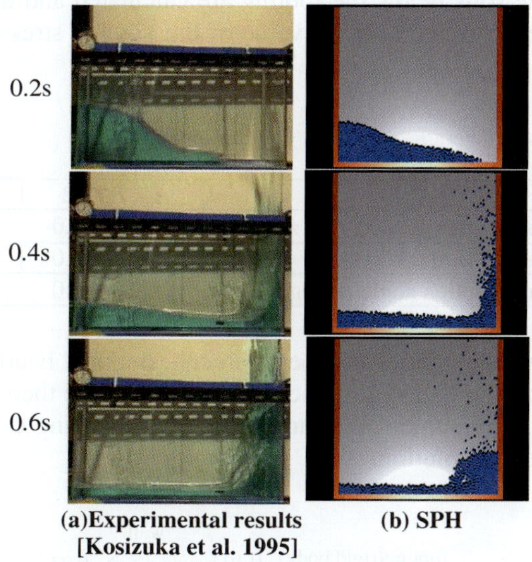

Fig. 9. Time histories of surface configuration

5.2 *Bearing Capacity Analysis of Cohesive Ground*

Here, we carry out the bearing capacity analysis of a cohesive ground. The simulated results are compared with the theoretical solution obtained from Prandtl's theory [18] as follows:

$$q_u = (2+\pi)c \cong 5.14c \tag{39}$$

where q_u is the ultimate bearing capacity, and c is the cohesion of soil. The theoretical solution in Eq. (39) has been verified numerically in the literature; see e.g. [19]. However, Eq. (39) involves some following assumptions:

- Pure cohesive material (c>0, ϕ 0)
- No friction between rigid body and ground
- Weightless material

Figure 10 shows the numerical model used in this simulation. A footing is placed on the ground and is assumed to be a rigid body in this simulation. The footing moves downward at a constant velocity. The constant value of 1.0×10^{-5} m/s is set in this simulation. In order to prevent the penetration, the constant velocity is set to a small value. The number of particles including the particles comprising the wall is 5,146. The initial interparticle distance is 0.01 m and the radius of influence domain is 0.02 m. Table 3 summarizes the material parameters. Three different values of cohesion are used in this simulation. No gravitational force is included in the simulation. Moreover the boundary condition between the bottom of the footing and the ground surface is described as a non-slip boundary. The vertical stresses in the soils below the footing are calculated and the bearing capacities are determined with the average value of the vertical stresses in the particles below the footing.

Table 3. Material parameters

Case	1	2	3
viscosity after yield η_0 [Pa· s]		1.0	
cohesion c[Pa]	1.0	2.0	3.0
internal friction angle ϕ [deg]		0.0	

Figure 11 shows the relationship between cohesion and bearing capacity at the different cohesion. The obtained numerical results and the theoretical solution are shown in Fig. 11. The numerical simulations give a bearing capacity value quite close to the theoretical solution.

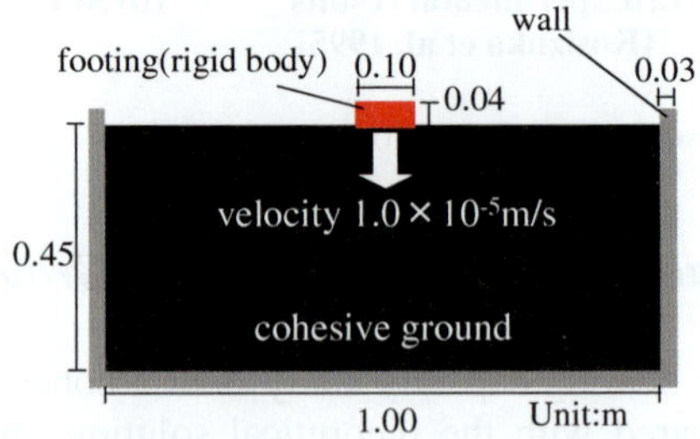

Fig. 10. Numerical model for bearing capacity problem

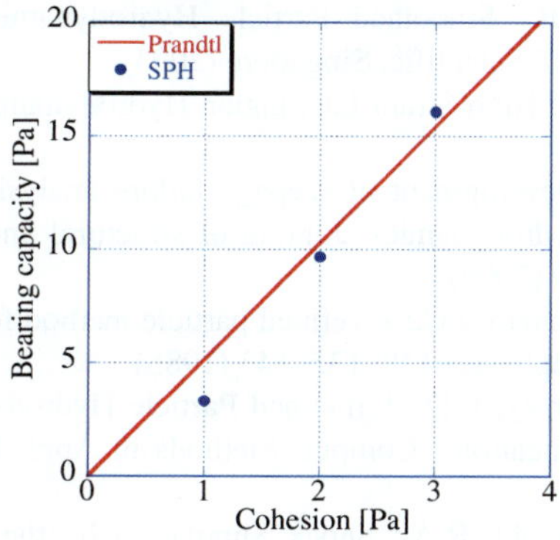

Fig. 11. Relationship between cohesion and bearing capacity

6 Conclusions

In this study, the application of SPH method for deformation analyses of geomaterials is discussed. From the solid analysis, it is confirmed that the numerical results for simple shear test using an elastic model and an elasto-plastic model are in good agreement with the theoretical solution as well as results of FE analysis. The fluid analysis of a Dam-break (breach) problem involving a Newtonian fluid also confirmed a good agreement between numerical results and existing experimental data. Geomaterials are assumed to be a fluid by introducing a Bingham model with shear strength of soils. It is confirmed that the bearing capacity of cohesive ground is given as a factor of the cohesion of soil which is obtained within a certain level of accuracy when compared to Prandtl's solution.

In the future, in order to take the influence of water into consideration, it is necessary to introduce a scheme for soil-water coupled analysis.

References

[1] Uzuoka, R.: Analytical study on the mechanical behavior and prediction of soil liquefaction and flow, Ph. D. Dissertation, Gifu University (2000) (in Japanese)
[2] Hadush, S.: Fluid dynamics based large deformation analysis in geomechanics with emphasis in liquefaction induced lateral spread, Ph. D. Dissertation, Gifu University (2002)
[3] Moriguchi, S.: CIP-based numerical analysis for large deformation of geomaterials, Ph. D. Dissertation, Gifu University (2005)
[4] Lucy, L.B.: A numerical approach to the testing of the fission hypothesis. Astron. J. 82, 1023–1024 (1977)
[5] Gingold, R.A., Monaghan, J.J.: Smoothed particle hydrodynamics: theory and application to non-spherical stars. Monthly Notices of the Royal Astronomical Society 181, 375–389 (1977)

[6] Liu, G.R., Liu, M.B.: Smoothed Particle Hydrodynamics: A Meshfree Particle Method, p. 449. World Scientific, Singapore (2003)
[7] Libersky, L.D., et al.: High Strain Lagrangian Hydrodynamics. J. Comput. Phys. 109, 67–75 (1993)
[8] Maeda, K., et al.: Development of seepage failure analysis method of ground with smoothed particle hydrodynamics. Journal of structural and earthquake engineering, JSCE 23(2), 307–319 (2006)
[9] Monaghan, J.J., Lattanzio, J.C.: A refined particle method for astrophysical problems. Astronomy and Astrophysics 149, 135–143 (1985)
[10] Randles, P.W., Libersky, L.D.: Smoothed Particle Hydrodynamics: Some recent improvements and applications. Comput. Methods in Appl. Mech. Eng. 138, 375–408 (1996)
[11] Monaghan, J.J., Gingold, R.A.: Shock Simulation by the Particle Method SPH. J. Comput. Phys. 52, 374–389 (1983)
[12] Schofield, A.N., Wroth, C.P.: Critical State Soil Mechanics, p. 310. Mcgraw-Hill, New York (1968)
[13] Amsdam, A.A., Harlow, F.H.: The SMAC Method: A Numerical Technique for Calculating Incompressible Fluid Flow. LA-4370. Los Alamos Scientic Laboratory (1970)
[14] Sakai, Y., et al.: Incompressible viscous flow analysis by SPH. Journal of the Japan Society of Mechanical Engineers. Series B 70(696), 1949–1956 (2004) (in Japanese)
[15] Cleary, W., Monaghan, J.J.: Conduction modeling using smoothed particle hydrodynamics. J. Comput. Phys. 148, 227–264 (1999)
[16] Martin, J.C., Moyce, W.J.: An Experimental Study of the Collapse of Liquid Columns on a Rigid Horizontal Plane. Philosophical Transactions of the Royal Society of London, Ser. A 244, 312–324 (1952)
[17] Koshizuka, S., et al.: Particle Method for Incompressible Viscous Flow with Fluid Fragmentation. Computational Fluid Dynamics Journal 4(1), 29–46 (1995)
[18] Prandtl, L.: Über die Härte plastischer Körper: Nachrichten von der Königlichen Gesellschaft der Wissenschaften zu Göttingen. Math. Phys. KI. 12, 74–85 (1920)
[19] Tamura, T., et al.: Limit analysis of soil structure by rigid plastic finite element method. Soils and Foundations 24(1), 34–42 (1984)

CIP-Based Numerical Simulation of Snow Avalanche

K. Oda[1], A. Yashima[1], K. Sawada[1], S. Moriguchi[1], A. Sato[2], and I. Kamiishi[2]

[1] Gifu University, Gifu, Japan
[2] National Research Institute for Earth Science and Disaster Prevention, Niigata, Japan

Abstract. In order to predict a flow area of snow avalanche, a numerical method based on fluid dynamics has been proposed. In the CIP–based numerical method, snow is modeled as a Bingham fluid with the consideration of the Mohr-Coulomb failure criterion. Therefore, the cohesion c and the internal friction angle ϕ are the material parameters for the flowing medium. In this study, using the numerical method with THINC in CIP, the snow avalanche model tests were simulated. In the model tests, snow avalanches were reproduced in a low-temperature room. In the tests, snow flowed on a model slope and the travel length of snow was measured. The parameters used in the simulations are obtained from a previous research. In order to investigate the applicability of the numerical method proposed in this study, the results of model tests were compared with simulation results.

1 Introduction

In Japan, there is a large number of areas where there are heavy snowfalls in winter. Damages caused by snow avalanches have to be predicted in those areas. Snow avalanches are seen to occur repeatedly at the same bare slopes due to adverse weather conditions, the amount of snowfall, and so on. It is, however, very difficult to predict the flow path and the travel length of snow avalanches. Various numerical methods to analyze snow avalanches have been proposed in the literature. For example, simulations of a snow avalanche using a Bingham model were carried out by Dent and Lang [7]. The model reproduced snow avalanches accurately. However, in order to obtain the travel length of snow avalanche, a value of the initial viscosity was determined arbitrarily. The model was not applicable for the condition in which the flow velocity of snow changes rapidly because the viscosity term was solved by an explicit calculation procedure. Other proposed methods have not been used in a practical setting because they were not able to reproduce the behavior of snow very well. In order to develop a reasonable simulation method for predicting the flow path and travel length of snow avalanches, a collaborative research has been carried out by Gifu University and Snow and Ice Research Center. In this study, the CIP–based numerical method based on fluid dynamics [1] is used for the simulation of a snow avalanche. In the numerical simulation, snow is modeled as a Bingham fluid with consideration of Mohr-Coulomb failure criterion for the solid part. It is difficult to define the interface

R. Wan et al. (Eds.): Bifurcations, Instabilities and Degradations in Geomaterials, SSGG, pp. 291–306.
springerlink.com

sharply between two materials in the simulation. To overcome this problem, the Tangent of Hyperbola for Interface Capturing technique (THINC) [2] is introduced in order to capture a free surface correctly. Using a numerical method with THINC technique implemented into CIP, available snow avalanche model tests were simulated. In the model tests, snow avalanches were reproduced in a low-temperature room. In the tests, snow flowed on a model slope and the travel length of snow was measured. In order to investigate the applicability of the numerical method proposed in this study, the results of model tests were compared with the simulation results.

2 Type of Snow Avalanche

Snow avalanches are divided into two types, full depth snow avalanche and surface snow avalanche. Figure 1 shows images of a full depth and surface snow avalanches. Additionally, the above types of avalanches are further divided into stream type avalanche and powder type avalanche. Figure 2 shows images of stream type and powder type, respectively. It is difficult to solve the behavior of powder type avalanche because the movement is turbulent. On the other hand, it is relatively easy to solve the behavior of a stream type avalanche because the movement is almost like water.

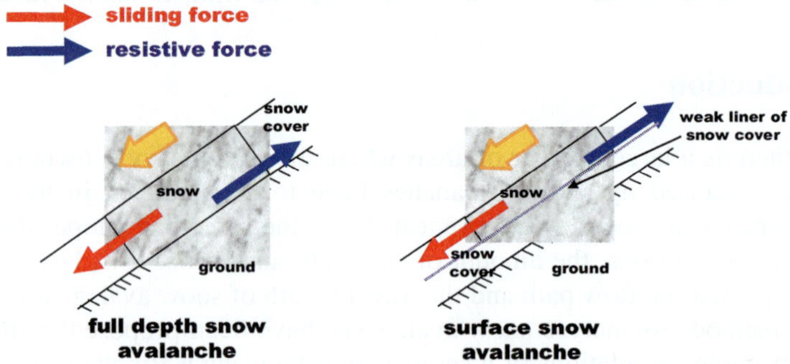

Fig. 1. Schematics of full depth snow avalanche and surface snow avalanche

Fig. 2. Examples of stream type and powder type avalanches

3 Experiment

To obtain the travel length of stream type snow avalanche, a series of experiments were carried out. Table 1 summarizes the classification of snow quality based on the density of snow [3]. It is preferable that a fresh snowfall is used in the experiments. However, the experiments were not carried out in the winter season. Therefore, we used snow stored in a low-temperature room at the temperature of -20 degrees C in Snow and Ice Research Center. This snow collected had fallen in Nagaoka City, Niigata Prefecture in 2008, and thus represents a sample of the type of snow fallen that year at that particular location. The snow quality is defined as fresh snow or granular in condition. It is, however, difficult to keep these conditions in a low-temperature room. Therefore, in this study lightly compacted snow was used because in these conditions, snow is very stable. All experiments were carried out at the same temperature conditions and in the same low-temperature room.

Table 1. Classification of snow quality by the density of snow

Classification	Density[kg/m^3]
fresh snow	50~150
lightly compacted snow	50~150 / 150~250
compacted snow	150~250 / 250~500
granular snow	250~500 / 300~500
lightly granular snow	300~500 / about 300
frost granular snow	about 300

3.1 Snow Avalanches Model Test

In this study, the model slope located in Snow and Ice Research Center was used. Photo 1 shows the model slope. Figure 3 shows the side view of the model slope and the initial position of the snow mass. The snow mass was 40cm in width, 70cm in length and 40cm in height. The bottom of the model slope has been covered with the frozen snow. In the experiments, the snow flowed on the model slope. Each test was run 2 times under same test conditions.

Photo 1. Model slope

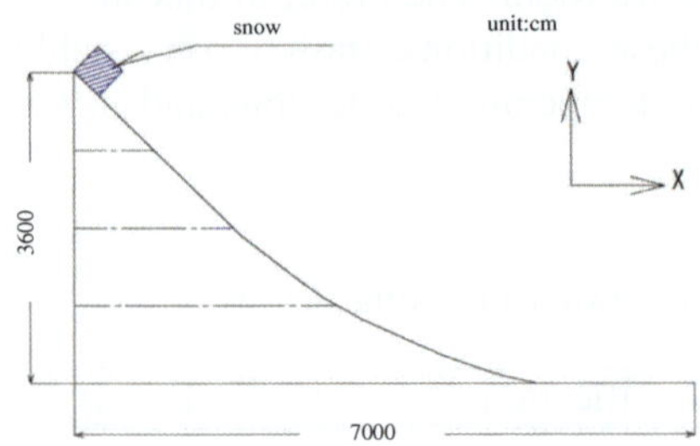

Fig. 3. Side view of the model slope and initial position of the snow mass

3.2 Model Test Result

Photos 2 and 3 show flowing snow at different times for first and second experiments, respectively. In first and second experiments, flowing snow stopped after 2.2 and 2.3 seconds, respectively. The travel length of snow was about 6.0m in both the first and second experiments. It was confirmed that the snow flowed like a block in the beginning which thereafter baroke around the middle of the slope. It was found that the solid snow mass changed into fluidized snow during flowing.

at 0.5s at 1.0s at 2.0s at 2.2s

Photo 2. First flowing snow test (at 0.5, 1.0, 2.0 and 2.2 second)

at 0.5s at 1.0s at 2.0s at 2.3s

Photo 3. Second flowing snow test (at 0.5, 1.0, 2.0 and 2.3 second)

4 Numerical Modeling

4.1 Constitutive Model for Flowing Snow

A number of rheological models have been proposed. The Bingham model has been recognized as one of the most promising models for the engineering simulation of lava flow [6] and snow avalanche [7]. A constitutive model used in this study is also based on the Bingham model. In one dimensional simple shear state, the Bingham model adopted in this study can be described as the linear expression between the shear stress and the shear strain rate as follows:

$$\tau = \eta_0 \dot{\gamma} + \tau_y \tag{1}$$

where τ is the shear stress, η_0 is the viscosity after yield, $\dot{\gamma}$ is the shear strain rate and τ_y is the yield shear strength. In order to describe both of cohesive behavior and frictional behavior, the Mohr–Coulomb criterion is introduced as the yield shear strength in the Bingham model, i.e.

$$\tau = \eta_0 \dot{\gamma} + c + p \tan\phi \tag{2}$$

where c is the cohesion, ϕ is the friction angle and p is the hydrostatic pressure. As it is difficult to solve Equation (2) directly, an equivalent viscosity η' obtained as in Eq. (3) is used [1]. The equivalent viscosity used for the Newtonian fluid is as follows:

$$\eta' = \frac{\tau}{\dot{\gamma}} = \eta_0 + \frac{c + p \tan\phi}{\dot{\gamma}} \tag{3}$$

The equivalent viscosity is used as the viscosity coefficient of a flowing material and its value is updated with the change of the shear strain rate.

4.2 Governing Equations

The flow material is assumed to be incompressible fluid. In this case, the following equations can be used as the governing equation:

$$\frac{\partial u_i}{\partial t} + u_j \frac{\partial u_i}{\partial x_j} = \frac{1}{\rho}\frac{\partial \sigma_{ij}}{\partial x_i} + g_i \tag{4}$$

$$\frac{\partial u_i}{\partial x_i} = 0 \tag{5}$$

where ρ is the density and g_i is the gravity acceleration vector. Equation (4) is the conservation equation of momentum and Eq. (5) is the equation of continuity that is derived from the conservation of mass. The conservation of momentum is represented as below:

$$\frac{\partial u_i}{\partial t} + u_j \frac{\partial u_i}{\partial x_j} = -\frac{1}{\rho}\frac{\partial p}{\partial x_i} + \frac{1}{\rho}\frac{\partial}{\partial x_j}\left[\eta'\left(\frac{\partial u_i}{\partial x_j} + \frac{\partial u_j}{\partial x_i}\right)\right] + g_i \tag{6}$$

For the case of a Newtonian fluid, the viscosity coefficient is treated as a constant value and its spatial derivative is not considered. However, as we can see from Eq. (6), the equivalent viscosity η' varies with location. Therefore, the effect of the spatial derivative should be taken into account in Eq. (6). For setting up the numerical solution, Eq. (6) is decomposed as follows:

$$\frac{u_i^{*} - u_i^{n}}{\partial t} = -u_j \frac{\partial u_i}{\partial x_j} \tag{7}$$

$$\frac{u_i^{**} - u_i^{*}}{\partial t} = g_i \tag{8}$$

$$\frac{u_i^{***} - u_i^{**}}{\partial t} = \frac{1}{\rho}\frac{\partial}{\partial x_j}\left[\eta'\left(\frac{\partial u_i}{\partial x_j} + \frac{\partial u_j}{\partial x_i}\right)\right] \tag{9}$$

$$\frac{u_i^{n+1} - u_i^{***}}{\partial t} = -\frac{1}{\rho}\frac{\partial p}{\partial x_i} \tag{10}$$

where subscripts n and $n+1$ indicate the quantities at each calculation time step and $*$ indicates the temporary quality. Equations (7), (8), (9) and (10) are the advection term, the external force term, viscous term and the pressure term, respectively. The advection term is solved by the THINC method whose details will be presented later. The non-advection terms (Eqs. 7, 9 and 9) are discretized by using the finite different method. The Poisson equation of the pressure is derived from the pressure term and Eq. (5). This equation is solved implicitly. The implicit procedure is also used for the viscous term. As previously mentioned, the equivalent viscosity η' depends on shear strain rate and its value changes severely in both time and space. Therefore, it is necessary to use the implicit time integration scheme for the viscous term.

4.3 Surface Treatment

Many surface capturing methods have been proposed [8], [9], [10] and [11]. In this study, THINC (Tangent of Hyperbola for Interface Capturing) technique proposed by Xiao [12] is used as a surface capturing method. In this method, a function φ, called VOF function, is defined at each calculation grid similar to the VOF method [8]. The VOF function has a value between 0 and 1 and it indicates the occupancy of fluid at each mesh. By solving the following advection equation of the VOF function, the movement and deformation of free surface can be captured, i.e.

$$\frac{\partial \varphi}{\partial t}+\frac{\partial (u_i \varphi)}{\partial x_i}-\varphi\frac{\partial u_i}{\partial x_i}=0 \tag{11}$$

The basic 1-D THINC scheme is devised for the one dimensional advection equation of the VOF function as

$$\frac{\partial \varphi}{\partial t}+\frac{\partial (u \varphi)}{\partial x}-\varphi\frac{\partial u}{\partial x}=0 \tag{12}$$

The calculation procedure of the THINC scheme is briefly illustrated in Fig. 4. In this figure, Δt describes the time increment at one time step, Δx is the grid interval, g is flux of the VOF function and subscript n indicates the value at time step n ($t = t^n$).

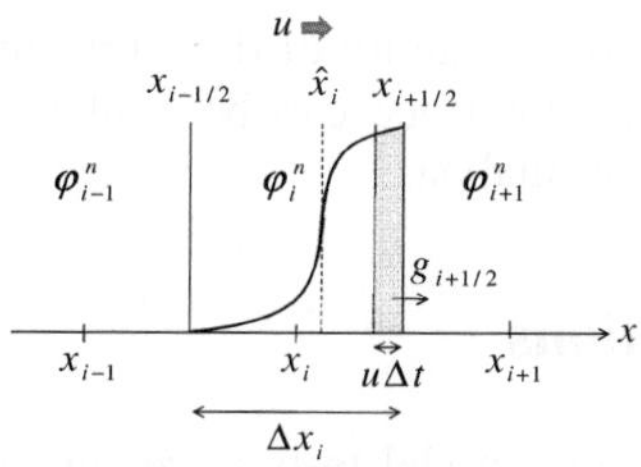

Fig. 4. Schematic image of one-dimensional VOF function

A piecewise modified hyperbolic tangent function is used as an interpolation function whose expression is given by

$$F_i = \frac{\alpha}{2}\left(1+\gamma \tanh\left(\beta\left(\frac{x - x_{i-1/2}}{\Delta x_i} - \hat{x}_i\right)\right)\right) \tag{13}$$

where F is the piecewise interpolation function, $\hat{x}$ is the middle point of the transition jump in the hyperbolic tangent function and the parameters α , β and γ are defined to adjust the shape and slope orientation. As mentioned by Xiao, values of the parameters α and γ can be automatically obtained from values of VOF function at x_{i-1} and x_{i+1} . The parameter β is an adjustable parameter which determines the steepness of the jump in the interpolation function. Xiao investigated the effect of the value of β . According to his investigation, a large value of β provides a sharp fluid interface and less numerical diffusion. However, a large β tends to wrinkle an interface which is parallel to the velocity direction. We use the value β =8.0 for all simulations conducted in this study because the fluid interface can be clearly defined with this β value. After α , β and γ are given, the only unknown in the interpolation function is the middle point of the transition jump $\hat{x}$. In order to obtain the value of $\hat{x}$, a constraint condition is used as follows:

$$\frac{1}{\Delta x_i}\int_{x_{i-1/2}}^{x_{i+1/2}} F_i(x)\,dx = \varphi_i^n \tag{14}$$

After the piecewise interpolation functions have been computed, the flux g is calculated by integrating the interpolation function. Finally, the VOF function φ is updated as below:

$$\varphi_i^{n+1} = \varphi_i^n - \frac{\left(g_{i+1/2} - g_{i-1/2}\right)}{\Delta x_i} + \varphi_i^n \frac{\left(u_{i+1/2} - u_{i-1/2}\right)}{\Delta x_i} \tag{15}$$

In the THINC method, the total quantity of the VOF function can be exactly conserved. In addition, the fluid interface can be kept sharp. These advantages are useful in the multi phase flow analysis.

5 Numerical Simulations

Simulation of snow avalanches model tests were carried out. Figure 5 shows the numerical model. The mesh size is 0.025m x 0.025m. The density of the snow was 500 kg/m^3. The other parameters (friction angle ϕ and cohesion c of snow) were obtained from the previous research [4]. Table 2 shows the material parameters obtained from the previous research [4]. It is found that, there is a wide variety for material parameters. In this study, two distinct snow types are considered: Type1: granular material (ϕ=10, 20 and 30 degree and c = 0 Pa) and Type2: cohesive material (ϕ=0 degree and c =30, 60 and 90 Pa).

Table 2. Material parameters of snow from previous research[4]

	Case1	Case2	Case3	Case4	Case5	Case6
Cohesion c [Pa]	25.5	85.3	28.5	60.8	34.3	12.8
Internal friction angle ϕ [degree]	23	23	24	24	29	32

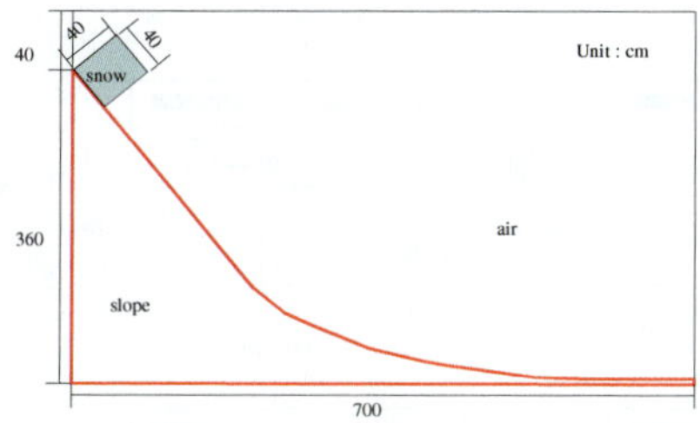

Fig. 5. Numerical model for snow avalanches model tests

5.1 Numerical Results

Figures 6 and 7 show simulated time histories of the surface configurations at different times for Type1 and Type2 snow materials, respectively. By using the THINC technique with the CIP method, it is possible to treat both air and snow together in the simulation. The density function (VOF function) is used to represent the volume of snow by the value between 0-1 (0: air, 1: snow). In both types, simulated results were different from experimental results. In the experiment, it took about 2 seconds until the snow stopped, whereas in the simulation, it took more than 2 seconds. It is considered that the numerical diffusion arose by the coarse mesh size used in these simulations. Therefore, the initial mesh size was changed from 140 x 85 to 280 x 170. Figures 8 and 9 show simulated results for finer mesh size cases. Numerical results for the granular materials are very different from the results of the experiment. On the other hand, the results for the cohesive materials are in good agreement with those of the experiment. Particularly, the case with the cohesion of 60 Pa gave the best fitting. Even if the cohesion was lower, the travel length of snow was not longer because the flowing snow mass became flat. On the other hand, when the cohesion was higher, the travel length of snow became larger because the flowing snow kept the block shape.

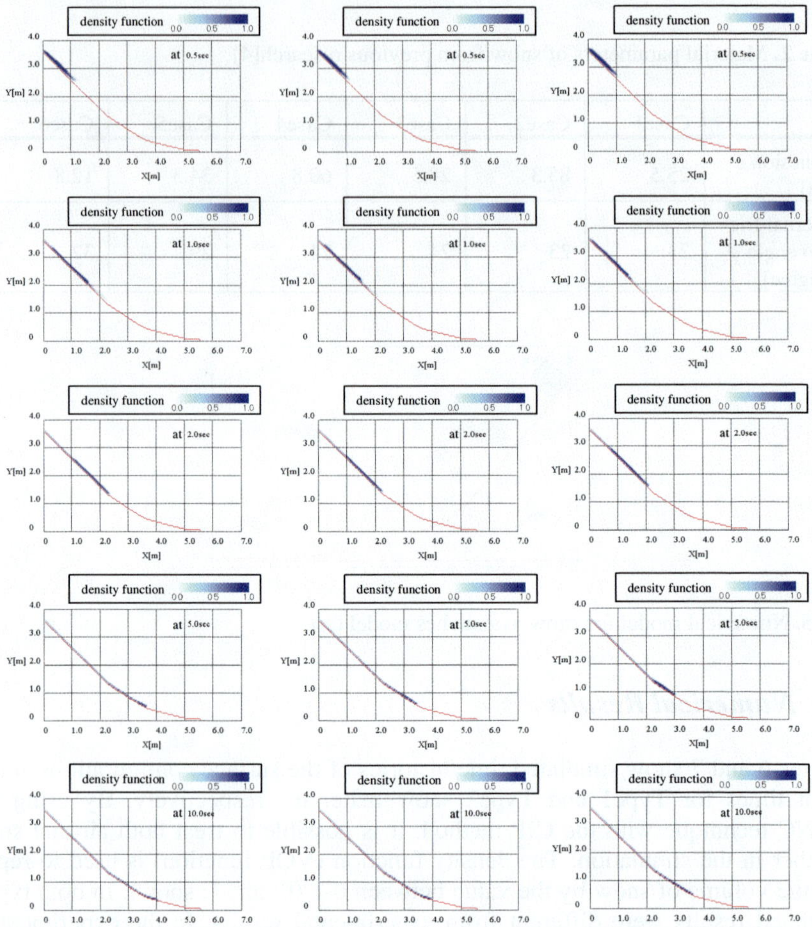

Fig. 6. Simulated time histories of the surface configurations at different times for Type1 snow (mesh size: 140 x 85)

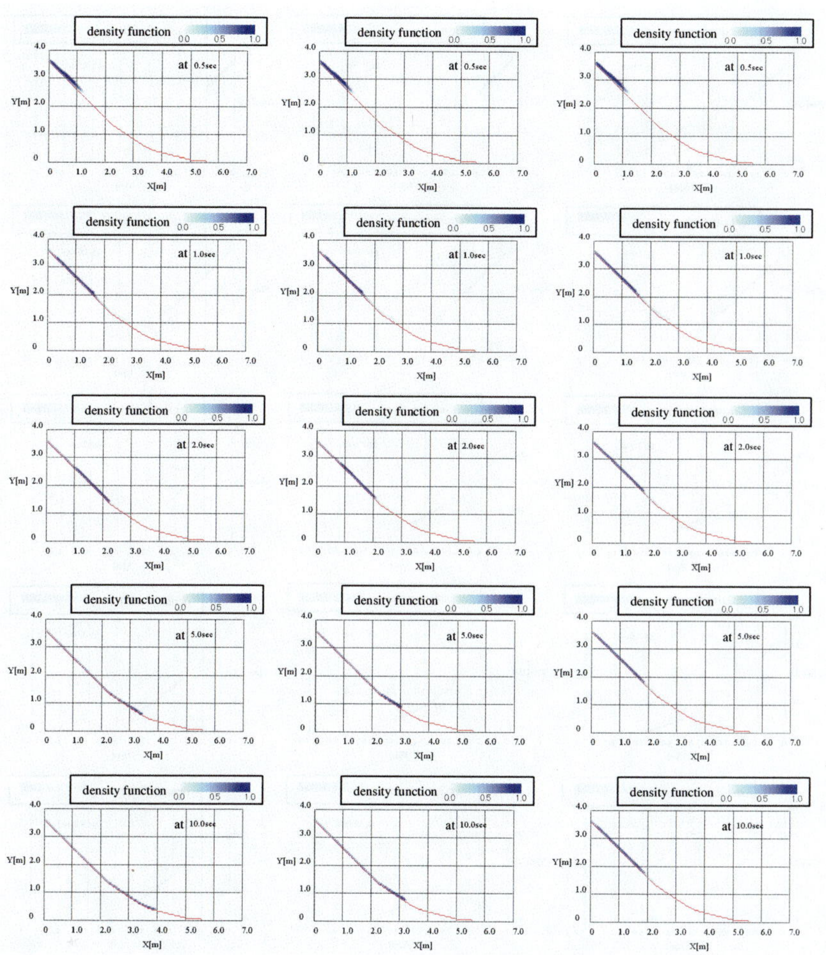

Fig. 7. Simulated time histories of the surface configurations at different times for Type2 snow (mesh size: 140 x 85)

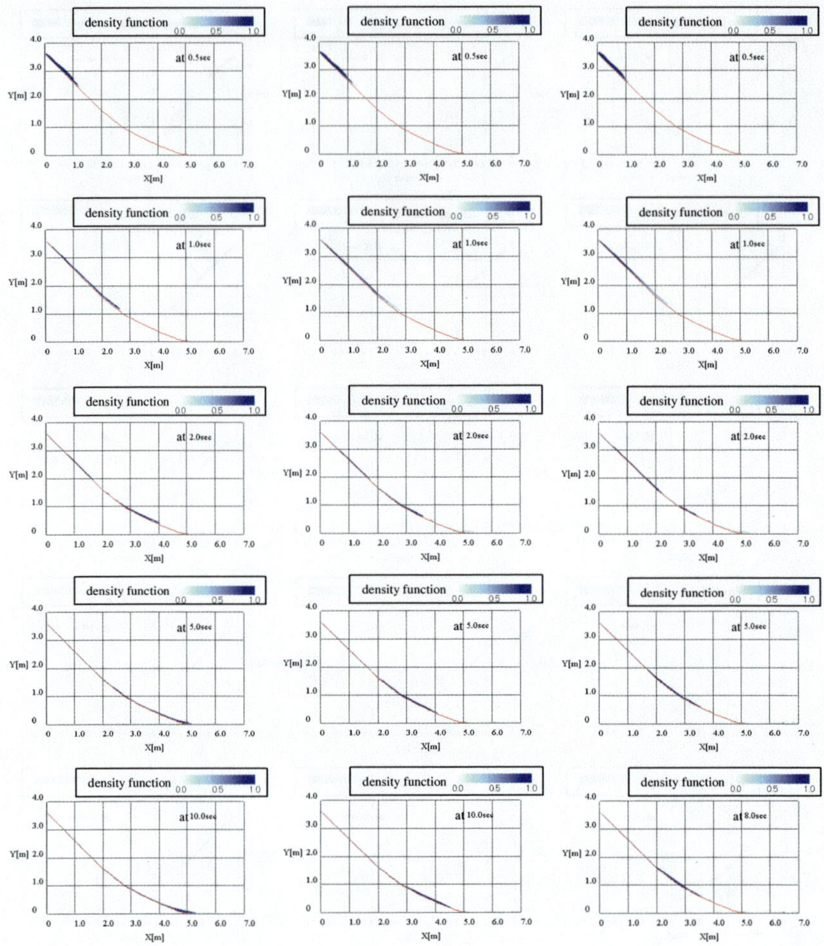

Fig. 8. Simulated time histories of the surface configurations at different times for Type1 snow (mesh size: 280 x 170)

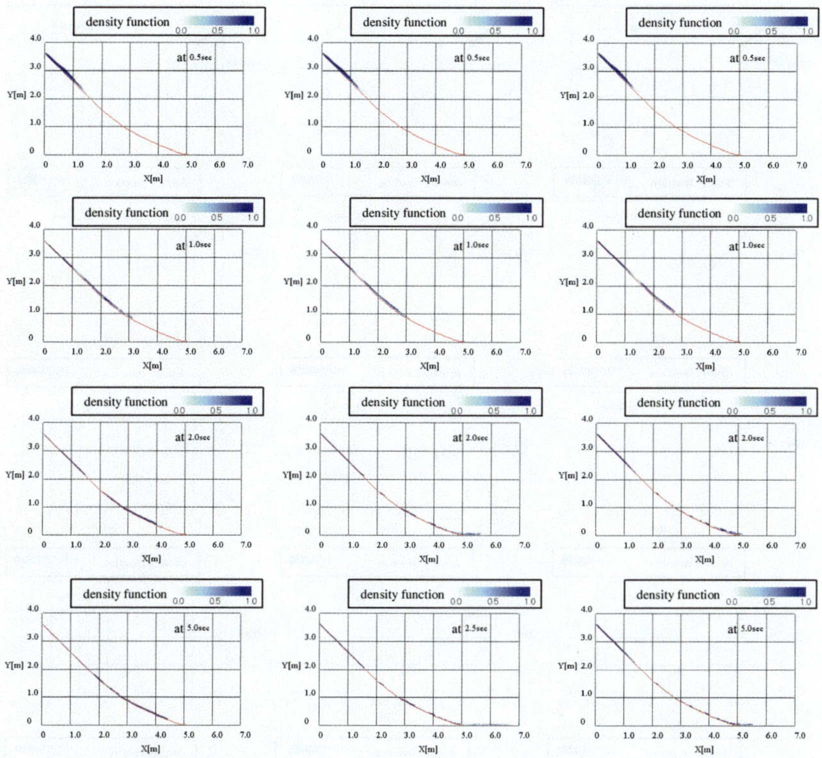

Fig. 9. Simulated time histories of the surface configurations at different times for Type2 snow (mesh size: 280 x 170)

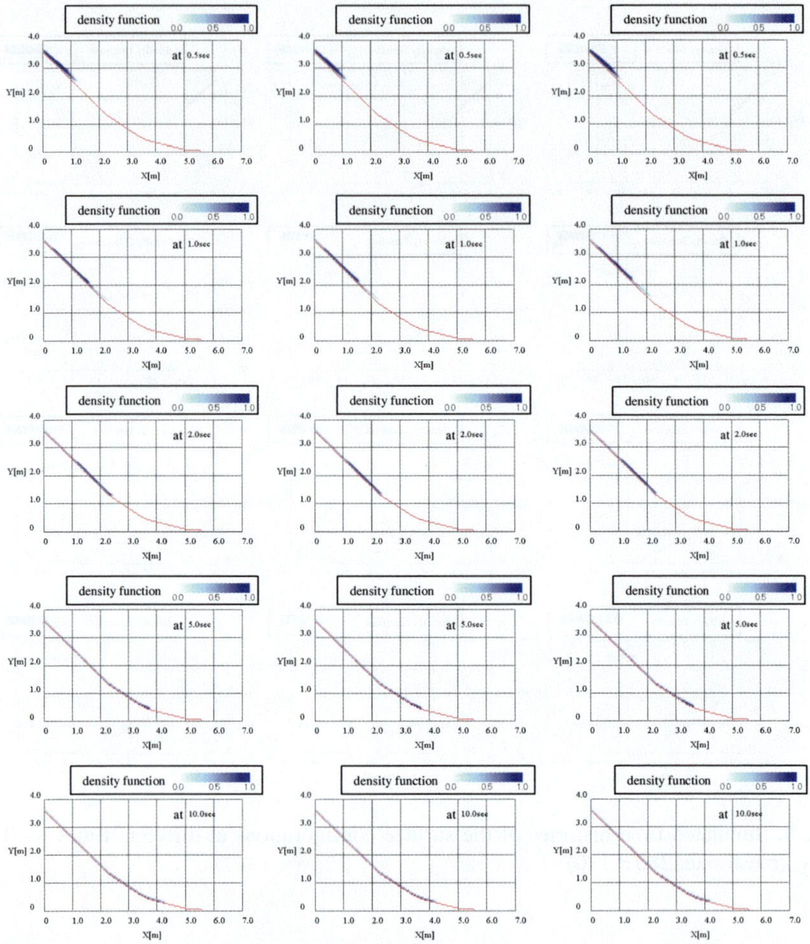

Fig. 10. Simulated time histories of the surface configurations at the different time for Type1 snow (mesh size: 140 x 85 and the value α_B =0.1)

However, when the cohesion was much higher, the travel length of snow became shorter again. In the experiment, the bottom of the model slope has been covered with the frozen snow. It is considered that the bottom friction angle is lower than the friction angle of flowing snow. In order to express the relation between bottom friction angle, the equivalent viscosity η' for the bottom material is changed as follows:

$$\eta' = \frac{\tau}{\dot{\gamma}} = \eta_0 + \frac{\alpha_A \times c + p \times \tan\phi \times \alpha_B}{\dot{\gamma}} \tag{16}$$

where α_A and α_B are values for controlling the bottom friction. We use the value α_A=0.0 for all simulations in this study, because the cohesion of the bottom is considered to be nearly zero. The ratio α_B is obtained from

$$\alpha_B = \frac{\tan\phi_b}{\tan\phi} \tag{17}$$

where ϕ_b is the bottom friction angle. The value of ϕ_b was 13 degrees as obtained from snow avalanche model tests. However, if the value ϕ_b=13 degrees is used, the value α_B should be high, which does not represent the conditions of the model slope. We used α_B=0.1 for all simulations conducted in this study. The simulation involving Type1 snow using a mesh size of 140 x 85 is carried out again. Figure 10 shows results of the simulation using a value α_B=0.1. We now find that the travel length of the snow is almost same for different ϕ values.

6 Conclusion

The laboratory experiments and the numerical simulations of snow avalanches were carried out. In the experiment, the travel length of snow was obtained. It was confirmed that the snow flowed like a block at the beginning and was broken into small pieces around the middle of the slope. It was found that solid snow mass changed into fluid snow during flowing. In the simulations, the travel length of snow flow was different from the one measured in the experiment. Therefore, the initial mesh size was changed from 140 x 85 to 280 x 170. As a result, numerical calculations for cohesive materials were in good agreement with experimental values. Particularly, the case with a cohesion of 60 Pa gave the best fitting. When the cohesion was lower, the travel length of snow was not larger because the flowing snow became flat. On the other hand, when the cohesion was made higher, the travel length of snow become larger because the flowing snow kept the block shape. However, when the cohesion was much higher, the travel length of snow became shorter again.

References

[1] Moriguchi, S.: CIP-Based Numerical Analysis for Large Deformation of Geomaterials. Ph. D. Dissertation. Gifu University (2005)

[2] Xiao, F., Honma, Y., And Kono, T.: A simple algebraic interface capturing scheme using hyperbolic tangent function. International Journal for Numerical Methods in Fluids 48, 1023–1040 (2005)

[3] Nishimura, K.: Studies on the Fluidized Snow Dynamics. Contributions from the Institute of Low Temperature Science A37, 1–55 (1991)

[4] Morikita Publishing Co., Ltd, Protection-against-snow engineering handbook. Japan Construction Mechanization Association, New edition. 96 (1988) (in Japanese)

[5] Casassa, G., Narita, H., Maeno, N.: Shear cell experiments of snow and ice friction. J. Appl. Phys. 69(6), 3745–3756 (1991)

[6] Dragoni, M., Bonafede, M., Boschi, E.: Downslope flow models of a Bingham liquid: implications for lava flows. Journal of Volcanology and Geothermal Research 30, 305–325 (1986)

[7] Dent, J.D., Lang, T.E.: A biviscous modified bingham model of snow avalanche motion. Annals Glaciology 4, 42–46 (1983)

[8] Hirt, C., Nichols, B.: Volume of Fluid (VOF) Method for the Dynamics of Free Boundaries. Journal of Computational Physics 39, 201–225 (1981)

[9] Osher, S., Sethian, J.: Fronts Propagating with Curvature Dependent Speed. Algorithms Based On Hamiliton-Jacobi Formulations. Journal of Computational Physics 79, 12–49 (1988)

[10] Yabe, T., Xiao, F.: Description of complex and sharp interface during Shock Wave Interaction with Liquid Drop. Journal of the Physical Society of Japan 62, 2537–2540 (1993)

[11] Tryggvason, G., Bunner, B.B., Esmaeeli, A., Juric, D., Al-Rawahi, N., Tauber, W., Han, J., Nas, S., Jan, Y.J.: A front-tracking method for the computations of multiphase flow. Journal of Computational Physics 169, 708 (2001)

[12] Xiao, F., Honma, Y., Kono, T.: A simple algebraic interface capturing scheme using hyperbolic tangent function. International Journal for Numerical Methods. Fluids 48, 1023–1040 (2005)

A Mesh Free Method to Simulate Earthmoving Operations in Fine-Grained Cohesive Soils

Joseph G. Gaidos and Mustafa I. Alsaleh

Caterpillar Inc, Peoria, IL, USA

Abstract. Gross distortion and eventual fragmentation of soil, which generally occur during earthmoving operations such as dozing and excavation, pose significant computational challenges to simulation by conventional Finite Element Methods (FEM). This deformation behavior in cohesive soils poses even greater challenges for simulation by the Discrete Element Method (DEM), since without the firm mathematical basis offered by continuum mechanics, DEM is heavily reliant on a mixed semi-analytical and empirical formulation. This paper focuses on the development of a 3D Mesh Free Method (MFM), specifically to extend the predictive capability of existing soil-machine interaction simulation tools to a variety of earthen materials important to earthmoving machines. This discretization method is seen as ideally suited for the prediction of implement forces and overall soil motion resulting from earthmoving operations in a fragmenting medium such as fine-grained cohesive soil. It is here, for simulations involving gross deformation and eventual fragmentation, that the absence of fixed connectivity (or "mesh" as the name implies) gives MFM great flexibility, while still retaining the highly desirable characteristics of a continuum mechanics based formulation. This work documents the theoretical aspects of the formulation, beginning with the MFM discretization of the governing partial differential equations. In addition, it covers the description of the coupled damage mechanics and plasticity constitutive model used to represent the soil, as well as, the details of the treatment of discrete fracture. The work also contains example results from 3D simulations of a blade cutting and a bucket excavating clay-type soil. These results depict a first attempt at capturing soil plasticity coupled with damage evolution, soil fragmentation at the end-state of damage and sustained contact of soil fragments with the earthmoving implementation and amongst the fragments themselves.

Keywords: mesh-free method, computational mechanics, soil dynamics, earthmoving simulation.

1 Introduction

This paper focuses on the development of a 3D Mesh Free Method (MFM), specifically to extend the predictive capability of existing soil-machine interaction simulation tools to a variety of earthen materials important to earthmoving

R. Wan et al. (Eds.): Bifurcations, Instabilities and Degradations in Geomaterials, SSGG, pp. 307–323.
springerlink.com

machines. It is here, for earthmoving simulations involving gross deformation and eventual fragmentation, that the absence of fixed connectivity (or "mesh" as the name implies) gives MFM great flexibility, while still retaining the highly desirable characteristics of a continuum mechanics based formulation.

In what follows, the Mesh Free Method - a continuum dynamics based numerical method - is seen as ideally suited for the prediction of implement forces and overall soil motion resulting from earthmoving operations in a fragmenting medium such as fine-grained cohesive soils. The photograph in Fig. 1, showing a track-type tractor performing a dozing operation, highlights many of the computationally challenging characteristics of earthmoving processes in general – they are dynamic in nature; they involve inelastic finite deformation; they involve gross distortion and severe fragmentation; they involve a large number of contacting fragments which continue to remain actively engaged in the process.

Fig. 1. A track-type tractor performing dozing operations

The classes of discretization methods, which define the approximation entirely in terms of neighboring points or nodes and hence do not rely on fixed connectivity to describe the field variables and the instantaneous topology of the domain, may be referred to collectively as meshless methods or Mesh Free Methods (MFM). First invented in 1977, by Lucy [1] and at the same time by Gingold and Monaghan [2], the then "smoothed particle hydrodynamics method" (now called standard-SPH) was originally applied to astrophysical and cosmological problems such as star and galaxy formation. Since [1,2], perhaps over twenty such methods have appeared in the literature. For the discretization of partial differential equations (PDEs) that describe a deforming medium and in particular, for problems involving gross deformation and eventual fragmentation, the absence of fixed connectivity (or mesh) is probably the most attractive general characteristic of the MFMs. These methods may be divided into two main categories based on how they discretize the balance laws - those that employ a variational (or weak formulation) and those that employ a collocation (or strong formulation). This work focuses on one of the collocation methods - the method of Corrected Smooth Particle Hydrodynamics (CSPH) [3,4] and how it may be adapted and applied to solving the partial differential equations that describe a deforming (and ultimately a fragmenting) medium. For the purposes of brevity and consistency of nomenclature, the now modified CSPH method will simply be referred to as the "MFM" in the remainder of this paper.

2 Governing Equations

As a continuum mechanics based method, MFM seeks to discretize and solve the differential forms of the balance laws of conservation of linear momentum and conservation of mass in a material reference frame. It seeks to simultaneously satisfy these balance laws, together with an appropriate constitutive model for the material, to ultimately obtain the instantaneous deformed configuration of the medium. In this setting, balance of linear momentum is solved in strong form (requiring continuity of the first derivative of Cauchy stress with-respect-to position) and hence, satisfied at discrete points.

The differential form of the balance of linear momentum, expressed in the spatial (or deformed) configuration, is commonly written as:

$$\vec{\nabla}\cdot\underset{\sim}{\sigma}+\rho\underline{b}=\rho\underline{\dot{v}} \tag{1}$$

Here, $\underset{\sim}{\sigma}$ is the Cauchy stress tensor, $\underline{b}$ is the body force vector per unit mass, $\underline{v}$ is the velocity vector, ρ is the mass density and of course, $\vec{\nabla}$ is the gradient vector-operator symbol for taking partial derivatives with respect to spatial coordinates. In the momentum equation, the gradient vector operator is applied as a scalar product to obtain the divergence of the Cauchy stress tensor. For proper formulation of the Initial Boundary Value Problem (IBVP), to this equation must be added both traction (natural) and velocity (essential) boundary conditions on their respective boundaries:

$$\underline{\hat{n}}\cdot\underset{\sim}{\sigma}=\underline{t}^0 \quad \text{on} \quad \Gamma_{\underline{t}} \tag{2}$$

$$\underline{v}=\underline{v}^0 \quad \text{on} \quad \Gamma_{\underline{v}} \tag{3}$$

Here, $\underline{\hat{n}}$ is the normal to the surface $\Gamma_{\underline{t}}$, on which traction boundary conditions are prescribed and $\underline{t}^0$ is the traction vector resulting from the operational product of the surface normal at the point on the boundary into the corresponding Cauchy stress tensor at the same point. Also here, $\underline{v}^0$ is the velocity prescribed on the portion of the boundary $\Gamma_{\underline{v}}$. The union of $\Gamma_{\underline{t}}$ and $\Gamma_{\underline{v}}$ comprises the entire boundary of the domain.

The differential form of the conservation of mass, expressed again in the spatial (or deformed) configuration, is commonly written as:

$$\dot{\rho}+\rho\vec{\nabla}\cdot\underline{v}=0 \tag{4}$$

Here again, the gradient vector operator symbol $\vec{\nabla}$ is applied as a scalar product, but in this case with the velocity $\underline{v}$, to obtain the divergence of the velocity field.

Together with these conservation laws, one also adds to the list of governing equations, the constitutive model which captures the internal constitution of the material and provides the link between deformation kinematics and internal forces. The constitutive behavior will be covered later in a separate section.

3 MFM Discretization

The above partial differential equations (PDEs) must now be discretized for algorithmic formulation and eventual solution. In this way, spatial and temporal derivatives are approximated by algebraic expressions and then the above set of PDEs (which now become algebraic expressions themselves) is solved incrementally. The first step to accomplishing this is the introduction of shape functions for interpolation of velocity and other field variables in the computational domain. The next step is to use these shape functions to construct a gradient vector operator for the approximation of spatial derivatives. The final step is the substitution of approximations for derivatives into the governing equations where the final MFM discrete forms emerge.

3.1 Shape Functions

The use of shape functions here, is similar to their use in the finite element method. They provide the means for locally interpolating field variables at any point in the domain from the corresponding field variables known at other spatial locations in the domain. These known field variables used in this interpolation are referred to as nodal parameters. As mentioned earlier, MFM employs no mesh or elements (fixed connectivity) in the formulation and consequently, simply uses points (or nodes) distributed throughout the domain (see Fig. 2) to both carry field

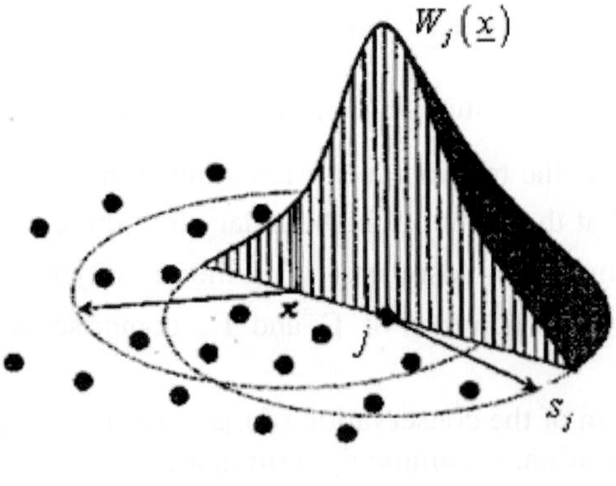

Fig. 2. An example of MFM interpolation points shown with local weighting or kernel function at point "j" (J. Bonet & T-S.L. Lok, 1999)

variables and describe the domain geometry. While in the literature, these points are loosely referred to as particles, they are really interpolation points within a continuous medium and not distinct, unconnected grain-like primitives as the particle label might imply.

The class of shape functions employed in this approach, are based in the use of moving-least-squares (MLS) interpolants [5,6]. These interpolants possess the property to exactly reproduce a desired set of basis functions. For generality, the MFM is formulated with basis functions sufficient to exactly interpolate a field variable up to either a constant or linear function of position. So, as the least-squares portion of the name implies, the interpolants result from the minimization of a weighted error functional but in this "moving" case, with respect to coefficients that vary with position. For example, the linear MFM basis functions in 3D are in general, expressed as a vector $\underline{p}$ of basis functions:

$$\underline{p}^T = [1, x_1, x_2, x_3] \tag{5}$$

The local weighting (or kernel) function used in this least-squares minimization is of compact support and can be viewed (as depicted in Fig. 2) as a smooth "bell" shaped function centered at and carried by each MFM particle. This weighting function ultimately brings smoothness and locality to the approximating functions. Since this kernel function is defined to be non-zero inside the support size of the MFM particle and zero outside of this dimension, it clearly represents a compact zone of diminishing influence for the MFM particle on other MFM particles in the local neighborhood. A cubic spline kernel function is used in this case [5] and is given as a function of the normalized distance measure $\bar{s} = \left\|\underline{x}_j - \underline{x}\right\| / s_j$:

$$W_j(\bar{s}) = \begin{Bmatrix} 2/3 - 4\bar{s}^2 + 4\bar{s}^3 & for\ \bar{s} \le \frac{1}{2} \\ 4/3 - 4\bar{s} + 4\bar{s}^2 - 4/3\bar{s}^3 & for\ \frac{1}{2} < \bar{s} \le 1 \\ 0 & for\ \ \bar{s} > 1 \end{Bmatrix} = W_j(\underline{x}) \tag{6}$$

Here, $W_j(\underline{x})$ denotes the local weighting (or kernel) function centered at the MFM particle "j", the quantity $\left\|\underline{x}_j - \underline{x}\right\|$ is simply the distance from the MFM particle "j" to the point "x" and s_j is the specified support size mentioned earlier.

The MLS interpolant resulting from the minimization process is expressed as:

$$\underline{v}(\underline{x}) = \sum_j \phi_j(\underline{x})\, \underline{v}_j \tag{7}$$

where, $\underline{v}(\underline{x})$ is the interpolated field variable at position $\underline{x}$, $\phi_j(\underline{x})$ are the position-dependent interpolation coefficients (or shape functions) associated with the MFM particles "j" found in the neighborhood of position $\underline{x}$, and $\underline{v}_j$ are associated field variables (or nodal parameters) carried by the MFM particle neighbors "j". The summation is over all MFM particle neighbors whose support dimension covers the position $\underline{x}$. The interpolation coefficients $\phi_j(\underline{x})$ (or shape functions) are computed as:

$$\phi_j(\underline{x}) = \underline{p}(\underline{x})^T \cdot A(\underline{x})^{-1} \cdot \underline{p}_j \cdot W_j(\underline{x}) \tag{8}$$

where the "moment" matrix $A(x)$ is given by

$$A(x) = \sum_j \underline{p}_j \cdot \underline{p}_j^T \cdot W_j(\underline{x}) \tag{9}$$

In this notation, $\underline{p}(\underline{x})$ is the vector of basis functions evaluated at the point "x" and $\underline{p}_j$ is the vector of basis functions evaluated at MFM particle neighbors "j".

3.2 Gradient Vector Operator

Having now developed expressions for first-order MLS shape functions, the next step in discretizing the governing PDEs is to arrive at expressions for derivatives of these shape functions and ultimately produce a gradient vector operator that will exactly recover the derivative of a linear field. This property of linear consistency was recognized early on, by many investigators [5,6,7], as crucial for stability and convergence of the MFM.

To compute shape function spatial derivatives at a position "x", one could simply apply the derivative directly to (8). However, in the present type of MFM collocation-based formulation (where MFM particles alone carry all of the field variables) direct derivative computation and use would result in the eventual computation of all field variables strictly from values known at the MFM particles. Belytschko et al [8] identified this type of direct derivative computation for the discrete gradient operator (or nodal integration if Galerkin weak form is used in place of collocation) as a source for instability in the general class of MFM methods. This instability is due, in part, to the vanishing derivative of the local weight functions $W_j(\underline{x})$ at the MFM particle on which it is centered.

To remove this cause of instability in the weak formulation, Chen et al [9] proposed Stabilized Conforming Nodal Integration (SCNI) and later Chen et al [10] proposed Stabilized Non-Conforming Nodal Integration (SNNI). These two techniques are essentially the same, differing only in the definition of the region of integration. In conforming integration, the integration process is carried out over non-overlapping integration cells such that any adjacent cells share a common surface. In non-conforming integration, overlap or space between integration cells is accepted. This weak form stabilization concept introduces strain smoothing by replacing the strain tensor with its smoothed counterpart obtained from application of the Mean Value Theorem and the resulting integration over a volume. The volume integration is then actually carried out as a surface integral using the Generalized Gauss Theorem and this, thus eliminates the use of derivatives of the local weight functions $W_j(\underline{x})$ in the weak form. In what follows, this same stabilization concept was adapted to fit with the strong formulation used in the present MFM and produce an explicitly non-local stabilized gradient vector operator for use whenever gradients are needed.

So to review, what is ultimately sought here is a stabilized gradient vector operator that will operate as a discrete "star" product (scalar product, cross product or tensor product) on any field variable $\mathbb{A}$ (scalar, vector or second order tensor) in the following manner:

$$\mathbb{A}(\underline{x}) * \overleftarrow{\nabla} = \sum_j \mathbb{A}_j * \overset{S}{\vec{\nabla}} \phi_j(\underline{x}) \tag{10}$$

Upon adapting SNNI in particular, to fit with the present strong formulation, the discrete form of the stabilized gradient vector operator emerges as:

$$\overset{S}{\vec{\nabla}} \phi_j(\underline{x}) = \underset{\sim}{L}(\underline{x}) \cdot \frac{1}{V} \int_\Gamma \varphi_j(\underline{x}) \underline{\hat{n}} d\Gamma \tag{11}$$

where, Γ is the surface of the closed volume of integration V , $\underline{\hat{n}}$ is the normal to the surface Γ, $\phi_j(\underline{x})$ are the MLS shape functions and $\underset{\sim}{L}(\underline{x})$ is a second order tensor. If first-order shape functions are used, they already possess the property of linear consistency and hence, the second order tensor $\underset{\sim}{L}(\underline{x})$ simply becomes the second order Identity tensor and (11) correctly recovers the gradient of a linear field. However, if zero-order MLS shape functions are used, they do not possess the property of linear consistency and hence their gradient must be corrected with the second order tensor $\underset{\sim}{L}(\underline{x})$, such that (11) again correctly recovers the gradient of a linear field. Figure 3 illustrates in 2D, an example of a rectangular region of integration centered at the MFM particle "i", presently at position "x". As shown, this MFM particle with support size "s" has other MFM particle neighbors "j".

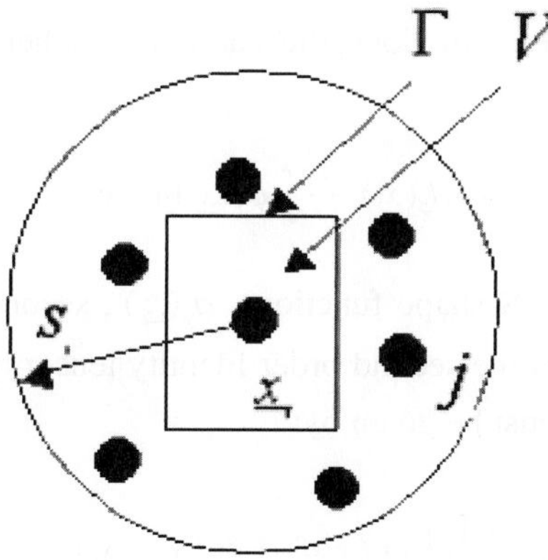

Fig. 3. An example of a rectangular region of integration used to compute the stabilized gradient vector operator

The surface integral over surface Γ of rectangular region V, may be numerically evaluated using a Gaussian Quadrature rule taken over each face. Since the shape of this region of integration is arbitrary in SNNI [9], a rectangular form is used here to simplify computation.

3.3 Summary of Discrete Form of MFM Governing Equations

Having now obtained the necessary expressions for shape functions and a discrete gradient vector operator, the final step in discretization of the governing PDEs is substitution of these discrete relations into the governing equations and final rearrangement of terms. In the following summary, the subscript "i" denotes field variables carried by particle "i" and summations are carried over all neighbors "j" that contain particle "i" in their support.

The balance of linear momentum then takes its final discrete form as:

$$\rho_i \dot{\underline{v}}_i = \sum_j \overset{S}{\vec{\nabla}} \phi_j(\underline{x}_i) \cdot \underset{\sim}{\sigma}_j \tag{12}$$

The conservation of mass takes the final discrete form as:

$$\dot{\rho}_i = -\rho_i \sum_j \underline{v}_j \cdot \overset{S}{\vec{\nabla}} \phi_j(\underline{x}_i) \tag{13}$$

It is also worth summarizing the expression for the discrete form of the velocity gradient:

$$\underline{v}_i \overleftarrow{\nabla} = \sum_j \underline{v}_j \otimes \overset{S}{\vec{\nabla}} \phi_j(\underline{x}_i) \tag{14}$$

In the above summary of expressions, the stabilized gradient vector operator is defined as:

$$\overset{S}{\vec{\nabla}} \phi_j(\underline{x}_i) = \underset{\sim}{L}(\underline{x}_i) \cdot \frac{1}{V} \int_\Gamma \varphi_j(\underline{x}_i) \underline{\hat{n}}\, d\Gamma \tag{15}$$

where, for first-order MLS shape functions $\phi_j(\underline{x}_i)$, second order correction tensor $\underset{\sim}{L}(\underline{x}_i)$ simply becomes the second order Identity tensor but for zero-order MLS shape functions, $\underset{\sim}{L}(\underline{x}_i)$ must be given by:

$$\underset{\sim}{L}(\underline{x}_i) = \left[\sum_j \frac{1}{V} \int_\Gamma \varphi_j(\underline{x}_i) \underline{\hat{n}} d\Gamma \otimes \left(\underline{x}_j - \underline{x}_i \right) \right]^{-1} \tag{16}$$

4 Constitutive Model

The constitutive model used to presently represent fine-grain cohesive soil in MFM, is adapted from one used to simulate damage evolution in concrete. This approach couples rate independent plasticity and damage mechanics in the operator-split formulation found in [11,12]. This coupling of plasticity with damage mechanics provides a convenient means to introduce "softening" behavior into the constitutive model to capture the degradation of material properties associated with micro-void initiation, growth and coalescence in real geomaterials. The presence of softening behavior is crucial for capturing the cohesive and ultimately fragmentary nature of fine-grain soils such as clay.

In this approach, algorithmic incremental updates of the relevant constitutive variables proceed first with an elastic damage predictor that is then followed by a plastic return-mapping corrector. The plastic corrector portion of this formulation is cast in strain space with the multi-surface yield condition consisting of a Drucker-Prager failure envelope and a hardening cap. This entire plasticity formulation is expressed in terms of the effective stress tensor $\bar{\underset{\sim}{\sigma}}$ such that it conforms to the damage mechanics Hypothesis of Strain Equivalence [11] where "the strain associated with the damaged state under the applied stress is equivalent to the strain associated with the undamaged state under the effective stress."

The Cauchy stress tensor is related to the effective stress tensor through the fourth order damage tensor $\underset{\approx}{D}$ as:

$$\underset{\sim}{\sigma} = \underset{\approx}{D} : \bar{\underset{\sim}{\sigma}} \tag{17}$$

Here, the formulation is quite general, as the damage tensor may be isotropic for ductile damage behavior or anisotropic for brittle damage behavior.

4.1 Damage Model

The governing equations for the isotropic elastic-damage evolution begin with the definition of an equivalent-strain based damage parameter $\bar{\tau}$ as:

$$\bar{\tau} = \sqrt{\underset{\sim}{\varepsilon} : \underset{\sim}{\varepsilon}} \tag{18}$$

where in this case $\underset{\sim}{\varepsilon}$ is the total strain tensor.

The strain-space criterion for damage g is then given as:

$$g = \bar{\tau} - r \leq 0 \tag{19}$$

where, the scalar r is the current threshold of equivalent-strain for active damage evolution.

The scalar damage variable d is introduced at this point, where d is defined to be in the interval $0 \leq d \leq 1$, such that $d = 0$ is undamaged and $d = 1$ is fully damaged. The general form of the governing equation for damage variable evolution is given as:

$$\dot{d} = \dot{\chi} H(\overline{\tau}, d) \tag{20}$$

where, $\dot{\chi}$ is the damage consistency parameter and H is the damage function. As presently embodied in this MFM, if the damage function H is taken as a simple but specific function of $\overline{\tau}$ alone, then with this function, (20) can be easily integrated directly to give:

$$d = \frac{\overline{\tau}_{\max}}{\overline{\tau}} \left(\frac{\overline{\tau} - \overline{\tau}_{\min}}{\overline{\tau}_{\max} - \overline{\tau}_{\min}} \right) \tag{21}$$

Here, $\overline{\tau}_{\min}$ is the initial threshold of equivalent-strain for damage initiation, $\overline{\tau}_{\max}$ is the equivalent-strain for damage completion, $\overline{\tau} \neq 0$ and $\overline{\tau}_{\min} < \overline{\tau}_{\max}$.

To these governing equations for damage are added the Kuhn-Tucker conditions for damage loading/unloading:

$$\dot{\chi} \geq 0; \quad g \leq 0; \quad \dot{\chi} g = 0 \tag{22}$$

and the general consistency condition:

$$\dot{\chi} \dot{g} = 0 \tag{23}$$

For the isotropic damage case discussed here, the fourth order damage tensor in (17) simply becomes $(1-d)$ times the fourth order identity tensor and (17) then simplifies to

$$\underset{\sim}{\sigma} = (1-d) \underset{\sim}{\overline{\sigma}} \tag{24}$$

4.2 *Plasticity Model*

The governing equations for the rate-independent Drucker-Prager plasticity model with elliptic cap begin with a general expression (cast in strain space) for the relationship between effective stress and total strain:

$$\underset{\sim}{\overline{\sigma}} = \underset{\approx}{C}_0 : \underset{\sim}{\varepsilon} - \underset{\sim}{\overline{\sigma}}^p \tag{25}$$

where $\underset{\approx}{C}_0$ is the isotropic (fourth order) linear elasticity tensor, $\underset{\sim}{\varepsilon}$ again is the total strain tensor and $\underset{\sim}{\overline{\sigma}}^p$ is plastic relaxation effective stress tensor.

While it is recognized that soils tend to be non-associative, for simplicity at this point in development of the MFM, an associative flow rule provides the necessary evolution equation for the rate of plastic relaxation effective stress and is given by:

$$\dot{\underset{\sim}{\overline{\sigma}}}^p = \dot{\lambda} \frac{\partial f}{\partial \underset{\sim}{\varepsilon}}$$

In this equation, $\dot{\lambda}$ is the familiar plasticity consistency parameter and f is the multi-surface yield condition.

An isotropic evolutionary hardening law is associated with expansion of the cap surface and takes the general form:

$$\dot{\kappa} = \dot{\lambda} h \tag{26}$$

where, κ, called the hardening parameter, appears next in the multi-surface yield condition and h is a hardening function related to plastic volume change [11].

The multi-surface yield condition consists of a Drucker-Prager failure envelope and an elliptic cap hardening surface as given by:

$$\begin{aligned} f_{env}(\bar{\underset{\sim}{\sigma}}) &= \left\{\bar{J}_2 - \bar{F}_{env}\left(\bar{J}_1\right)\right\} \\ f_{cap}(\bar{\underset{\sim}{\sigma}}, \kappa) &= \left\{\bar{J}_2 - \bar{F}_{cap}\left(\bar{J}_1, L(\kappa)\right)\right\} \end{aligned} \tag{27}$$

Here, $\bar{J}_1$ is the first invariant of the effective stress tensor and $\bar{J}_2$ is the second invariant of the effective stress deviator.

To complete the list of governing equations for the constitutive model, one must add the Kuhn-Tucker conditions for plastic loading/unloading:

$$f \leq 0; \quad \dot{\lambda} \geq 0; \quad \dot{\lambda} f = 0 \tag{28}$$

and the well known consistency condition:

$$\dot{\lambda} \dot{f} = 0 \tag{29}$$

To these six governing equations ((25) to (29)), which now fully define the plasticity portion of the constitutive model, we can now apply a generalized midpoint time integration rule to obtain the discrete return-mapping algorithm as discussed in detail in [12]. The algorithmic approach in [12] correctly treats both the tension cut-off and the "corner region" intersection of the failure envelope and the elliptic hardening cap surfaces.

4.3 Fracture Model

In the context of the present MFM, degradation of material properties associated with micro-void initiation, growth and coalescence in geomaterials (damage) evolves as described in the Constitutive Model section and may culminate in the separation or fracturing of material. Discrete fracture may thus be characterized as the end-state of damage whereby new surfaces evolve and the boundary value problem is irreversibly altered. It is here, for problems involving gross deformation and eventual fragmentation that the absence of fixed connectivity (or mesh) gives MFM great flexibility.

The fracture model chosen for the present MFM is adapted from the kernel function attenuation method proposed in [13] to treat the large number of cracks which evolve in a fragmenting medium, rather than crack tip basis function enrichment methods often used to accurately treat a single crack. In this case, for two neighboring MFM particles that satisfy certain damage criteria, the attenuation strategy essentially nullifies the contributions that the MFM particle neighbors make to each other's MLS shape functions. The discrete fracture process is then reflected in the irreversible conversion of MFM interior particles to boundary particles, with the accompanying presence of newly established surface normals, which ultimately leads to the formation of entirely new boundary surfaces associated with the fragmenting soil.

4.4 Discrete Fracture

As mentioned earlier, to represent discrete fracture at MFM particle neighbors that now satisfy fracture criteria, the MLS shape functions are reconstructed with the contributions of the fractured neighbors nullified. The gradient vector operator associated with each of these fractured neighbors is then recomputed to again recover linear consistency. This decoupling of fractured MFM particle neighbors is easily accomplished by introducing a scalar attenuation factor δ_{ij} into MLS shape function construction by replacing $W_j(\underline{x}_i)$ with $W_j(\underline{x}_i)\cdot\delta_{ij}$ in (8) and (9) respectively, as follows:

$$\phi_j(\underline{x}_i) = \underline{p}(\underline{x}_i)^T \cdot A(\underline{x}_i)^{-1} \cdot \underline{p}_j \cdot W_j(\underline{x}_i)\cdot\delta_{ij} \tag{30}$$

$$A(x_i) = \sum_j \underline{p}_j \cdot \underline{p}_j^T \cdot W_j(\underline{x}_i)\cdot\delta_{ij} \tag{31}$$

Here the generalized expressions (8) and (9) have been specialized to relate the MFM particle "i" to its neighbor "j" and the scalar attenuation factor δ_{ij} takes on the values:

$$\delta_{ij} = \begin{Bmatrix} 1 & \text{for no fracture} \\ 0 & \text{for fracture} \end{Bmatrix} \tag{32}$$

5 MFM Temporal Update

Time stepping in the present MFM proceeds using a predictor-corrector leap-frog explicit scheme as found in [14]. In this time integration scheme, MFM particle velocities and velocity gradients are computed at half-time points and all other quantities are updated at full time points.

6 MFM Earthmoving Simulation Examples

In the two examples of MFM simulation of earthmoving operations presented next, the earthmoving implements traverse predefined paths. These examples are not intended to show validation of the method, but rather to illustrate the present level of development of MFM for simulating soil-machine interaction. The first example is a MFM simulation of a track-type tractor size blade (half scale) cutting a clay-type material. The second is a MFM simulation of a wheel loader size bucket (half scale) excavating the same type material. Both simulations (containing approximately 100,000 particles each) help to highlight the capability of MFM to capture gross deformation and severe fragmentation. These results capture soil plasticity coupled with damage evolution, soil fragmentation at the end-state of damage and sustained contact of soil fragments with the earthmoving implement and amongst the fragments themselves.

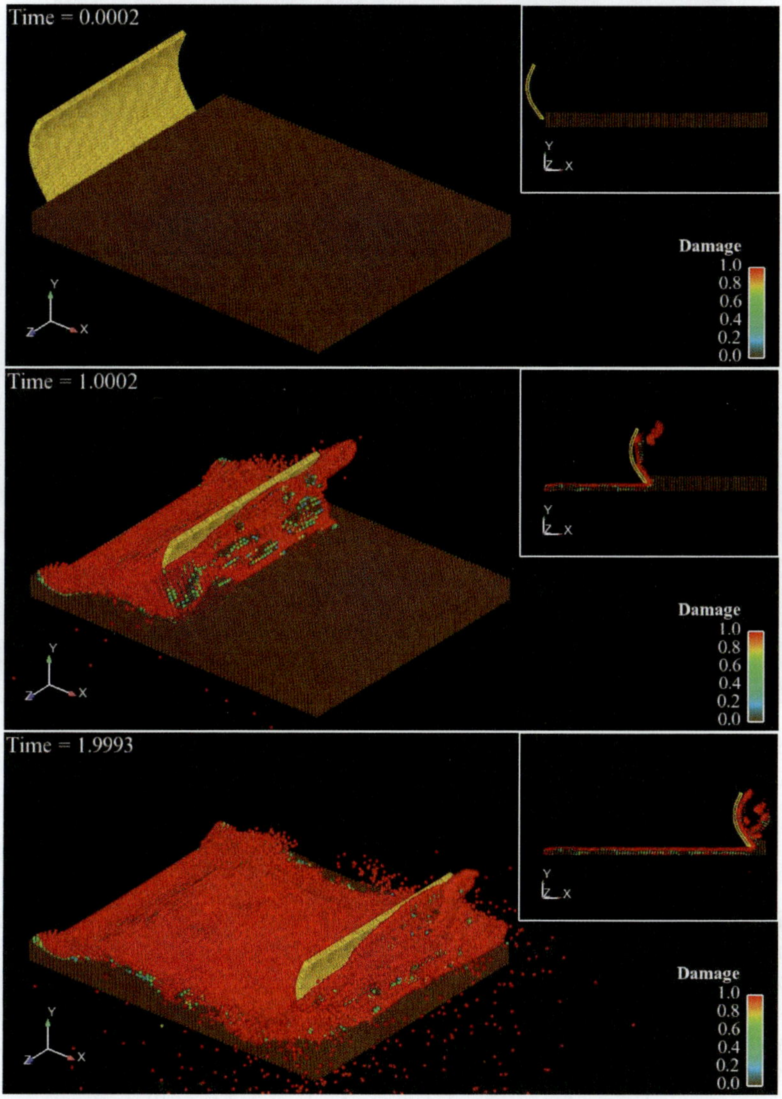

Fig. 4. Time sequence of isometric views for blade cutting example

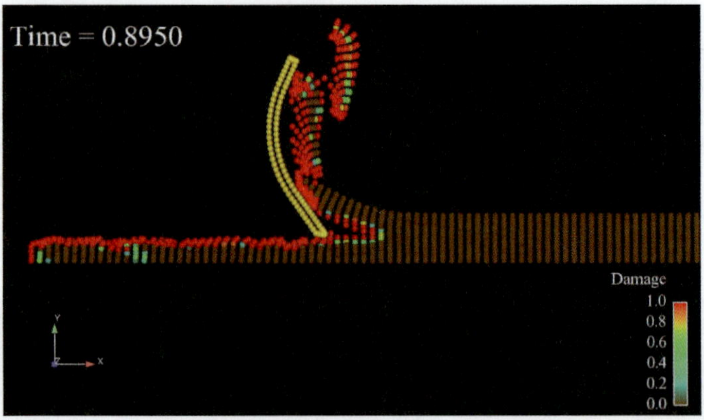

Fig. 5. Clipping plane at centerline of MFM domain

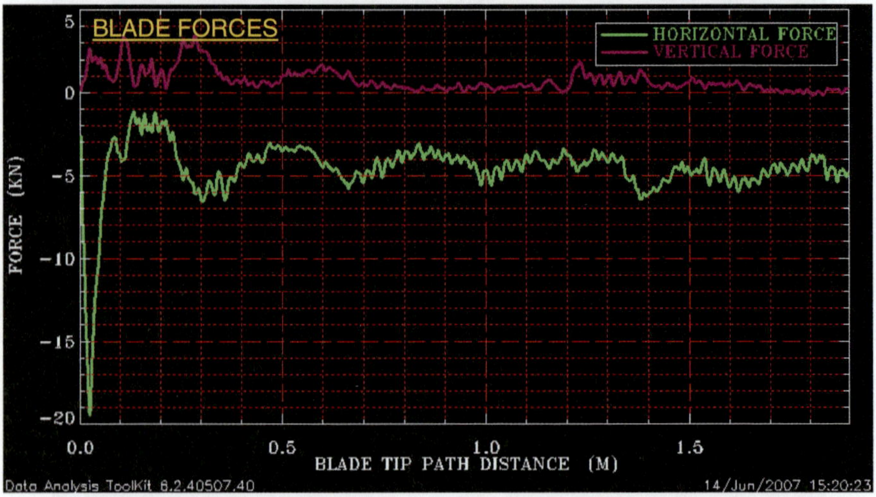

Fig. 6. Horizontal and vertical blade force as a function of distance for blade cutting example

6.1 Soil Cutting MFM Simulation

In this example, a half scale dozer blade was set to cut at a constant depth of 50mm while ramping-up to and then traveling at a constant horizontal velocity of approximately 1m/sec. The high cohesion soil in this example had a measured cohesion of 57.0 kPa and a friction angle of 20 degrees. In Figure 4 and the following figure, rigid particles used to describe the blade geometry are color coded yellow and MFM particles are color coded based on the level of the damage variable they carry – brown for minimal damage to deep red for full damage.

Figure 4 depicts a sequence of isometric views of the general flow of material during the cutting process. The blade has traveled approximately 2.0m in the final frame. Figure 5, resulting from a clipping plane at the domain centerline, highlights the resolution in the field ahead of the blade and illustrates the "tensile" mode of failure as one of four general modes of soil failure identified by [15]. Figure 6 gives a plot of both horizontal and vertical forces on the blade in this example, as a function of distance traveled.

6.2 Soil Excavation MFM Simulation

In this example, a half scale loader bucket was set to enter the soil, rack and then lift by moving along a predefined path. The high cohesion soil in this example

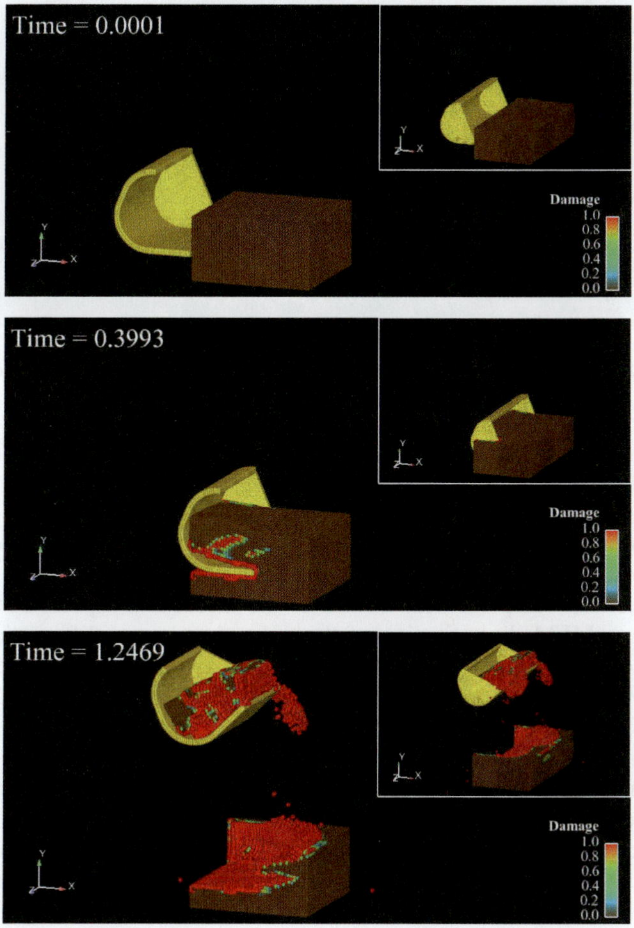

Fig. 7. Time sequence of isometric views for bucket excavating example

again had a measured cohesion of 57.0 kPa and a friction angle of 20 degrees. In Fig. 7, rigid particles used to describe the bucket geometry are color coded yellow and MFM particles are color coded, based on the level of the damage variable they carry – brown for minimal damage to deep red for full damage. While the entire bucket was present during the simulation, Fig. 7 depicts a cut-away view to reveal soil behavior inside the bucket. Figure 8 gives a plot of both horizontal and vertical forces on the bucket in this example, as a function of distance traveled.

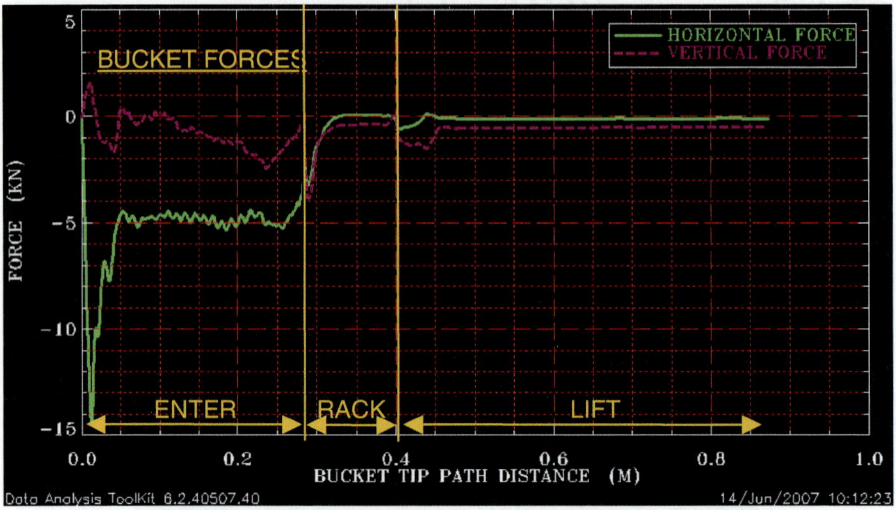

Fig. 8. Horizontal and vertical forces as a function of distance for bucket excavation example

7 Summary and Conclusions

The development of a 3D Mesh Free Method (MFM), intended to extend the predictive capability of existing soil-machine interaction simulation tools to a variety of earthen materials, was presented. Details of the theoretical aspects of the formulation, including MFM discretization of the governing partial differential equations, were provided. In addition, the description of the coupled damage mechanics and plasticity constitutive model used to represent the soil, as well as, the details of the treatment of discrete fracture were covered. Example results from 3D simulations of a blade cutting and a bucket excavating clay-type soil were also given. These earthmoving simulation examples highlight the present level of development of this MFM in providing predictive capability for soil-machine interaction studies. The lengthy process of validation of the MFM, as applied to soil cutting in a variety of natural and artificial soils, is underway and may be the subject of a future publication. Preliminary indications, however, from comparisons of MFM simulations of soil cutting with experimental evidence in both low and high cohesion soils, indicate

MFM is able to capture essential aspects of the fragmentation pattern and the approximate magnitudes of soil cutting forces.

For earthmoving simulations, involving gross deformation and eventual fragmentation of cohesive soils, the absence of fixed connectivity in the discretization, gives MFM great flexibility while still retaining the highly desirable characteristics of a continuum mechanics based formulation.

References

[1] Lucy, L.B.: A Numerical approach to the testing of the fission hypothesis. Astron. J. 82, 1013 (1977)
[2] Gingold, R.A., Monaghan, J.J.: Smoothed Particle Hydrodynamics: Theory and applications to non-spherical stars. Mon. Not. R. Astr. Soc. 181, 375 (1977)
[3] Bonet, J., Lok, T.-S.L.: Variational and momentum preserving aspects of smooth particle hydrodynamic formulations. Comput. Meth. Appl. Mech. Engrg. 180, 97–115 (1999)
[4] Bonet, J., Kulasegaram, S.: Correction and stabilization of smooth particle hydrodynamics methods with applications in metal forming. IJNME 52, 1203–1220 (2001)
[5] Belytschko, T., Krongauz, Y., Organ, D., Fleming, M.: Meshless methods: An overview and recent developments. Comput. Meth. Appl. Mech. Engrg. 139, 3–47 (1996)
[6] Dilts, G.A.: Moving-least-squares-particle hydrodynamics – I. Consistency and Stability. IJNME 44, 1115–1155 (1999)
[7] Chen, J.-S., Pan, C., Wu, C.-T., Liu, W.K.: Reproducing kernel particle methods for large deformation analysis of non-linear structures. Comput. Meth. Appl. Mech. Engrg. 139, 195–227 (1996)
[8] Belytschko, T., Guo, Y., Liu, W.K., Xiao, S.P.: A unified stability analysis of meshless particle methods. IJNME 48, 1359–1400 (2000)
[9] Chen, J.-S., Wu, C.-T., Yoon, S., You, Y.: A stabilized conforming nodal integration for Galerkin mesh-free methods. IJNME 50, 435–466 (2001)
[10] Chen, J.-S., Wang, D., Yoon, S., You, Y.: Accelerated and adaptive meshfree method for earthmoving simulation, Progress Report To Caterpillar Inc. (2001)
[11] Simo, J.C., Ju, W.J.: Strain and stress based continuum damage models - I. Formulation. Int. J. Solids Structures 23, 821–840 (1987)
[12] Simo, J.C., Ju, W.J.: Strain and stress based continuum damage models - II Computational aspects. Int. J. Solids Structures 23, 841–869 (1987)
[13] Randles, P.W., Libersky, L.D.: Smoothed particle hydrodynamics: some recent improvements and applications. Comput. Meth. Appl. Mech. Engrg. 139, 375–408 (1996)
[14] Randles, P.W., Petschek, A.G., Libersky, L.D., Dyka, C.T.: Stability of DPD and SPH. In: Griebel, M., Schweitzer, A. (eds.) Proceedings, Meshfree Methods for Partial Differential Equations, Bonn LNCSE, pp. 340–357. Springer, Heidelberg (2001)
[15] Elijah, D.L., Weber, J.A.: Soil failure and pressure patterns for flat cutting blades. Transactions of the ASAE, 781–785 (1971)

Analysis of Deformation and Damage Processes in Soil-Tool Interaction Problems

Jan Maciejewski[1,2] and Zenon Mróz[1]

[1] Institute of Construction Machinery Engineering, Warsaw University of Technology, Narbutta Str. 84, 02-524 Warsaw
e-mail: jmaciejewski@simr.pw.edu.pl
[2] Institute of Fundamental Technological Research, ul. Świętorzyska 21, 00- 049 Warsaw, Poland
e-mail: zmroz@ippt.gov.pl

Abstract. The processes of interaction of earth-working machines with soils are related to large inelastic deformation inducing soil structure variation with dilatancy, compaction and critical states developing during the process. The effects of hardening, softening, strain localization in shear or tensile rupture bands accompany the machine tools operation in the cohesive soil. The aim of this contribution is to provide simplified incremental analysis of some typical processes, such as soil cutting, digging, filling, and compaction among others, by applying constitutive models relevant to the type of process. The aim of analysis is to predict the deformation modes, forces interaction and energy required for the process, and also to generate optimal process control in order to minimize some parameters of soil-tool interaction

1 Introduction

The classical soil plasticity is based on a perfectly plastic model with the Coulomb or Drucker-Prager yield condition. The limit analysis theorems valid for the associated flow rules then provide the foundation for different methods of assessment of limit loads and safety factors of geotechnical structures [1,3]. However, for soil tool interaction processes, more refined models are needed accounting for large localized deformation of soil and configurations changes [4]. The models of material softening, hardening and critical state are introduced and applied in the analysis of soil deformation. The incremental equilibrium analysis is applied with account for softening and hardening effects and generation of periodic flow mechanism occurring during tool motion. For the analysis of soil compaction, the multisurface hardening model has been applied with account for material memory effects.

The theoretical predictions are confronted with ample experimental data obtained in laboratory testing of soil-tool interaction problems. The typical processes considered in the paper are:

R. Wan et al. (Eds.): Bifurcations, Instabilities and Degradations in Geomaterials, SSGG, pp. 325–345.
springerlink.com

- monotonic tool motion inducing progressive soil deformation: soil cutting, excavation, digging, wedge, punch and pile penetration;
- controlled tool motion with the objective to minimize energy dissipation required for the process execution, the wear of tools and its effect on analyzed deformation mode is clarified;
- soil compaction induced by a moving roller with the analysis of cyclic compaction and related soil cracking effects.

The constitutive models used in the analysis are presented in Fig. 1

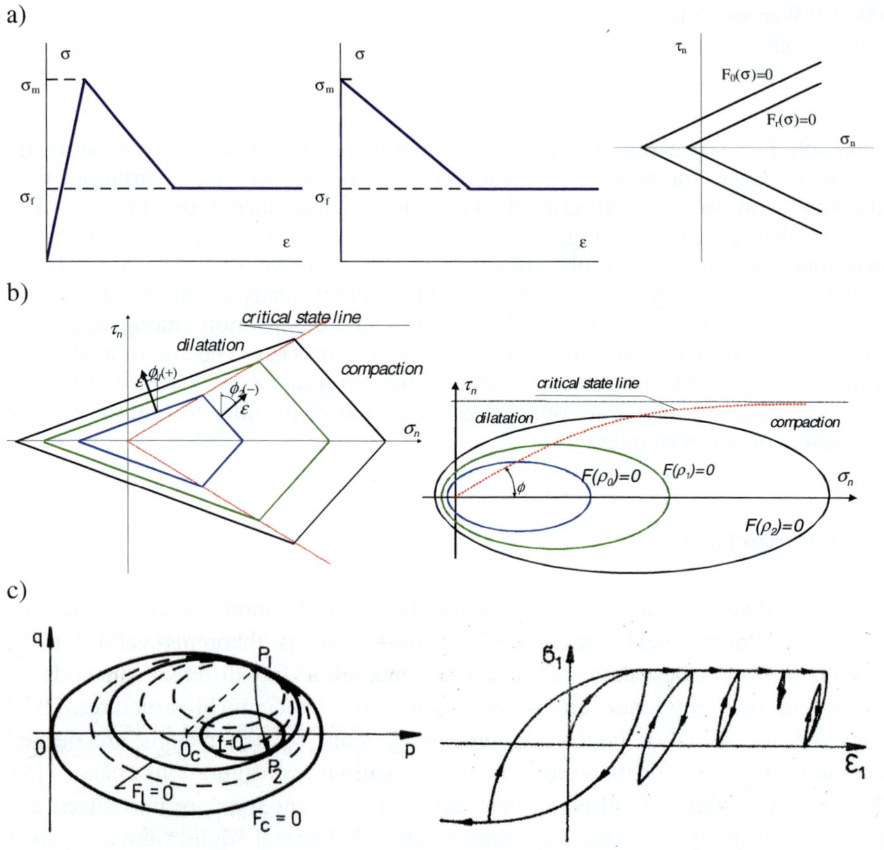

Fig. 1. Constitutive models used in analysis

The elastic or rigid-plastic softening model is shown in Fig. 1a. For low mean pressure regimes and compacted soils the softening response is most important in the analysis of deformation modes in soil cutting processes. The initial high yield stress σ_m , residual stress σ_f and the softening modulus K_s characterize the material response. The friction angle is assumed as constant and cohesion varying from the initial to residual values. The elastic stiffness moduli are additional parameters for the elastic-plastic analysis. Figure 1b presents the density hardening models with both softening and hardening regimes, separated by the critical state regime with vanishing dilatancy. The model can be used for both low and high mean pressure states occurring in problems of deep foundation or pile penetration. Figure 1c presents the multisurface hardening model with sequential loading surfaces, able to simulate cyclic response of soils. This model application is presented by considering compaction of cohesive soils by rolling cylinder where loading unloading-reloading events follow after each roller pass. The details of model formulation are presented in separate papers [6, 18].

2 Soil Cutting Process: Shear Band Patterning Effects

During the cohesive soil-rigid tool interaction process a characteristic deformation pattern consisting of several rigid blocks separated by shear bands is observed [4-12]. This phenomenon is correlated with the material softening occurring within the zone of shear band and local degradation of material parameters, as well as the global drop of the force acting on the rigid tool. During progressive motion of the tool the initial failure mechanism evolves. Some of the shear bands are material planes, while remaining bands move with respect to the material in order to constitute a kinematically admissible failure mode. A characteristic effect of switching to new failure modes occurs at particular states of the deformation process, resulting in an oscillatory force-displacement response.

The effects of varying density on granular material were studied in [13], and strain localization effect in flowing material through hopper was analyzed in [14]. The shear band pattern generation behind an elastic soil supporting wall was discussed in [15-17] presenting both experimental data and analytical model.

The phenomenon of shear band generation becomes the most important factor for the optimisation of the soil cutting process combined with the machine tool filling process. The idea of such a machine tool control to displace the tip of the tool along the previously generated shear surfaces with residual strength was introduced and developed in papers [5,7,8,12]. The specific energy for the soil excavating process depends on several parameters, among which the most important are: material behaviour (softening), tool shape, the tool trajectory and free boundary configuration of the excavated material.

An experimental program was executed on the specialized laboratory stand equipped with a soil bin located at the Technical University of Kielce, Poland. As the bin, the stand and sample preparation were precisely described in [7,8], only the most important details will be presented here. The bin dimensions were 2 m long x 0.6 m wide x 1.2 m deep. The scheme of the stand is presented in Fig. 2. The tool (1) was moved inside the container by means of three hydraulic cylinders, whose motion was fully computer-controlled through electric-hydraulic proportional valves. The horizontal motion was executed by a hydraulic cylinder (2), pushing the front cart (5), while the rear cart (6) was fixed to the rail. A

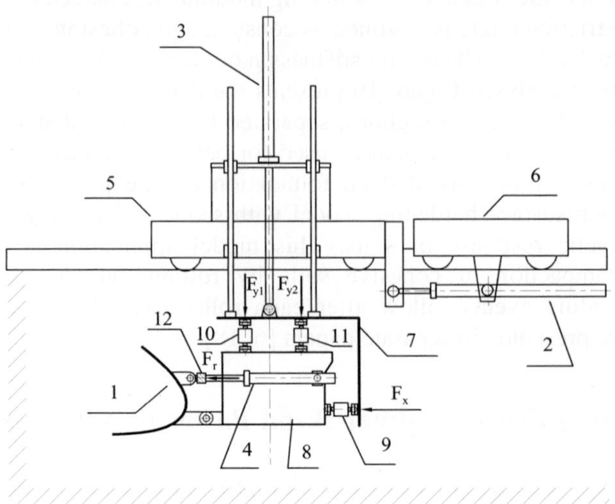

Fig. 2. Scheme of the laboratory stand: (1) model tool; (2), (3), (4) hydraulic cylinders; (5) front cart; (6) rear cart; (7) rigid frame; (8) rigid element; (9-12) load cells

vertical motion device was mounted on the front cart. It consisted of a rigid frame (7) driven by a hydraulic cylinder (3). Another hydraulic cylinder (4) responsible for the rotation of the tool was mounted to the rigid element (8) supporting various tool models (1). However, as all tests were performed in the quasi-static conditions the velocity of the tip of the tool was kept close to 10 mm/s. The set of load cells (9-12) equipped with strain gauges with analogue-numeric transducer allowed for the force measurement and on-line data storage in the computer. The load cell (9) recorded the horizontal force F_x, the load cells (10) and (11) were measuring the vertical force components F_{y1} and F_{y2}, and a loading cell (12) was mounted on the axis of the rotational cylinder (4) to allow for direct measurement of the force F_r . Thanks to this cell the part of work consumed for rotation of the tool was directly calculated.

A mixture of cement, bentonite, sand and white vaseline was used to imitate a clayey soil. No water was added. Use of white vaseline as one of the components resulted with generation of a cohesive soil, whose parameters were not affected by air humidity and liquid. The soil sample was prepared in the soil bin applying layer by layer method. The compaction of each layer was obtained by means of a rigid stamp, covering the free surface during several passes. In this procedure the pressure exerted by the stamp and the subsequent horizontal movement of the stamp were computer controlled. The soil sample prepared according to this procedure can be described by the Coulomb model with the following initial parameters: loose soil: specific density γ=16.2 kN/m^3, internal friction angle ϕ=27°, cohesion c=~15 kPa, medium dense soil: γ=16.8 kN/m^3, ϕ=27° and c=~30 kPa, and dense soil: γ=17.2 kN/m^3, ϕ=27° and c=~45 kPa, with parameters similar to those of sandy clay.

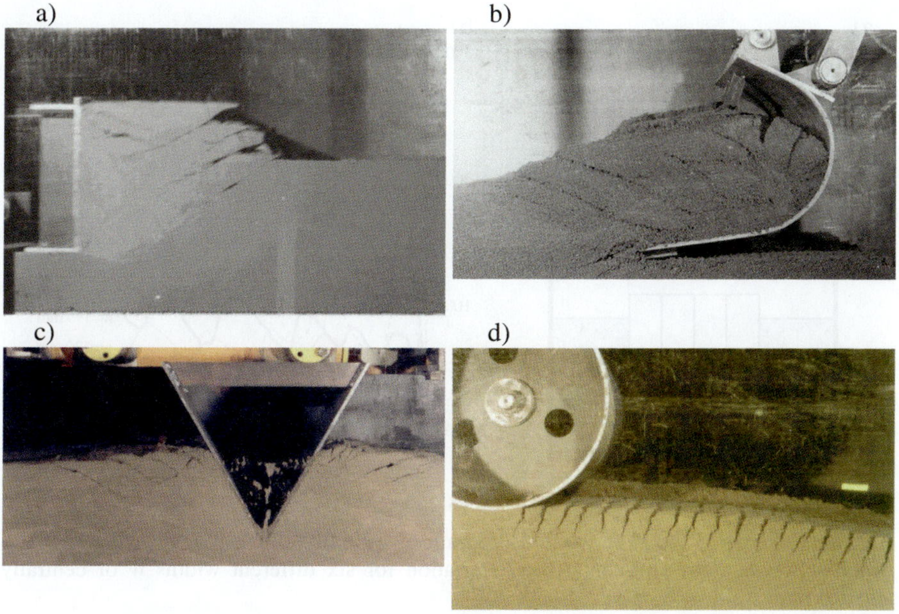

Fig. 3. Typical tool motion inducing progressive soil deformation a) the rigid wall pushing process [4], b) soil excavation using the model of the excavator bucket [12], c) wedge penetration [4], d) soil compaction under towed rigid cylinder [6]. Note the cracking pattern after passage of towed cylinder.

Figure 3 presents several different types of soil-tool interaction processes inducing progressive soil deformation. The cyclic repeatability of the process was observed with the developing displacement of the tool. The deformation pattern consisted of several rigid blocks sliding on the consecutive shear bands where material softening was observed. As a consequence, reduction of the horizontal force and abrupt changes of the deformation mechanisms were noticed. These effects were observed for various tools shapes. As an example, Fig. 4b presents the force variation for the monotonic horizontal motion of a rigid wall with varying width (Fig. 4a). The results of the force variation for each of the six centrally situated rigid walls are plotted for 400 mm displacement of the tool. The result for horizontal movement of the vertical rigid wall of the width equal to that of the soil bin s is additionally plotted in this figure. Precise description of this soil-tool interaction was given in [8].

For a continuous deformation process, the initial soil configuration changes and the mode of flow evolves. With a dilatant and softening cohesive soil the failure mode is usually composed of localised deformation within shear bands. Some of these bands are material planes, while other bands move with respect to the material in order to ensure a kinematically admissible failure mode.

A characteristic effect of switching to the new failure mode occurs at particular stages of the deformation process, resulting in an oscillatory force-displacement response. Examples of subsequent stages of the deformation pattern observed by

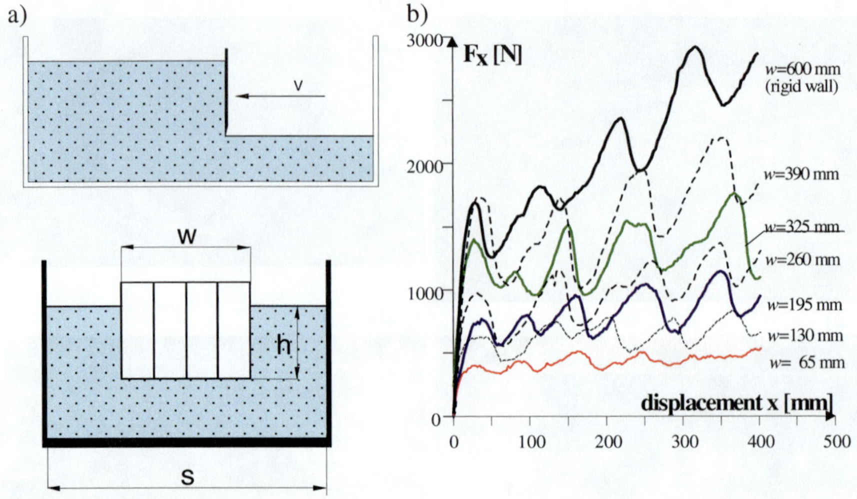

Fig. 4a. Scheme of the experimental program of centrally situated rigid wall tools with different width; **4b)** Horizontal force variation for six different widths *w* of centrally situated rigid walls

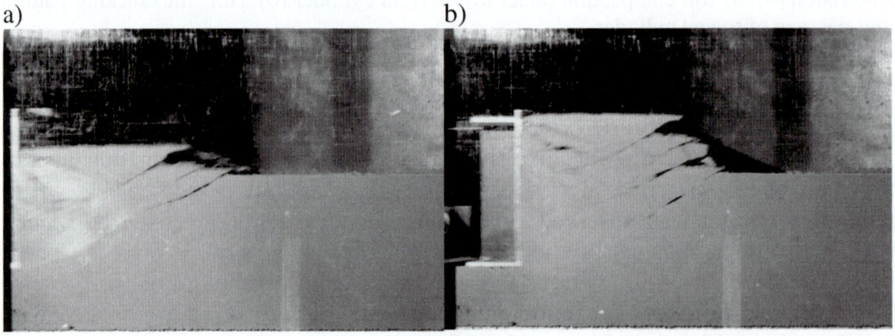

Fig. 5. Subsequent stages of the deformation for the rigid wall pushing process

the authors for the typical experiment during which the vertical rigid wall was acting on the cohesive soil sample are presented in Fig.5.

The results were obtained for the medium dense soil. The width of the rigid wall was equal to the width of the bin (b=0.6 m) and the cutting depth was 0.18 m. Photos presented in Figs. 5a and 5b were taken after the displacement of the wall: 150mm, 250mm, respectively. The first and the second, and the third well developed shear bands are clearly represented in these photos. The characteristic oscillatory force-displacement diagram recorded during the experiment is presented in Fig. 6. The instants at which photos were taken are marked in this

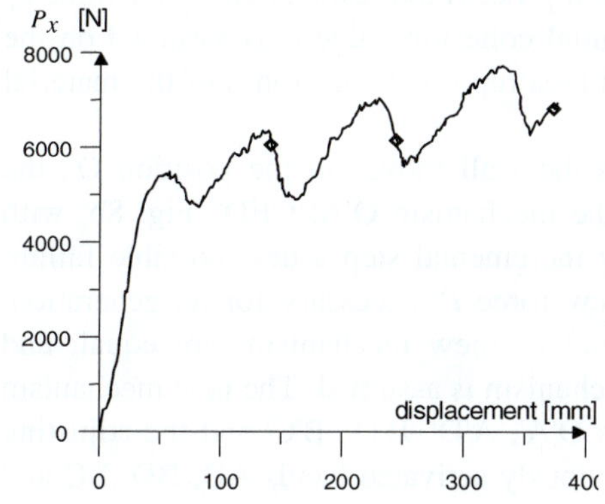

Fig. 6. The force-displacement diagram recorded for the rigid wall pushing process

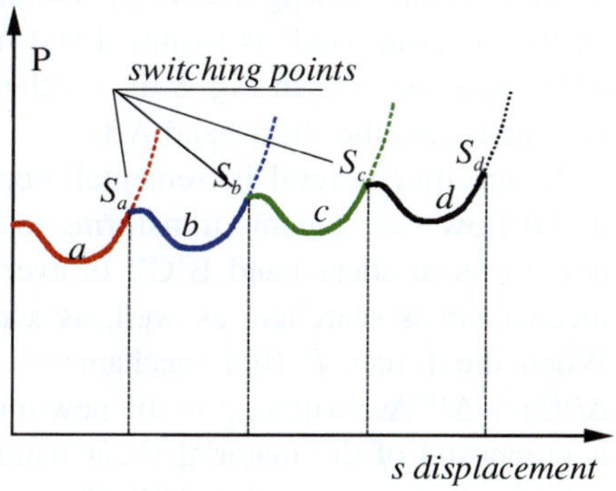

Fig. 7. Schematic force variation with generation new failure mode

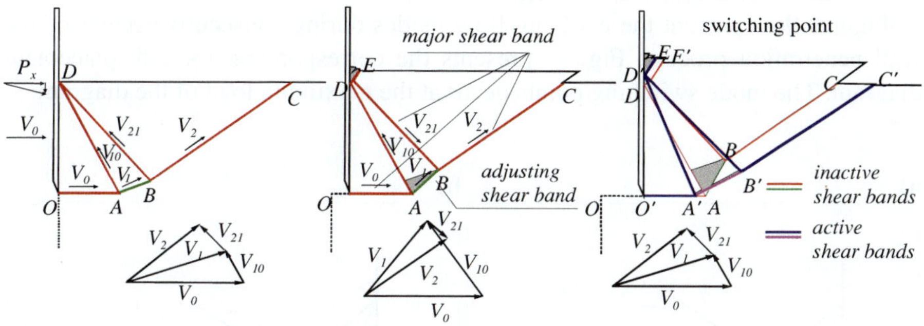

Fig. 8. Evolution of the deformation pattern with consecutive material shear band BC, B'C' developing during the process

figure using additional squares. The generation of each shear band approximately corresponds to the subsequent peak values of the plate driving force.

Let us discuss the incremental procedure for the case of a rigid wall penetration into the soil. The modified Coulomb yield condition, with linear softening rule and a non-associated flow rule is applied, Fig. 1a. The kinematically admissible flow mode is considered with the associated force equilibrium conditions of rigid blocks. This procedure does not correspond to the upper bound solution of limit analysis.

Consider the flow mode shown in Fig. 8. At the initial state, three rigid blocks OAD, ABD and BCD are separated by velocity discontinuities lines (shear bands). The incremental force ΔP inducing small displacement Δs of the wall is specified with account for configuration changes. It is assumed that BC, OA, DA and BD are material shear bands and AB is an adjusting band moving with respect to the material. During the wall penetration process the residual cohesion value c_r on the material bands is assumed and initial cohesion value c_0 is assumed on the adjusting band AB. In Fig. 8 the hatched area represents the zones of the material moving across the shear band AB.

When after several incremental steps the wall moves to the position O', the initial flow mechanisms transforms to the mechanism O'ABCED', Fig. 8b, with new material shear band B'C'. In every incremental step a new possible failure mechanism is searched, as well, as a new force P' necessary for its generation. When the forces P (old mechanism) and P' (new mechanism) are equal, and $\Delta P/\Delta s > \Delta P'/\Delta s$ switching to the new mechanism is assumed. The new mechanism is composed of the material shear bands O'A', A'D', B'D', B'C' and the adjusting velocity discontinuity line A'B'. The previously activated bands AD, BD, BC and the line AB are no longer active and move with the material. When the new mode is activated the load decreases because of the softening along the material bands. Next, due to configuration changes, the load starts to increase until the new consecutive switching point is reached (Fig. 7).

Figures 9a-c present the evolving flow modes during consecutive stages of the wall penetration process. Fig. 9c presents the corresponding force-displacement diagram. The mode switching points occur at the maximum load of the diagram.

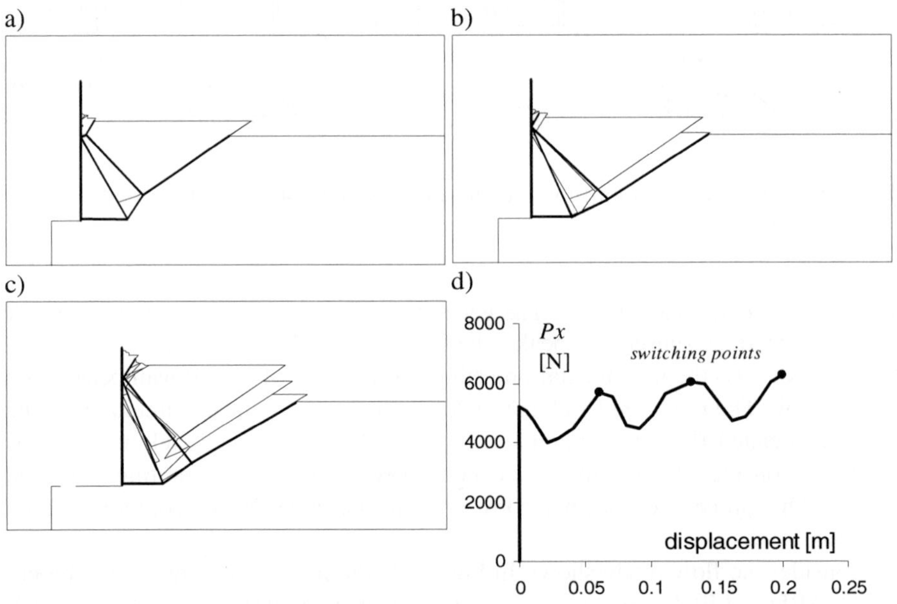

Fig. 9. a)-c) Computer simulation of the soil cutting process due to rigid wall motion; d) Evolution of the penetration load versus the rigid wall displacement

The material parameters were selected as follows:

c_0=15 kPa, St_0=7.5 kPa, c_r= 5 kPa, St_r= 2.5 kPa, ϕ=27°, δ= 15°, γ=16.2 kN/m^3, s_0 = 0.01 m, h= 0.18 m, b= 0.6 m,

where δ denotes the friction angle between the soil and the wall, s_0 is the softening parameter corresponding to the transition from initial to residual state, γ denotes the soil density, h is the depth of the penetration, and c_0, St_0, c_r, St_r denote initial and residual values of cohesion and tensile strength.

Although the presented theoretical solution is based on a kinematically admissible failure mode, the predicted horizontal force (Fig.9d) reaches lower values than those experimentally recorded (Fig.6). This phenomenon is caused by the friction effect between the soil and vertical walls of the soil bin. The influence of friction on laboratory tests performed in soil bins and range of the zone of the plane strain deformation within the soil sample was discussed by Maciejewski *et al* [8,12].

The presented method does not take into account the elastic properties of soil. For this reason the horizontal force on the predicted force-displacement diagram (Fig. 9d) reaches the first limit value at zero displacement. However, due to elastic deformation of the soil sample and the stand the first limit load was observed in our experiments at the displacement of the wall close to 50 mm (Fig. 6).

3 Process Optimization: Specification of Tool Trajectory

Based on Jarzębowski et al. conclusions [5] that there are optimal soil cutting trajectories in laboratory conditions, an experimental program was developed to replicate tests where the soil free boundary before and after the experiments was similar. It is assumed, that the whole excavation task consists of several replicate cycles (for example in long trench digging with automated excavators), when the computer aided machine may execute any prescribed tool trajectory and the optimization of the single cycle plays significant role in the energetic efficiency of the whole earth-working task (Fig. 10). A simplified model of the bucket of the K-111 produced by Waryński excavator company, Warsaw, Poland, without teeth was used in this experimental program.

All tests were performed on the same bin and with the same modeling material. As in the first series of experiments the force acting on the tool was measured and recorded and the total work W_f was calculated after each experiment. To allow direct comparison of results obtained for various tests the total work W_f was corrected according to the initial position of the center of mass of the dug out volume:

$$W_c = W_f - (y_c - y_i)\, Q, \tag{1}$$

were W_c is the corrected value of the total work of the process, Q is the weight of the material which remained in the bucket and y_i and y_c denote the vertical components of the positions of the centers of mass of the dug out material in the

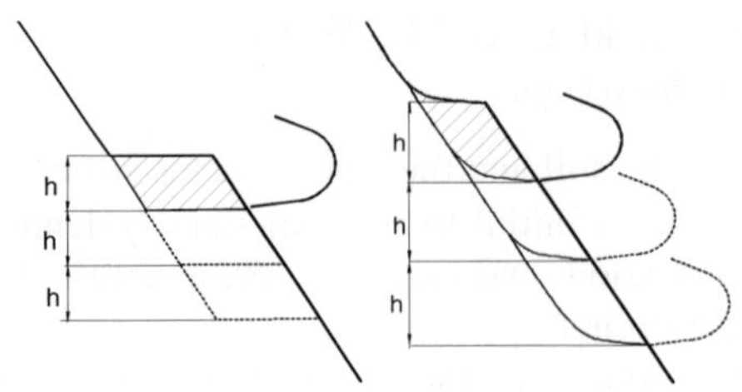

Fig. 10. Scheme of sample preparation in the soil bin

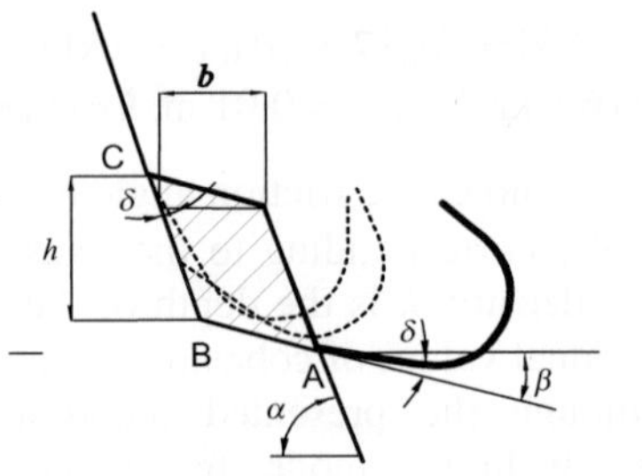

Fig. 11. Scheme of the single cycle and its parameters

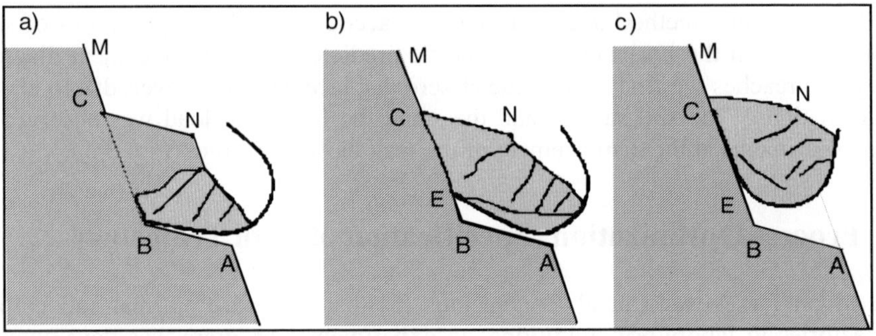

Fig. 12. Subsequent phases of the single cycle of digging process

initial and the final positions of the process, respectively. The value of y_C was found basing on a photographic documentation of each test. The value of y_i was determined from the cross-section of the dug out material before the test.

Piecewise-linear trajectories were used in this experimental program. Typical piece-wise linear trajectories for which soil configurations before and after each test were similar are presented in Fig. 11. To describe the single cycle of the working process the following parameters were used: α - inclination of soil slope (equal to the inclination of the lifting phase), β - inclination of the initial phase of trajectory, δ- tool inclination with respect to β, h - height of the excavated material, b - width of the excavated material.

For each experimental cycle of the working process (Fig. 12) several characteristic phases may be distinguished. At the initial phase of the movement the translation displacement of the tool along the straight line AB was executed.

The second phase of the trajectory started at point B. The tool changed its direction passing to the lifting phase. During the change of the direction of the tool motion the shear band BEC (Fig. 12b) was created. It is worth to point out, that the shear band was generated from the point B to the point C (i.e., the corner of the free boundary MCN), independently of the slope inclination, as well, as of other parameters of the working cycle. The inclination of the new generated shear band BEC was close to the slope inclination of the original soil. In the final phase of the trajectory (Fig. 12c) the translation of the bucket was executed to conduct the tip of the tool along the shear band EC and proceeded to point M.

Some of the experimental results are plotted in the Figs. 13 and14. The influence of the inclination of the initial part of trajectory β on the specific unit energy of the process was investigated for the slope equal to 70° and constant ratio of b and h equal to 2:3. The scheme of trajectories used in these tests and the results are presented in Fig. 13. The lowest value of the specific energy was obtained for the initial lateral part of the trajectory (β=0°).

The influence of the ratio b:h on the specific unit energy of the process is presented in Fig.14. Two different values of the slope inclination were used: α=50° and α=70°. The existence of the optimal value of the ratio b:h was observed for the ratio b:h equal to 2:3. It was shown [7] , that the soil cutting strategy based on the cutting of a horizontal or vertical thin layer was not reasonable. Neither specific unit energy nor exerted forces were optimal for this method of soil cutting.

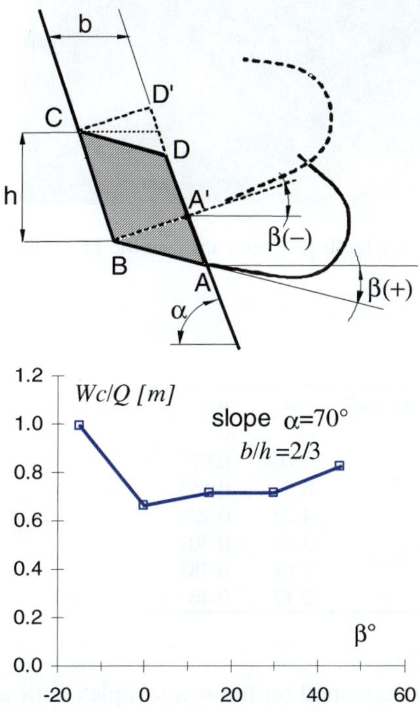

Fig. 13. Specific excavation energy versus the inclination of the initial phase of trajectory

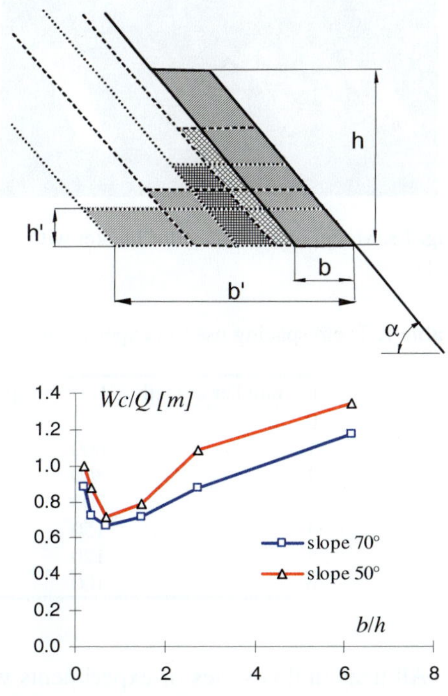

Fig.14. Specific excavation energy versus the width b

4 Test Procedure and Results for the Bucket with Teeth

Now some of the tests of excavation process using the model of an excavator's bucket with teeth will be presented, Fig. 15. The bucket was equipped with teeth, which were easily mounted/dismounted directly on the inside lip without any adapters. The shape and size of each tooth corresponded to that which is produced for the small bucket excavator K-111 by Waryński Company (Poland). They stand out 95 mm in front of the inside lip. Thanks to simple construction of each tooth any number of teeth could be mounted on the lip. The model with 1, 2, 3, 4, 5 and 6 symmetrically situated teeth was used, as well, as the model without any teeth. In the case of model equipped with one tooth it was situated centrally (teeth spacing 600 mm). For the multi-teeth bucket, the teeth spacing l was 300 mm, 200 mm, 150 mm, 120 mm and 100 mm. The values of teeth spacing according to the number of teeth for the performed tests are given in Table. I, where w = 46 mm is the width of a single tooth. The distance from the middle of the side teeth to the sidewalls of the bin was equal to half of the distance between teeth.

Fig. 15. Model of the excavator bucket with teeth, the teeth geometry used in the laboratory tests

Table 1. Teeth spacing used in experiments

n – number of teeth	l – teeth spacing [mm]	l/w	w/l
0	∞	∞	0
1	600	13.04	0.077
2	300	6.52	0.153
3	200	4.35	0.230
4	150	3.26	0.307
5	120	2.61	0.383
6	100	2.17	0.461

All tests in this series of experiments were executed on the soil samples with a horizontal free surface. The same trajectory, consisting of two linear stages was used. During the first stage the bucket moved horizontally with various cutting depths 100, 150 and 200 mm until it was filled up with the soil. The distance of the horizontal motion was equal for each number of teeth. The bucket inclination was 3° to avoid friction between the outer surface of the bottom of the bucket and the remaining soil surface. In the second stage of the trajectory the bucket was lifted without rotation along the straight line inclined at 45° to the horizontal direction.

In Fig. 16, results of the horizontal force variation for the number of teeth 1, 3 and 5 are plotted. For the tool equipped with one tooth, initially, the 3-dimensional failure zone was observed leading to small oscillation of horizontal force prior to the peak value associated with the generation of the first shear band. Subsequently, when the lip of the bucket reached the soil sample the superposition of the plane strain mechanism with the 3-dimensional failure mode occurred. For the buckets with the number of teeth greater or equal to 4 (*l*/*w* less then 3.26), the first and subsequent shear bands were not generated from inside lip but from the line connecting the tips of the teeth. For the number of teeth exceeding 3 the line connecting the tips of the teeth played the role of the inside lip. The teeth did not act as separate three-dimensional objects but as one wide tool built up from several modules. Thus, the deformation pattern in front of such an assembly of teeth was again the plane strain deformation pattern, which was observed by authors for tools of the width equal to the width of the soil bin (Fig. 2b).

The total work generated by the external horizontal force for horizontal displacement to 150 mm was calculated for each test. The obtained values of work denoted W_{150}, were plotted in Fig. 17. The total work was higher for the tools with teeth compared to the tool without teeth. This work increased with increasing number of teeth (increasing *w*/*l* ratio in Fig. 17), and tended to stabilize for higher numbers of teeth. For the initial values of parameter *w*/*l*, corresponding to number of teeth 1 and 2, the value of work increased linearly. It was caused by simple addition of earth working mechanisms generated by one and two teeth. For values of *w*/*l* greater then 0.3 (more than 4 teeth) the differences in work values disappeared, especially for more advanced stages of the process. For values of ratio *w*/*l* greater then 0.15 and smaller than 0.3 a transition between described-above cases was observed.

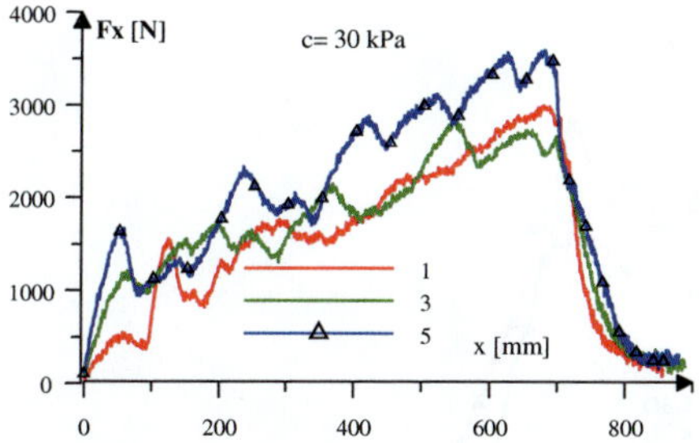

Fig. 16. Horizontal force variation for the bucket with different number of teeth

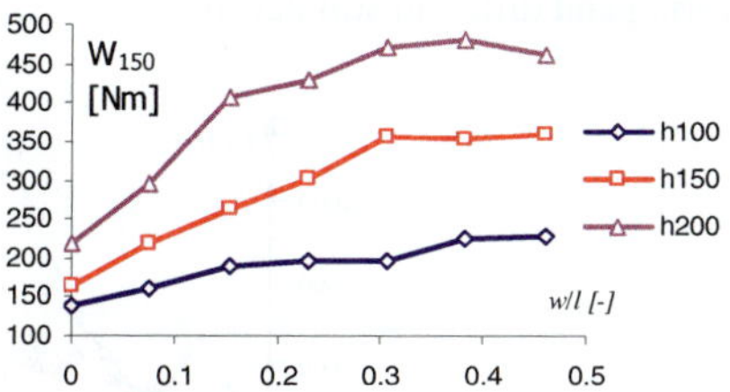

Fig. 17. Work W150 for tools with different number of teeth

The influence of the teeth shape is presented on Figs. 18 and 19. Four different teeth were used: "sharp tooth" (2 mm worn-out), 10 mm worn-out and 20 mm worn-out, and teeth with three dimensional shape. In Fig. 18 results of the horizontal force variation for the cutting depth 175 mm and the tool equipped with five teeth are presented. Three different sets of teeth were used: sharp teeth, 10 mm worn-out and 20 mm worn-out. The results are presented for two different soil samples, characterised by the cohesion c_1 =~15 kPa (Fig. 18a) and c_2 =~45 kPa (Fig. 19b).

According to previous results for the soil cutting problem the initial failure mechanism consists of several rigid zones separated by shear bands. When the plastic deformation of the cohesive soil is associated with the softening behaviour, the evolution of the failure mechanism occurs. During progressive movement of the tool, the initial failure mechanism is modified with some shear bands becoming material interfaces and other bands moving with respect to the material. A characteristic effect of switching to new failure modes occurs at particular states of the deformation process.

The value of force at the moment of generation of new shear bands is highly influenced by the degree of teeth wearing, whereas the evolution of the already existing failure mechanisms is not sensitive to this wearing state.

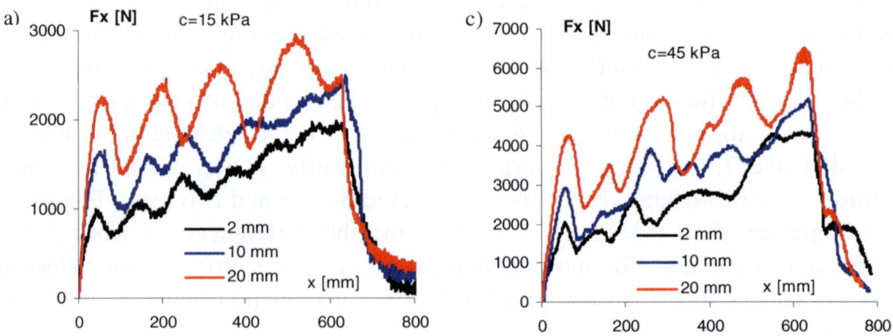

Fig. 18. Horizontal force variations for the tool with 5 teeth and different stages of teeth wearing and different soil density

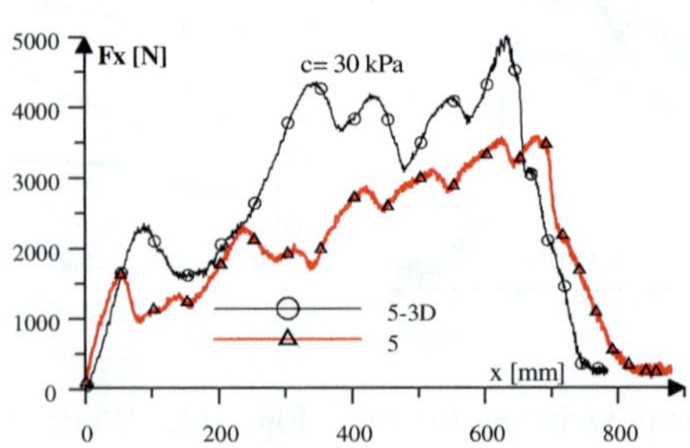

Fig. 19. Horizontal force variations for the tool with 5 plate teeth and 5 three dimensional teeth

The horizontal force component was much higher for the tests performed on dense soil (c_2=~45 kPa). In addition, for the loose soil (c_1 =~15 kPa) the material was compacted within the tool frontal zone what was not observed for the dense material. As a result the intervals between subsequent shear bands were higher for the dense soil than for the loose one.

Figure 19 presents horizontal force variations for the excavator bucket with 5 plate sharp teeth and 5 three dimensional teeth (compare with Fig. 15). It is clearly seen that the three dimensional shape is inefficient. In this case the sets of 3D teeth did not act as a one wide tool, but as a several separate three-dimensional objects.

5 Compaction of Cohesive Soil by the Rolling Cylinder

In this section the compaction process of the cohesive soil layer is presented. The laboratory tests were carried out simulating single static cylinder working process under plane strain conditions. The laboratory stand (Fig. 2) was equipped in the rolling cylinder, with independent control of rigid motion of the cylinder and its rotation.

The sample was prepared in the soil bin applying layer by layer method. Each layer consisted of equal amount of loose material. In order to analyse the deformation of soil a set of markers was used. The markers were introduced into the soil sample by putting them on each layer, Fig. 20. Samples consisted of two, three, four and five layers according to the experimental program.

Fig. 20. Subsequent stages of the compaction soil layer under rolling cylinder

Tests were executed in two stages: sinkage and rolling. During the first stage the rigid cylinder was pushed down until certain assumed vertical force was achieved. Next, the rolling stage with prescribed traction parameters was executed. Tests for various slip values s were performed, where slip was defined as:

$$s = (\omega R - V)/\omega R, \tag{2}$$

where, R denotes the radius of the cylinder, ω denotes its rotational velocity and V denotes the velocity of the cylinder centre system. The towed cylinder tests were performed by dismounting independent driving of the cylinder.

In Fig. 21 the values of drawbar-pull over the travelling distance for the driven and the towed cylinder tests are presented. At the beginning of the rolling stage of the driven cylinder test the drawbar-pull increased until the value of travelling distance close to 300 mm was reached. Then, the value of drawbar-pull stabilised. For the towed cylinder tests negative values of drawbar-pull were obtained. In the case of driven cylinder tests the uniform deformation field was observed behind the cylinder

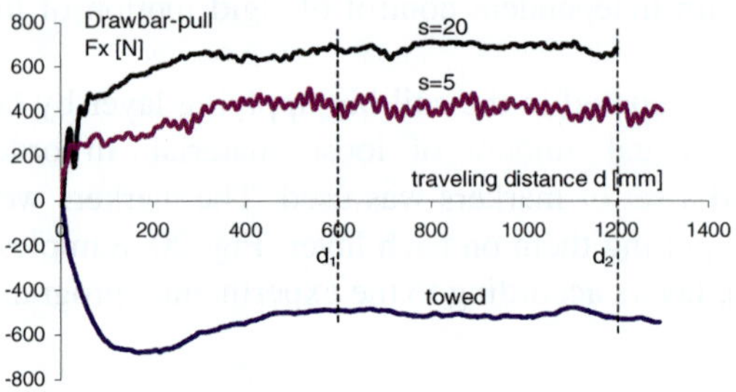

Fig. 21. Drawbar-pull versus slip for subsequent passes

In Fig. 23 the particle paths are presented for each layer of the soil sample. The trajectories obey a characteristic loop shape. All loops are positioned within the first and the fourth quarters of the graphs, what means that particles were pushed forward and downward by the cylinder. The compaction effect was observed as a final effect after the test. The precise description of the compaction soil layer under static cylinder is presented in papers [6, 11, 12].

In Fig. 22 the drawbar-pull versus slip is presented for subsequent passes of the rubber-coated cylinder. Starting from the second pass the drawbar-pull diagram exhibits maximum. Test results for subsequent passes of the towed cylinder (negative drawbar-pull points) were additionally plotted in this figure. As the soil sample compacts, the value of slip increases for each subsequent pass of the towed cylinder from -24% in the first pass, to the value of -2% in the tenth pass.

In Fig. 24 the final compaction effect for subsequent passes of the cylinder is presented. For the slip within the range of +10% to +20% the strongest final compaction was obtained. This optimal slip value is connected with the maximal drawbar-pull force value, Fig 22.

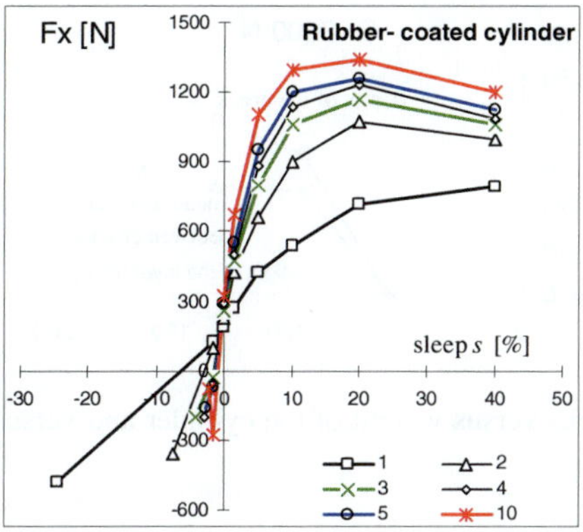

Fig. 22. Drawbar-pull versus slip for subsequent passes

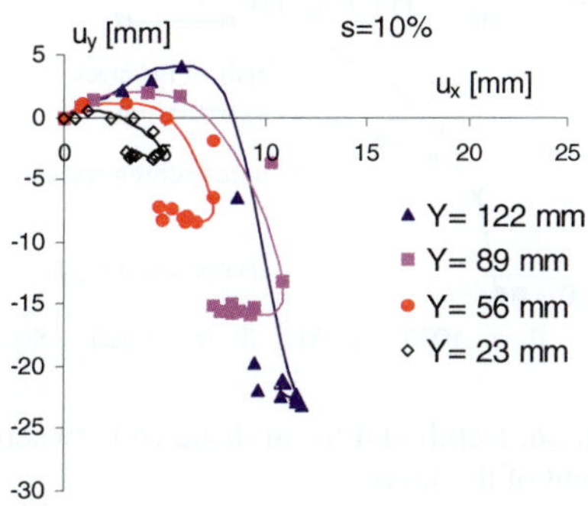

Fig. 23. Particle paths. for the rubber-coated cylinder rotating with 5 % slip, Fy=2000 N, h=122 mm

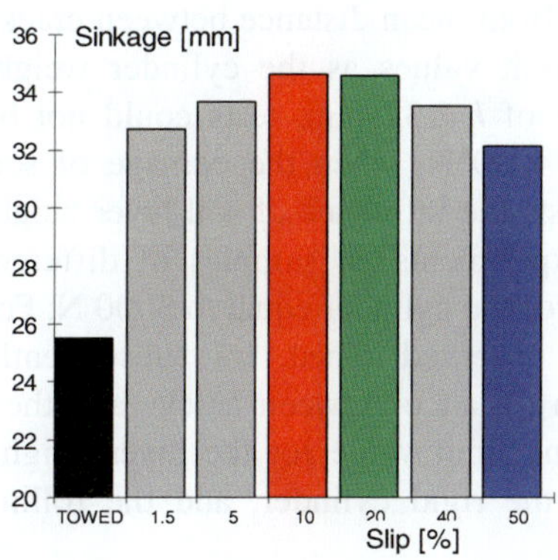

Fig. 24. The final compaction effect for different slip values after 10 passes of the cylinder

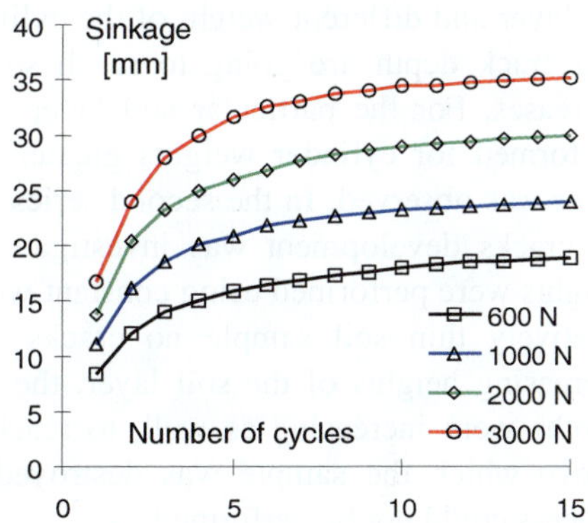

Fig. 25. Compaction effect for different cylinder weight for subsequent passes of the towed cylinder

The compaction effect after several subsequent passes of the towed cylinder of various weights is presented in Fig. 25 for the initial height of the soil layer equal to 132 mm. Obviously, the final compaction effect depends on the weight of the cylinder and tends to stabilise for numerous passes of the cylinder on the same sample.

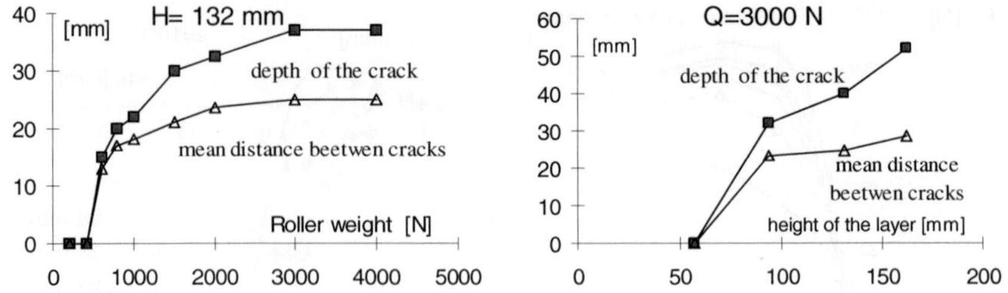

Fig. 26. Depth and mean distance between cracks versus weight of the cylinder and versus height of the layer

During the towed cylinder tests in the zone prior to the cylinder the material was pushed, whereas in the zone behind the cylinder the soil deformation was localised along nearly vertical lines separated by blocks of material. Those lines (cracks) are presented in Fig. 2 for an example a towed cylinder test.

Figures 26 present the mean distance between cracks and their depth for two groups of tests. In the first one, the tests were conducted with constant height of the layer and different weight of the cylinder. Both mean distance between cracks and crack depth are going to reach some limit values as the cylinder weight increases. For the particular soil layer height of h=132 mm, tests could not be performed for cylinder weights greater than 4000 N, when the damage of soil layer was observed. In the second series of tests the influence of soil layer height on cracks development was investigated. Experiments on samples of different heights were performed using constant weight of the cylinder equal to 3000 N. For relatively thin soil sample no cracks were observed. Then, for subsequently increasing heights of the soil layer, the mean distance between cracks and their depths were increasing as well, to reach some limit value for the layer height, above which the sample was destroyed by the rigid cylinder, and the rolling process could not be performed.

For the prediction of cohesive soil layer under rolling cylinder the numerical analysis by FEM was applied. For the description of soil the multisurface hardening model (Fig. 1), developed by Jarzębowski and Mróz [18] was used. This model allows for proper description of hardening and softening, dilatancy and compaction and cyclic loading behavior. Prediction of the sinkage stage is presented in Fig. 27. Figure 28 presents the drawbar pull and sinkage variation during the rolling stage for different slip value. Comparison of experimental data and numerical prediction was performed for towed cylinder tests for two different initial densities. Figures 29 and 30 present drawbar-pull and sinkage variation versus horizontal cylinder displacement for both initial densities. It is seen that multisurface model of cyclic plasticity is able to describe layer compaction phenomena. The numerical analysis is presented in more details in papers [6, 12].

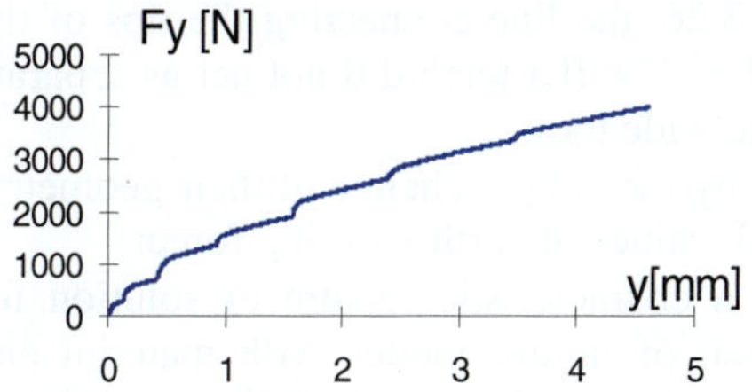

Fig. 27. Vertical force variation during the sinkage stage

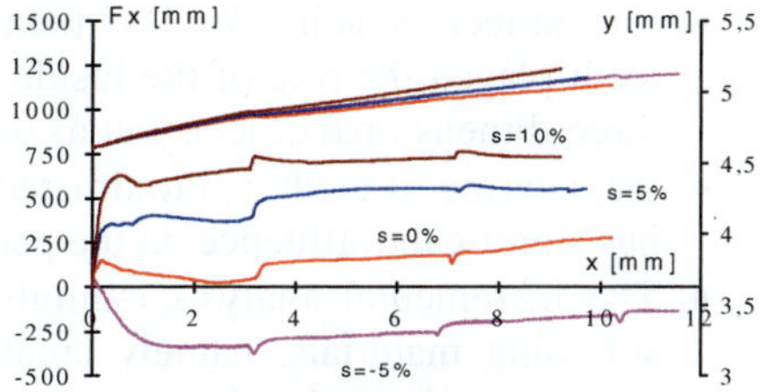

Fig. 28 . Drawbar pull and sinkage variation during the rolling stage for different slip value

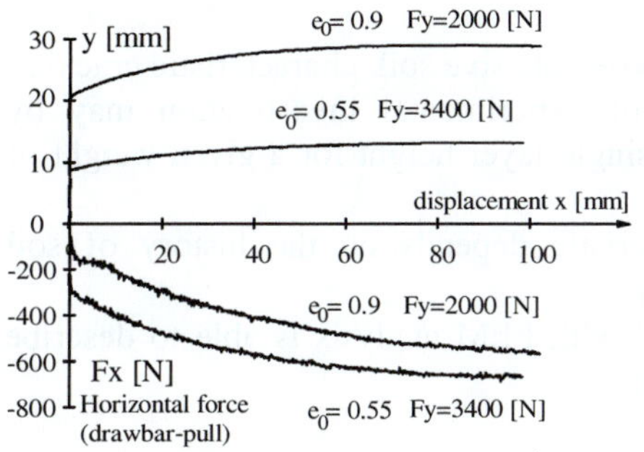

Fig. 29. Towed cylinder experiments for different initial void ratios

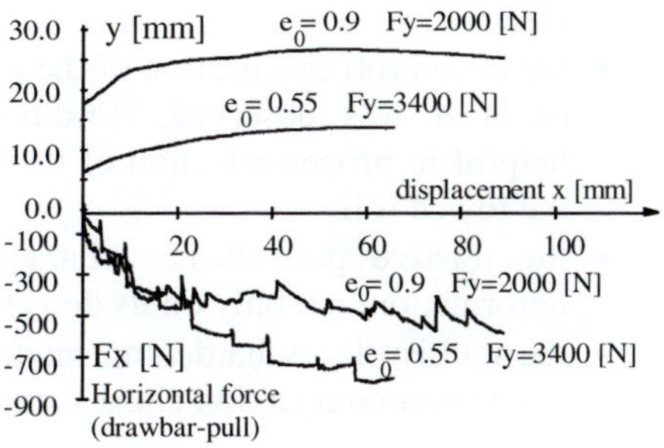

Fig. 30. Numerical prediction for different initial void ratios

6 Conclusion Remarks

The following conclusion can be drawn from the presented analysis of soil tool interaction problems:

1. Progressive tool motion

- periodic character of the digging process (observed for the cohesive soil with softening) manifested in the oscillatory character of force components and generation of the subsequent shear bands is independent of the shape of the tool;
- for the cohesive soil the energetically most efficient trajectory consists of the indentation phase and the withdraw phase. For the withdraw phase the tip of the tool should follow the shear band generated during the indentation phase;
- for the repeatable excavation cycles and the particular cohesive soil the following parameters of the single cycle, turned out to be most efficient: α=50°, β =0°, δ =0° and b/h=2/3;

- for the teeth spacing *l/w* less then 3.26, the line connecting the tips of the teeth played the role of the inside lip. The flat teeth did not act as separate three-dimensional objects but as one wide tool;
- the wearing of teeth, approximated by the simple change of their geometry, has significant influence on the peak values of earth-working force;
- The incremental analysis exhibits a characteristic feature of solution for softening materials, namely creation of failure modes with material and adjusting shear bands, mode switching effect and oscillatory force-displacement diagram.

2. Soil compaction by the rolling cylinder

- for driven roll tests the optimal slip value giving maximum compaction was observed;
- for towed roll compaction of the loose cohesive soil characteristic cracking of layer was observed. Results of experimental investigation may by helpful in proper selection of the single layer height for a given weight of the towed roll;
- the tractive performance characteristic depends on the history of soil deformation, not only on its density;
- the multisurfaces hardening model with FEM analysis is able to describe most layer compaction phenomena;

References

[1] Harr, M.E.: Foundations of Theoretical Soil Mechanics. McGraw-Hill, New York (1966)

[2] Chen, W.F.: Limit analysis and soil plasticity. In: Developments in Geotechnical Engineering, vol. 7, Elsevier Scientific Publishing Co., Amsterdam (1975)

[3] Izbicki, R., Mróz, Z.: Developments in Geotechnical Engineering. In: Rock and Soil Mechanics, vol. 48, pp. 423–638. Elsevier Scientific Publishing Co-PWN, Warsaw (1989)

[4] Mróz, Z., Maciejewski, J.: Post-critical response of soils and shear band evolution. In: Proc. 3rd Int. Workshop On Localisation and Bifurcation Theory for Soils and Rocks, Grenoble (Aussois), France, Balkema, pp. 19–32 (1994)

[5] Jarzębowski, A., Maciejewski, J., Szyba, D., Trąmpczyński, W.: On the energetically most efficient trajectories for heavy machine shoving process. Engng. Trans. 43, 169–182 (1995)

[6] Jarzębowski, A., Maciejewski, J.: Localisation effects in a compaction process under the rigid rolling cylinder. In: Proc. 4th Int. Workshop Localisation and Bifurcation Theory for Soils and Socks, Gifu, Japan, Balkema, pp. 107–116 (1998)

[7] Maciejewski, J., Jarzębowski, A.: Laboratory optimization of the soil digging process. Journal of Terramechanics 39(3), 161–179 (2002)

[8] Maciejewski, J., Jarzębowski, A., Trąmpczyński, W.: Study On The Efficiency Of The Digging Process Using The Model Of Excavator Bucket. J. of Terramechanics 40(4), 221–233 (2004)

[9] Maciejewski, J., Jarzębowski, A., Trąmpczyński, W.: The Influence Of Teeth Wear On The Digging Process. In: Proceedings of the 9th European Conference of the ISTVS, Harper Adams, UK, pp. 273–282 (2003)
[10] Maciejewski, J., Jarzębowski, A.: Application of kinematically admissible solutions to passive earth pressure problems. Int. J. of Geomechanics 4(2), 127–136 (2004)
[11] Maciejewski, J., Jarzębowski, A.: Experimental analysis of soil deformation below a rolling rigid cylinder. Journal of Terramechanics 41(4), 223–241 (2004)
[12] Maciejewski, J.: Soil-tool interaction in the cohesive soil (in Polish). IPPT Reports, January 2008, pp. 262, Warsaw (2008)
[13] Drescher, A., Michalowski, R.L.: Density variation in pseudo-steady plastic flow of granular media. Geotechnique 34(1), 1–10 (1984)
[14] Michalowski, R.L.: Strain localization and periodic fluctuations in granular flow processes from hoppers. Geotechnique 40(3), 389–403 (1990)
[15] Leśniewska, D., Mróz, Z.: Limit equilibrium approach to study evolution of shear band systems in soils. Geotechnique 50, 521–536 (2000)
[16] Leśniewska, D., Mróz, Z.: Study of evolution of shear band systems in sand retained by flexible wall. Int. Journ. Num. Anal. Meth. Geomechanics 25, 909–932 (2001)
[17] Leśniewska, D., Mróz, Z.: Shear bands in deformation processes. In: Labuz, J.F., Drescher, A. (eds.) Bifurcations and Instabilities in Geomechanics, Balkema, pp. 109–121 (2003)
[18] Jarzębowski, A., Mróz, Z.: A constitutive model for sands and its application to monotonic and cyclic loadings. In: Saada, Bianchini (eds.) Constitutive Equations for Granular Non-Cohesive Soils, Balkema, pp. 307–323 (1988)

Modeling Excavator-Soil Interaction

M.G. Lipsett and R. Yousefi Moghaddam

Department of Mechanical Engineering,
University of Alberta, Edmonton, AB Canada

Abstract. This paper reviews models of how ground-engaging tools interact with soils, the rigid-body dynamics of excavating machines, and how to combine these models to estimate soil parameters or to find faults in machines from anomalous dynamic behaviour. Soil-tool interaction models rely primarily on assumptions of homogeneous, isotropic soil properties, and tools that have simple geometries and steady motion through the soil. In many cases, these are reasonable assumptions, provided that the strain rate of the soil is not extreme. Parametric formulations are discussed for soil failure under stress from a non-deformable tool; but finite-element and distinct-element methods are not considered. The formulation of governing equations for the rigid-body dynamics of the machine that carries the tool are discussed. By using parametric equations for the combined system of machine and soil, it is possible to estimate the parameters of the system by measuring machine motions and interaction forces at the tool and base. Possible sources of error are discussed for this approach, with recommendations for how to determine the parameters of the system.

1 Introduction

Model-based analysis of the dynamics of earthmoving can be applied to equipment design, system identification, performance monitoring, simulation, and control. The knowledge of the forces encountered by a tool in earthmoving operations is important for tool and machine design. Moreover, it may also be required for machinery automation (Singh, 1997; Hemami, 1994) even though some approaches to automate excavation do not make an explicit use of soil-interaction forces (Lever & Wang, 1995; Seward, Bradley, Mann, & Goodwin, 1992). Since the forces during earthmoving activities are significant, improved machine control could be achieved through improved understanding of the external forces imparted on the machine. The goal in soil-tool interaction modeling is a mathematical formulation of the resistive force of a medium on a tool during a general excavation task, in terms of a small number of representative parameters of the medium, the tool, and the tool motion. This contribution reviews methods for modeling the forces encountered during work on a medium, specifically the forces of penetration into and cutting the earth, and lumped-parameter dynamic modeling of excavators with single buckets.

R. Wan et al. (Eds.): Bifurcations, Instabilities and Degradations in Geomaterials, SSGG, pp. 347–366.
springerlink.com

2 Soil-Tool Interaction Models

A tool has different interactions with a medium depending on whether the action is penetration, cutting, and excavation. Penetration is the insertion of a single solid body of a fixed orientation into a medium. The tool normally has a longitudinal shape, and the depth of penetration is the controlled variable. The depth of material ahead of the penetrating element is considered to be infinite. Cutting is a lateral motion into a medium made by a single, solid, blade-like body, which is inclined and usually maintained at a constant depth with respect to the surface of the medium. The speed of operation and the rake angle (inclination) are constant. Unlike penetration, the medium in front of a cutting tool is considered to be semi-infinite. Side effects are related to the blade width, and so cutting models can be defined for families of narrow, wide, and infinitely wide blades, where a blade is considered wide when its width is at least twice its depth of operation (Gill & Vanden Berg, 1968) and it becomes effectively infinitely wide when the width-to-depth ratio is greater than six (Osman, 1964).

Excavation is defined here in the restricted sense of loading a bucket or a shovel with soil or fragmented rock. The action comprises a combination of penetration, cutting and breaking of material, followed by scooping to remove material for loading (Hemami, Goulet, & Aubertin, 1994). Since a bucket has sidewalls almost no material moves in the direction perpendicular to the motion of the tool, which is analogous to an infinitely wide blade (McKyes, 1985). Thus, the tool and its trajectory can be defined, especially if incorporated into a dynamic model of the machine that moves the tool.

A parametric model of an earthmoving tool interacting with a medium will include tool geometry and trajectory (including angle of attack) and parameters of the forces and flows of the deformable medium and how the medium fragments under load. Various materials have been classified into a small number of categories based on functional definitions to simplify analysis (McCarthy, 1993). Medium classifications include clay, particulate, soil, and combinations. Clay is a finely grained earthy material that exhibits internal cohesion (shear strength) and plasticity (nonlinear strain to shear ratio and irreversibilities). Particulate is a coarse-grained material, generally assumed to have neither cohesion nor adhesion to tools; particulate media range from crushed rock to dry sand. Soil is a mixture of loose sediments (clay, sand, silt, etc.) with consistency determined by water content and mixture ratio. Interfacial forces between water and mineral particles result in negative pore pressure, and thus cohesion and shear strength. Freezing also affects the strength of a medium containing water.

In earthmoving modeling, two assumptions are generally made about the medium: that it is homogeneous, and it has continuity. The assumption of homogeneity implies that any property is assumed to be constant anywhere in the volume being considered (Gill & Vanden Berg, 1968; Thakur & Godwin, 1990; Willman & Boles, 1996; Zelenin, Balovnev, & Kerov, 1985). In reality, a tool experiences variation in resistance during any earthmoving task because soil characteristics can vary significantly (Bernold, 1993). Different facies of material combinations can be found in a given depositional environment, and so the soil mechanics properties of a medium in place can be heterogeneous and anisotropic. The only materials in a natural state that are close to being homogeneous are rock and sand (Fowkes, Frisque, & Pariseau, 1973), in many mines and quarries, however, the variability is fairly low, and ore bodies can be assumed to be fairly homogeneous. Once the material has been fragmented and classified or blended, the classified material is more homogeneous than the original ore. Mined ore is usually particulate, with some exceptions, such as oilsands, which are unconsolidated, heterogeneous sedimentary deposits with some lithofacies that may be highly cohesive with high friction due to sand particle subangularity.

The second assumption is that of medium continuity, which is assumed to be present throughout a task (Vesic, 1977; Korzen, 1985; Zelenin, Balovnev, & Kerov, 1985). In earthmoving actions such as soil cutting, the medium usually experiences discontinuous strains of large magnitude accompanied by restructuring and alteration of the volume of voids (McKyes, 1985). If the medium grain size is much smaller than the tool size (e.g. sand and spade), or if the particle size is much larger than the tool (for example, inserting a rod into a pile of broken rock), then the medium exhibits a continuous-like behaviour (Fowkes, Frisque, & Pariseau, 1973). The third assumption is isotropic behaviour of the medium (Korzen, 1985; Makarov, 1992; Thakur & Godwin, 1990; Willman & Boles, 1996). This property implies that the reaction of the material to an external stress does not depend on the direction of application.

While physical properties such as density, friction, cohesion, and adhesion characterize a medium, density plays a particularly important role in tool-medium interaction, including the state of compaction and the stress-history.

Friction is often an influential parameter in cutting or penetrating a medium. For two sliding bodies of the same material, the friction is internal. If one of the bodies is the tool, the friction is termed external. The external friction characterizes the interface properties between tool and medium, while the internal friction (including intergranular friction) is considered as an inherent characteristic of the medium. In the specific case of a granulated medium in a stable pile, the angle of repose gives an accurate average value of the internal angle of friction of the uncompacted material.

Cohesion is a third parameter of importance when characterizing a medium. Cohesion can be thought of as the resistance to separation amongst a group of particles, usually due to the negative pore pressure of interstitial liquid.

Adhesion characterizes the tool-medium interfacial tension, which depends on the tool material, the water content, and external friction.

These properties are assumed to be constant throughout the earthmoving action. Soil tillage experiments have demonstrated that the shear rate does not affect any of the friction parameters (internal and external), cohesion, or adhesion (Swick & Perumpral, 1988). Variations do occur: when water is added to a medium, adhesion undergoes a logarithmic increase with increasing medium-tool sliding speed (McKyes, 1985; Yao & Zeng, 1990). There is a limit to this effect; once the moisture content rises beyond a threshold, liquefaction can occur; in that case, the shear strength of the medium reduces drastically and the material behaves more like a Bingham fluid than a solid. This effect shows the sensitivity of adhesion to moisture content.

Practical models for soil-tool interaction are not based on complex overall inter-particular behaviour, but rather on characterization of the medium by its global behaviour (McCarthy, 1993; Yong & Warkentin, 1966). The stress-strain relation provides relevant information by defining the reaction of the medium in the presence of an external force. Elasticity and plasticity theory can be used to model a medium exhibiting a complex combination of elastic and plastic behaviour when a tool interacts with it for cutting, penetration, or excavation; but such a formulation requires continuity of the medium (McKyes, 1985). For a penetration task performed over a particulate medium whose particle size is equivalent to the tool size, plasticity theory does not explain the phenomenon because the continuity assumption is invalid (Fowkes, Frisque, & Pariseau, 1973).

Rheological modeling can be used for soil-tool interactions; viscosity, elasticity, and plasticity in a medium can be associated with a corresponding lumped-parameter element, with a spring as an ideal Hookean elastic model, a damper as an ideal Newtonian viscous model, and a dry frictional contact element as a plasticity model. The rheological approach can be used if the medium is a continuum and behaves according to a set of elasticity and plasticity assumptions, with the parameters representing the rheological elements being single-valued to satisfy the homogeneity assumption. Compared to other earthmoving expressions, rheological models can be considered as a simple and effective approach for modeling earthmoving forces.

The interaction between an earthmoving tool and the medium can be modeled as a set of interconnected mass-spring-dashpot systems, illustrated in Figure 1, which is sometimes referred to as impedance modeling (Salcudean, Tafazoli, Lawrence, & Chau, 1997). The challenge of a rheological model is estimating lumped parameters that emulate the behaviour of the medium.

3 Excavation Modeling

The key aspect of modeling an excavation activity is the tool-soil interaction. A complex action such as cutting or excavation involves a failure of the medium. Among the various types of failure, shear is the mechanism of failure concerned with the interaction of a cutting tool and a medium exhibiting a plastic behaviour (Gill & Vanden Berg, 1968). For the specific case of a non-cohesive medium, the interaction between contacting particles determines the shear strength. The magnitude of the force required to cause the failure is a function of the shear strength of the medium and the dimension of the internal rupture surface. This failure surface is usually considered to be a curved (i.e., logarithmic) or a flat shape, with the latter being commonly adopted as an approximation of the former, as shown in Figure 2. A curved surface is characterized by a large set of parameters derived graphically or empirically, described below, while a flat surface is described by variables with physical meaning: speed v, shear plane angle ρ, rake (or cutting) angle β, operating depth d, surcharge q (if present), the blade width w and submerged length of the blade l as illustrated in Figure 3.

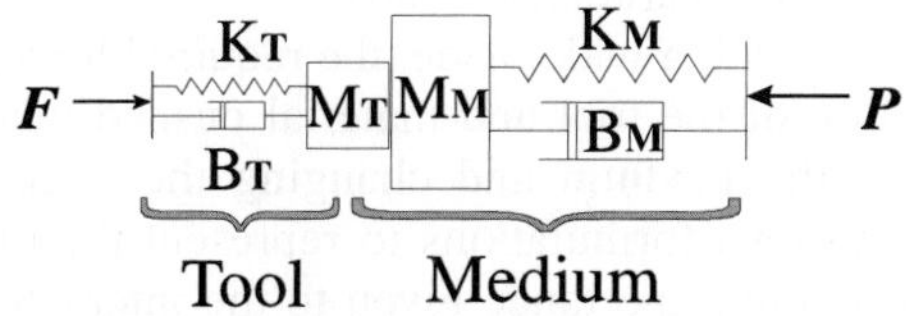

Fig. 1. Rheological model of interacting tool and medium

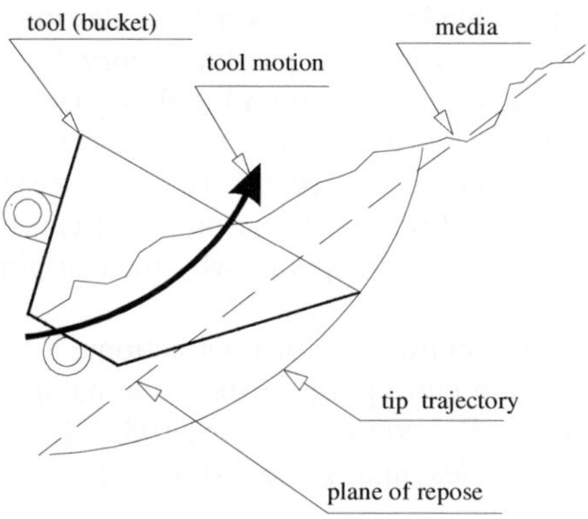

Fig. 2. A two-dimensional view of excavation

Friction and cohesion principally determine the shear strength of the medium. Two kinds of friction interactions take place: internal friction within the medium characterized by friction angle φ and cohesion c, and external friction between the medium and tool, appearing at angle $\delta$with adhesion C_a). From these expressions, shear strengths may be defined as

$$S_{mm} = c + \sigma_n tan(\varphi)$$

$$S_{mt} = C_a + \sigma_n tan(\delta) \tag{1}$$

where σ_n is the normal pressure acting perpendicularly on the shearing surface, and S_{mm} and S_{mt} are the shear strengths in terms of the internal medium failure and the medium-tool interface, respectively.

Coulomb considered the failure surface to be a flat plane, an assumption also adopted for the wedge (trial wedge) theory (McKyes, 1985). Mohr introduced a method to determine stresses at different failure planes within a (homogeneous) medium at an equilibrium state. Coulomb's law and Mohr's method can be used commonly to identify the failure strength of a medium only if the failure occurs at a constant speed (McKyes, 1985). In an excavation task, the medium failure speed cannot be controlled as easily as in cutting, and so combining Coulomb's law and the Mohr's method is more difficult, as shown in experiments performed with a particulate medium (Yong & Warkentin, 1966).

During a cutting action with a wide blade, the required force is attributed to two mechanisms: acceleration of the tool and material pushed (and lifted) as the tool moves, and loading of the medium and changing the strength of the medium (McKyes, 1985). The various formulations to represent the mechanics of cutting or excavating depend on the importance given to the interactive dynamic effects. Indeed, dynamics is often omitted due to the relatively small influence of acceleration forces in specific applications. For cases in which dynamics cannot be ignored, an analogy is usually made with a static process to simplify the analysis. For instance, a wide cutting blade moving perpendicular to its depth may be modeled (in two dimensions) by using the passive theory for large retaining walls (Osman, 1964). For the same reason, even a highly dynamic process such as pile driving is modeled as a static process (McCarthy, 1993). The static analysis of earthmoving forces may either be based on passive pressure on large retaining walls, passive or limiting equilibrium, or earthmoving equations. All these originate from foundation theory, which explains why their original form depicts a zero-speed process.

Although penetration and cutting are distinct actions, it has been noted that resistive forces observed while cutting are of the same nature as those encountered during penetration (Zelenin, Balovnev, & Kerov, 1985). Figures 3b and 3c provide generic diagrams for variables and forces encountered in the cutting models presented below.

Osman, one of the pioneering researchers on cutting models, divided the cutting resistance into two components: one component considers a frictional and heavy medium (without cohesion and surcharge) and the second component considers a (weightless) frictional medium with cohesion and surcharge effects (Osman, 1964). Osman is one of the few authors to consider a logarithmic failure plane. This distinction, however, brings unique parameters that make comparison with other models more difficult. The resultant cutting force T experienced by the tool is

$$\begin{aligned} T = 0.5\, w\gamma t^2 \tan^2(45+0.5\,\rho)\, d_1 d_3^{-1} + \frac{e^{2w' \tan\rho}{}_{-1}}{4\tan\rho} r_0^2 \gamma w d_2 d_3^{-1} + \\ qt \tan^2(45+0.5\rho)\, d_4 d_3^{-1} w + 0.5(r_1^2 - r_0^2)\frac{c}{\tan\rho} d_6^{-1} w + 2\,ct\, \tan(45+0.5\rho) \\ +2\,ct\, \tan(45+0.5\rho)\, d_4 d_6^{-1} w + \frac{qt}{sin(45-0.5\rho)} d_5 d_6^{-1} w + C_a\, l d_7 d_6^{-1} w \end{aligned} \quad (2)$$

where variables $\gamma, t, w', (r_0, r_1)$ and $d_1, \ldots, d_7$ represent respectively the bulk density, the Rankine passive zone depth, the polar angle, the radii of the logarithmic spiral and distances found graphically. The horizontal force H is given by

$$H = T\sin(\beta + \delta) \quad (3)$$

corresponds to the projection forces T along the $\boldsymbol{x}$ axis, as shown in Figure 3c. As pointed out by (Reece, 1965), the inertia effects are not present in the previous formulation, but it could be included by integration of the cutting velocity.

Gill and Vanden Berg proposed the following general expression for the horizontal force

$$H = N_0 \sin\beta + \delta\, N_0 \cos\beta + kw \tag{4}$$

where $\mathrm{N_0}$ represents the normal load on the inclined tool, and $\mathbf{k}$ is the pure cutting resistance of the medium. The parameter $\mathbf{k}$ is important only in presence of an obstacle (or significant wear). Therefore, for a homogeneous medium the previous equation can be rewritten as

$$H^* = H - kw = N_0 \sin\beta + \delta\, N_0 \cos\beta \tag{5}$$

on more precisely

$$H^* = \left[\gamma \frac{\sin(\beta+\rho)}{\sin\rho}\left(l + \frac{d\cos\beta + \rho}{\sin\rho} + \frac{d\sin(\beta+\rho)\tan\beta}{2\sin\rho}\right) + \frac{c}{\sin\rho\,(\sin\rho + \varphi\cos\rho)}\right.$$

$$\left. + \frac{\gamma v^2 \sin\beta}{g\sin(\beta+\rho)(\sin\rho+\varphi\cos\rho)}\right] \frac{wd(\sin\beta + \delta\cos\beta)(\sin\rho + \varphi\cos\rho)}{\sin(\beta+\rho)(1-\varphi\delta) + \cos(\rho+\beta)(\varphi-\delta)} \tag{6}$$

with g the gravitational force. Unlike the model derived by Osman, Gill and Vanden Berg obtained an expression containing terms for the weight and inertia effects.

Swick and Perumpral developed an equation for the resultant force T encountered by a tool in the form

$$T = \frac{wd}{\sin(\beta+\varphi+\rho+\delta)}\left[\frac{-C_a\cos(\beta+\varphi+\rho)}{\sin\beta} + \frac{\gamma d^2}{4}(\cot\beta + \cot\rho)\times\right.$$

$$\left. \sin(\varphi+\rho) + \frac{qd}{2}(\cot\beta + \cot\rho)\sin(\varphi+\rho) + \frac{c\cos\varphi}{\sin\rho} + \frac{\gamma v^2\sin\beta\cos\varphi}{g\sin(\beta+\rho)}\right] \tag{7}$$

from which the horizontal force can be obtained (Swick & Perumpral, 1988). In contrast with Gill and Vanden Berg's model, this expression includes the effects of adhesion, weight, surcharge, cohesion, and inertia.

The fundamental equation of earthmoving was first introduced by Reece (1965) and summarized by McKyes (1985). The most complete form of the earthmoving equation contains terms representing weight N_γ, cohesion N_c, adhesion N_{ca}, surcharge N_q, and inertia force N_a (which was not in Reece's formulation) as:

$$T = w\left(\gamma g d^2 N_\gamma + c d N_c + C_a d N_{ca} + q d N_q + \gamma v^2 d N_a\right) \tag{8}$$

which expands as

$$T = \frac{wd}{\cos(\beta+\delta) + \sin(\beta+\delta)\cot(\rho+\varphi)}\left[\frac{\gamma g d(\cot\beta + \cot\rho)}{2} + q(\cot\beta + \cot\rho) + c(1 + \cot\rho\cos(\rho+\varphi)) + C_a(1 - \cot\beta\cot(\rho+\varphi)) + \frac{\gamma v^2(\tan\rho + \cot(\rho+\varphi))}{1 + \tan\rho\cot\beta}\right] \tag{9}$$

from which the horizontal force can be obtained as

$$H = T\sin(\beta+\delta) \tag{10}$$

The earthmoving equations have been applied in a range of application as diverse as blade cutting, wire cutting, pile driving, and chisel plough wing cutting, simply by modifying the terms of the earthmoving equation. For pile driving (McCarthy, 1993; Vesic, 1977) and wire cutting (Thakur & Godwin, 1990), the only additional requirement was inclusion of cohesion and surcharge factors. An operation with a chisel plough wing (Fielke & Riley, 1991) required a model with the influence of weight and adhesion. Luengo et al. (1998) proposed a variation of the earthmoving equation for an inclined shape of the terrain (Luengo, Singh, & Cannon, 1998).

Three-dimensional models are extensions of two-dimensional models to consider side effects from bucket walls. There appear to be few interactions between the blade and the walls of a bucket; and so the action of cutting with an infinitely wide blade can be associated with excavation. Boccafogli et al. (1992), McKyes (1985) and Swick and Perumpral (1988) offer three-dimensional versions of their two-dimensional models. Willman and Boles (1996) performed an experimental investigation on various three-dimensional cutting models.

When modeling cutting with a blade, the cutting force is most often separated into two components, normal and tangential to the blade surface. These components originate from the presence of surcharge, cohesion, etc. When a bucket is considered, the addition of the sides and the bottom to transform a blade into a bucket leads to other forces that must be brought into account. Some researchers refute the validity of an existing analytical expression for an excavation model because of the difficulty in determining the plastic/elastic predominance associated with the tool action, or simply due to the lack of a rigid theoretical approach (Bernold, 1993; Zelenin, Balovnev, & Kerov, 1985). Since excavation can be seen as a combination of cutting and penetration, this subsection considers the complete excavation process as well as the analysis of each proposal for penetration-cutting force formulation. Figure 4 illustrates the force components.

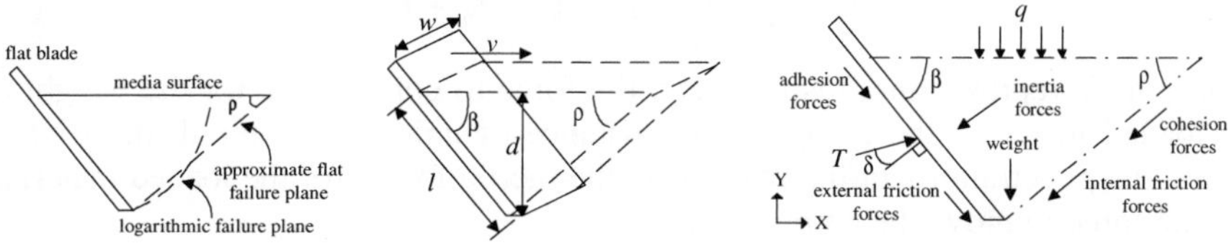

a) Cutting failure planes b) Cutting operation variables c) Cutting forces

Fig. 3. Force components during a cutting task

Alekseeva et al. (1985) suggested a formulation for the tangential components of the resistance force as a set of vectors:

$$F = P + R_A + P_A \tag{11}$$

where R_A denotes frictional forces, P_A is the sum of three forces (force to move the filled bucket, drag prism resistance, and compression resistance), and P represents the penetration-cutting force. The component P is a dominant force component in many excavation tasks. Similar to blade cutting models, there are both analytical and empirical expressions for the magnitude of P. Alekseeva et al. (1985) proposed an empirical expression

$$\|P\| = dwk_2 \tag{12}$$

where k_2 corresponds to an empirically determined cutting resistance.

Zelenin et al. (1985) regrouped the forces so that the total force F to be supplied by an excavating machine for loading a bucket is expressed as

$$F = P + R_Z + P_Z \tag{13}$$

where P is the penetration-cutting force, R_Z represents the total friction force experienced by the bucket with the medium, the force associated to the drag prism in front of the bucket and the resistance of the chip to transverse compression inside the bucket, and P_Z is the force necessary to move the filled bucket including its own weight. Zelenin et al. (1985) developed an empirical model by varying the parameters in an individual manner to determine their respective influence, resulting in the following expression for the cutting force experienced by a bucket (with a cutting and no teeth):

$$P = 10C_0 d^{1.35}(1 + 2.6w)(1 + 0.0075\,\beta)(1 + 0.03s)e_z k_z \tag{14}$$

where C_0, s, e_z and k_z represent the compactness and cutting resistance index, the cutting edge index, the tool plate thickness and the index for the type of cutting.

Hemami (1994) developed a formulation with six force components in the following form

$$F = f_1 + f_2 + f_3 + f_4 + f_5 + f_6 \tag{15}$$

with f_1 to f_6 representing, respectively, the force to compensate for the weight of the loaded material, the compacting resistance of unloaded material, the friction forces, the penetration-cutting resistance, the inertia force (for the loaded material) and the force to move the empty bucket.

The augmented number of force components enabled Hemami to relate to the expressions developed by Zelenin et al. (1985) and Alekseeva et al. (1985) as:

$$P = f_4,\ R_Z = f_1 + f_2 + f_3, P_Z = f_5 + f_6 \tag{16}$$

and

$$P = f_4,\ R_A = f_3, P_A = f_1 + f_2 + f_5 + f_6 \tag{17}$$

The excavation forces shown in Figure 4 are for an arbitrary point in time. Except for f_1, which represents the weight of the load inside the bucket and has its direction always oriented vertically, both the direction and the magnitude of the other components of the excavation force F change during the course of bucket motion. All the force components are functions of properties of the material, the tool, and the motion of the tool. Whereas the magnitude of the forces f_2 to f_6 are not directly bucket dependent, their directions change with the orientation of bucket. As the bucket rotates during loading, all the forces except f_1 rotate with the bucket. This observation is useful when loading on a known slope, in which case the gravity force f_1 can be decomposed into two components. The component normal to the terrain has the same role as f_1 in Figure 4, while the component parallel to the slope is in line with the forward forces to be provided by the machine crawlers. Separating out f_1 would allow for adjustment of the parameters in an automatic control scheme.

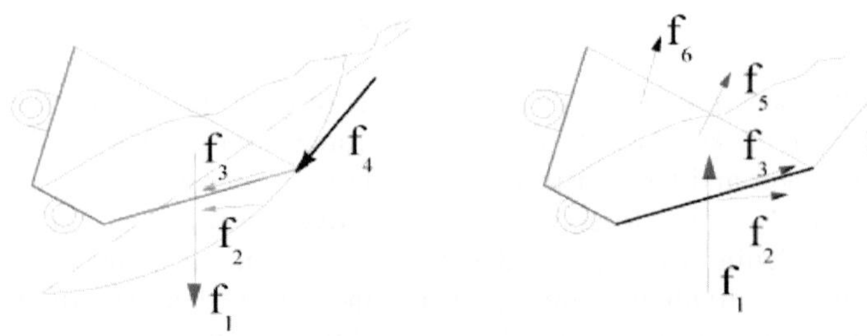

Fig. 4. Excavation forces

Balovnev (Balovnev, 1983) developed an analytical expression by extending passive pressure theory for large retaining walls to a bucket, by dividing the bucket into its constitutive parts (blade, sides, etc.) and then adding their individual influences. Balovnev's model is expressed in terms of the structured formulation of Hemami (1994) as

$$P = f_4 = P_1 + P_2 + P_3 \text{ and } f_3 = P_4 \tag{18}$$

where P_1 is the cutting resistance of the flat trenching blade with a sharp edge, P_2 is the additional resistance due to the wear of the edge, P_3 represents the resistance offered by the two sides, and P_4 is the resistance due to friction of the sides. The first three force vectors can be obtained by

$$\begin{bmatrix} P_1 \\ P_2 \\ P_3 \end{bmatrix} = \begin{bmatrix} wd\left(1+\frac{\tan\delta}{\tan\beta}\right)A_1 & 0 & 0 \\ 0 & we_b\left(1+\frac{\tan\delta}{\tan\beta}\right)A_2 & 0 \\ 0 & 0 & 2sdA_3 \end{bmatrix} \begin{bmatrix} d & 1 & d-l\sin\beta \\ e_b & 1 & d \\ d & 1 & d-l\sin\beta \end{bmatrix} \begin{bmatrix} \frac{\gamma}{2} \\ \frac{c}{\tan\varphi}+q \\ \gamma\frac{1-\sin\varphi}{1+\sin\varphi} \end{bmatrix}$$

with

$$A_1 = A(\beta), \qquad A_2 = A(\alpha_b), \qquad A_3 = A\left(\frac{\pi}{2}\right)$$

and

$$A(\beta) = \frac{1-\sin\varphi\cos(2\beta)}{1-\sin\varphi} \qquad \text{if } \beta < 0.5\left[\sin^{-1}\left(\frac{\sin\delta}{\sin\varphi}\right)-\delta\right]$$

$$A(\beta) = \frac{\cos\delta\left(\cos\delta+\sqrt{\sin^2\varphi-\sin^2\delta}\right)}{1-\sin\varphi} e^{\left(2\beta-\pi+\delta+\sin^{-1}\left(\frac{\sin\delta}{\sin\varphi}\right)-\delta\right)\tan\varphi}$$

$$\text{if } \beta \geq 0.5\left[\sin^{-1}\left(\frac{\sin\delta}{\sin\varphi}\right)-\delta\right]$$

where e_b and α_b represent the blunt edge height and the angle of worm surface with the cut surface. Superposition of the individual effect of bucket components may not be a valid assumption in all cases. Experiments of driving two parallel piles have shown that if their distance is small they mutually influence each other (Vesic, 1977). The same influence may exist with the bucket sides (depending on the bucket width), because the trenching plate travelling and loosening the medium facilitates the insertion of the sides within the medium.

Korzen (1985) developed a semi-empirical methodology to obtain excavation models for a heterogeneous medium, using a probabilistic model to describe the cutting force at each instant is a sum of the microreaction forces *F* with their respective statistics and some disturbances *e*, the deviation from an expected value. The large number of variables makes the method difficult to implement in practice.

Excavation experiments have shown that friction accounts for almost half of the total resistance along the direction of motion (Bernold, 1993).

3.1 Modeling Excavator Machine Dynamics

Lumped-parameter, rigid-body dynamic models are appropriate for mechanisms and machines such as industrial robots, which are firmly mounted on solid foundations. In large part, the superstructure and shovel attachment of an excavator can be modeled as rigid links connected by lower-pair joints, that is, joints that can only move with one degree of freedom, for example hinged pin joints.

Bernold (1993), Vähä (Vähä & Skibniewski, 1993), Koivo *et .al* (1996), Koivo (1994), Frimprong *et. al.*(2005, 2008) modeled a hydraulic front shovel and a cable shovel using a Newton-Euler formulation, with the assumption that there was no soft ground and no compliance in ropes or other parts of the system. Salcudean *et. al.* (1997) developed a dynamic model including an impedance model of the interaction with the soil.

There are several limitations to current models. Excavator dynamic models are formulated with a crawler frame that is assumed to be rigidly connected to the ground. This is not in fact realistic: excavators continually reposition their digging location, and often sit on uneven or poorly compacted ground. An extension to the machine model can be added to include the dynamic effects of soft ground contact. An example of such model is illustrated in Figure 5, where two sets of spring-dashpot are present to model the rear and front sinkage. Assuming that this results in rotation only, and not vertical motion, this soft foundation would add one degree-of-freedom to the system. Sepehri (Abo-Shanab & Sepehri, 2002; Abo-shanab & Sepehri, 2005) developed a dynamic model with consideration of the ground compliance.

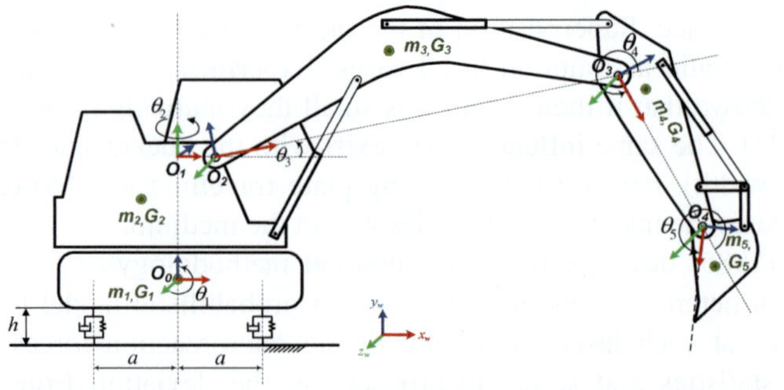

Fig. 5. Hydraulic Shovel Model and Coordinate Systems Described with 5 DOF. One Degree of Freedom Takes Rear and Front Sinkage into Account.

Table 1 gives the kinematic parameters for an excavator operating on soft ground, using the Denavit-Hartenberg convention.

Table 1. Denavit-Hartenberg Parameters for the Excavator System with 5 DOF

Link i	d_i	a_i	α_i	θ_i
1	0	0	-90	θ_1
2	$\overline{O_0O_1}$	$\overline{O_1O_2}$	90	θ_2
3	0	$\overline{O_2O_3}$	0	θ_3
4	0	$\overline{O_3O_4}$	0	θ_4
5	0	$\overline{O_4O_5}$	0	θ_5

4 Soil Parameter Estimation

Soil-tool interaction - and specifically cutting force prediction for shovels - has been discussed in various works (Flores *et. al.*, 2007; Frimpong & Hu, 2004). A common way to measure the resistive force is through the use of an instrumented flat plate moved by a manipulator to dig through a pile of soil (Dechao & Yusu, 1992; Singh, 1995; Hong, 2001). The displacement of the plate develops a failure surface between the two layers of moving and stationary soil.

Prediction of the soil-tool interaction force using a parametric formulation depends on accurate estimates of soil parameters. Luengo *et. al.* (1998) used a set of methods to estimate soil parameters from measured force data, applying exhaustive search, efficient gradient descent, stochastic search, and a combination of all three estimators. They also presented a reformulated version of the classical fundamental equation of earthmoving. Similarly, Hong (Hong, 2001) employed a robotic manipulator and a flat plate to estimate soil parameters for the perfectly plastic behavior of cohesionless soils, in loose and dense states, by applying various earth pressure models. It should be noted, having the soil parameters, one can predict the cutting force.

Following this work, other researchers proposed different methods for soil estimation, including a scheme based on the Newton-Raphson method (Tan *et. al.*, 2005; Tan *et. al.*, 2005) and the application of dissipated energy during excavation (Vahed *et. al.*, 2007, 2008).

4.1 Parameters Estimation from Measured Force Data

As the first step toward estimation through the measured force data, a soil model is required to characterize the interaction between the shovel blade and the cutting material. There are various earth pressure models that can be applied to represent the interaction, with the Coulomb-Mohr model being a simple and useful formulation.

To calculate the force required to fail the cutting material, the required soil parameters are: soil density γ, soil-tool interface (external) friction angle δ, soil-soil internal friction angle φ, and soil cohesion c. Geometrical required properties of the tool include: height of the blade H, the rake angle of the blade with respect to horizontal axis α, and angle of soil surface β. (These parameters have been illustrated above in Figure 3c.) According to the Coulomb earth pressure theory, cutting force can be computed from the fundamental equation of earthmoving (FEE) (Reece, 1965). The Mohr-Coulomb model assumes that the soil or substance fails on a flat plane in a wedge-like shape. This approximation works well for compacted particulate material with low cohesion. Thus, cohesionless material can be used to show how the estimation works. Knowing the density of the material, estimation can be done for the remaining two unknown parameters of external and internal friction angles: δ and φ.

The failure force is the maximum horizontal measured force during a dig (Tan C. P., Zweiri, Althoefer, & Seneviratne, 2005). The forces at the point of failure can be modeled as

$$F = F_{Passive} - F_{Active} = \frac{1}{2}\gamma H^2\left(K_P - K_A + \frac{4c}{\gamma H}\left(\sqrt{K_P} + \sqrt{K_A}\right)\right). \tag{19}$$

K_P and K_A, are passive and active coefficients. If the soil overburden is pushed with the blade horizontally, the soil constantly fails (Singh, 1995). For each unknown parameter that is to be estimated, a unique function is required. Although there is only one equation used to describe the system, the Mohr-Coulomb equation, this can be circumvented by obtaining force measurements at different rake angles. This creates multiple independent equations which can be used to obtain the desired variables.

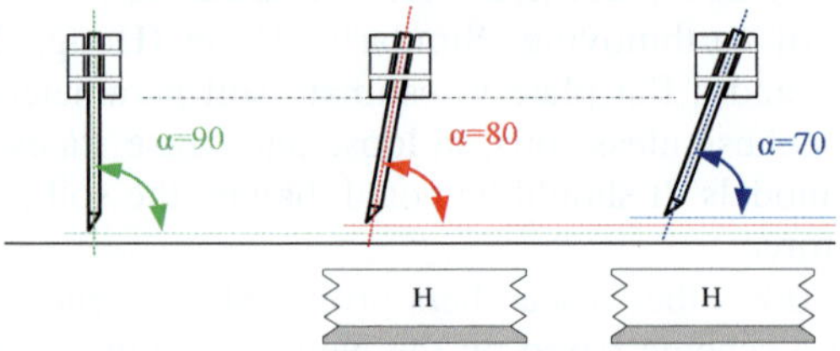

Fig. 6. Attack angle needs to be set at $\alpha = 80^\circ$ and 70°, while the depth is fixed

Since the Mohr-Coulomb equation is non-linear with respect to the desired parameters, a solving method must be applied to obtain the friction angles. The Newton-Raphson iterative method can be used to estimate the values of φ and δ. This method is chosen as it provides a single solution with relatively rapid convergence for a set of equations

$$J\begin{pmatrix}\Delta x_1\\ \Delta x_2\\ \cdots\\ \Delta x_n\end{pmatrix} = \begin{pmatrix}-f_1\\ -f_2\\ \cdots\\ -f_n\end{pmatrix}$$

where J is the Jacobian of the system, Δx_i is the incremental improvement of the desired parameter, and f_i is the function, in this case, the Mohr-Coulomb equation. For this parameter estimation, the functions used to estimate φ and δ were:

$$f_1 = M(\alpha_1, \delta_{es}, \varphi_{es}) = F_1$$
$$f_2 = M(\alpha_2, \delta_{es}, \varphi_{es}) = F_2$$

where M is the Mohr-Coulomb function evaluated at the rake angle α_i for a specific trial, using the estimated values of δ_{es} and φ_{es} and F is the force recorded from a given trial. The change in each desired parameter, then, can be written as:

$$\Delta\delta = \frac{-f_1\left(\frac{\partial f_2}{\partial \varphi}\right) + f_2\left(\frac{\partial f_1}{\partial \varphi}\right)}{J(f_1, f_2)}$$

$$\Delta\varphi = \frac{-f_2\left(\frac{\partial f_1}{\partial \delta}\right) + f_1\left(\frac{\partial f_2}{\partial \delta}\right)}{J(f_1, f_2)}$$

The Newton-Raphson iterative method operates by making a series of guesses to improve an initial estimate. This estimate is then refined until the difference between the current and previous estimation falls below a convergence criterion.

The goal of the parameter estimation is to take force measurements from the blade, along with the density of the material being cut γ, and obtain information about the material.

5 Limitations in Excavator Models Due to Machine Faults

Excavation entails fragmenting and loading rock and soil, which requires a high amount of interaction between the excavator and the medium being cut and moved. The forces of interaction between the machine and the soil depend on the tool geometry, the material, the operating conditions, and the soil characteristics (density, degree of compaction, cohesion, friction against the tool, moisture content, etc.) (Karmakar *et. al.*, 2009). The time-varying nature of the soil-tool interaction forces causes stresses in the components of earthmoving machines, which may result in damage.

Faults change the behavior of a system such that the system can no longer deliver the required level of performance. Failure modes are often followed by a change in the dynamics of the machine, which in turn affects the structure of the dynamical model (Khoshzaban-Zavarehi, 1997). A limitation of most earthmoving machine models is that any damage to the machine structure and drive transmission elements is not included. Cracks reduce the structural integrity; looseness in gearboxes introduces backlash in the power transmission system.

A common method for compensating for a faulty component is to reject undesired dynamics as disturbances. While this strategy would correct errors from the desired output trajectory, the power required for disturbance rejection may also aggravate the fault condition and accelerate the rate of damage accumulation.

Avoiding fault propagation in the system depends on identifying and isolating the faulty component in the early stages of failure. Most failures develop over time, in which cases there is some sort of degradation process that shifts the state of the system from normal to abnormal over a period of time, and this variation is identifiable in most cases using fault detection techniques. Even catastrophic failures, such as brittle failure of a structural element in cold weather, have root cause features that should be identifiable by inspection.

Many studies have been conducted to monitor the health of earthmoving machines as part of condition-based maintenance programs. Khoshzaban (1997) applied a model-based methodology for automatic generation of fault symptoms in a laboratory-scale hydraulic test rig, using on-line processing of low-quality raw sensor data. This work applied novel state-space models to estimate difficult-to-measure state variables. Allen and Sundermeyer (2005) developed an information management system for earthmoving machines with a focus on structural health. They analyzed sensor data so as to determine the machine's structural loading, assess its remaining structural life, and identify structural failures early, before damage became severe. Yin (2007) studied the fatigue failure of heavy mining equipments through the field monitoring of a cable shovel boom with a history of chronic fatigue cracking, using a finite-element model of the boom to predict the stress and strain histories with the objective of predicting the fatigue life of the shovel.

Timusk (Timusk, Lipsett, & Mechefske, 2008) studied on-line detection of faults in swing machinery of an electromechanical excavator, using vibration analysis. This study was performed in transient operating conditions where no historical fault data were available, employing an anomaly detection scheme with feature extraction and signal processing methods for monitoring the system and identifying anomalous situations (with possible faults) that were outside the expected variability of normal operation.

6 Conclusions

Soil-tool interaction models rely primarily on assumptions of homogeneous, isotropic soil properties, and tools that have simple geometries and steady motion through the soil. In many cases, these are reasonable assumptions, provided that the strain rate of the soil is not extreme. By modeling the rigid-body dynamics of the machine carrying the tool, and measuring its motions and interaction forces, it is possible to estimate the soil parameters. Heterogeneity in the soil and machine faults are possible sources of error in this approach; and care should be taken to collect data during the digging mode and not during other machine actions.

Acknowledgment. The contributions of S. Blouin and A. Hemami to the review of earthmoving models are greatly appreciated, as some parts of this paper originally appeared in S. Blouin, A. Hemami, M.G. Lipsett, "Review of Resistive Force Models for Earthmoving Processes." *ASCE Journal of Aerospace Engineering,* Vol. 14, No. 3, July 2001.

Nomenclature

a Acceleration;

C_a Adhesion;

C_o Compactness and cutting index;

c Cohesion;

d Tool depth;

d_{mean} Average particle size;

$d_1, \ldots, d_7$ Graphical distances;

E Elastic modulus;

e_b Blunt edge height;

e_z Cutting edge sharpness index;

$\mathbf{F}$ Tangential excavation force;

$\mathbf{f}_1, \ldots, \mathbf{f}_6$ Excavation force vectors;

g Gravitational force;

H Horizontal force;

J Clay fraction;

$k_1, k_2, \ldots, k_z$ Cutting resistance indices;

l Tool length;

m Failure plane parameters;

N_c, N_{ca} Cohesion, adhesion coefficients;

N_q, N_a, N_{g} Surcharge, inertia, weight coefficients;

N_o Normal load on inclined tool;

$\mathbf{P}$ Penetration-cutting force;

$\mathbf{P}_1, \ldots, \mathbf{P}_4, \mathbf{P}_A, \mathbf{P}_Z$ Excavation forces;

q Surcharge;

R_c Compressive strength;

$\mathbf{R}_A$, $\mathbf{R}_Z$ Excavation forces;

r_0, r_1 Curvature radius;

S_{mm}, S_{mt} Internal, external friction strength;

S_z Compacting ratio;

s Tool plate thickness;

T Resultant cutting force;

t Depth of Rankine zone;

v Tool speed;

w Tool width;

w' Polar angle;

a Failure plane parameters;

a_b Worm angle;

b Rake or cutting angle;

g Specific weight;

d External friction angle;

ε Porosity;

h Absolute viscosity;

n Poisson's ratio;

r Shear plane angle;

s_n Normal pressure to shear surface;

s, t Stress tensor field;

f Internal friction angle; and

v Relative moisture.

References

Abo-shanab, R.F., Sepehri, N.: Dynamic modeling of tip-over stability of mobile manipulators considering the friction effects. Robotica 23(2), 189–196 (2005)

Abo-Shanab, R.F., Sepehri, N.: Effect of base compliance on the dynamic stability of mobile manipulators. Robotica 20(6), 607–613 (2002)

Alekseeva, T.V., Artem'ev, K.A., Bromberg, A.A., Voitsekhovskii, R.I., Ul'yanov, N.A.: Machines for earthmoving work, theory and calculation. Intl Public Service, Balkema (1985)

Allen, W., Sundermeyer, J.: A structural health monitoring system for earthmoving machines. In: IEEE International Conference on Electro Information Technology, pp. 1–5 (2005)

Balovnev, V.I.: New methods for calculating resistance to cutting of soil. Amerind Pub. Co, New Delhi (1983)

Bernold, L.E.: Motion and path control for robotic excavation. J. Aerosp. Engrg. ASCE 6(1), 1–18 (1993)

Boccafogli, A., Busatti, G., Gherardi, F., Malaguti, F., Paoluzzi, R.: Experimental evaluation of cutting dynamic models in soil bin facility. Journal of Terramechanics 29(1), 95–105 (1992)

Dechao, Z., Yusu, Y.: A dynamic model for soil cutting by blade and tine. Journal of Terramechanics 29(3), 317–327 (1992)

Fielke, J.M., Riley, T.W.: The universal earthmoving equation applied to chisel plough wings. J. Terramechanics 28(1), 11–19 (1991)

Flores, F., Kecskeméthy, A., Pöttker, A.: Workspace Analysis and Maximal Force Calculation of a Face-Shovel Excavator using Kinematical Transformers. In: 12th IFToMM World Congress on the Theory of Machines and Mechanisms, Besancon (2007)

Fowkes, R.S., Frisque, D.E., Pariseau, W.G.: Materials handling re-search: Penetration of selected/granular materials by wedgeshaped tools. Washington, D.C.: Rep. of Investigations 7739, Bureau of Mines, U.S.D.I (1973)

Frimpong, S., Hu, Y.: Intelligent Cable Shovel Excavation Modeling and Simulation. International Journal of Geomechanics 8(2), 2–10 (2008)

Frimpong, S., Hu, Y.: Parametric Simulation of Shovel-Oil Sands Interactions During Excavation. International Journal of Mining, Reclamation and Environment 18(3), 205–219 (2004)

Frimpong, S., Hu, Y., Awuah-Offei, K.: Mechanics of cable shovel-formation interactions in surface mining excavations. Journal of Terramechanics 42(1), 15–33 (2005)

Frimpong, S., Hu, Y., Inyang, H.: Dynamic Modeling of Hydraulic Shovel Excavators for Geomaterials. International Journal of Geomechanics 8(2), 20–29 (2008)

Gill, W.R., Vanden Berg, G.E.: Agriculture handbook: Soil dynamics in tillage and traction Vol.316, Agricultural Research Service, U.S. Department of Agriculture, Washington, D.C. (1968)

Hemami, A.: Analysis and preliminary studies for automatic scooping. J. Adv. Robotics 8(5), 511–529 (1994)

Hemami, A., Goulet, S., Aubertin, M.: On the resistance of particulate media to bucket loading. In: Proc. 6th Can. Symp. on Min. Automation, pp. 171–178. Canadian Institute of Mining, Montreal (1994)

Hong, W.: Modeling, estimation, and control of robot-soil interactions. Massachusetts Institute of Technology. Dept. of Mechanical Engineering (2001)

Karmakar, S., Ashrafizadeh, S., Kushwaha, R.: Experimental validation of computational fluid dynamics modeling for narrow tillage tool draft. Journal of Terramechanics 46(6), 277–283 (2009)

Khoshzaban-Zavarehi, M.: On-line Condition Monitoring and Fault Diagnosis in Hydraulic System Components using Parameter Estimation and Pattern Classification. The Univeristy of British Columbia (1997)

Koivo, A.J.: Kinematics of excavators backhoes for transferring surface material. J. Aerosp. Eng. 7(1), 17–32 (1994)

Koivo, A.J., Thoma, M., Kocaoglan, E., Andrade-Cetto, J.: Modeling and Control of Excavator Dynamics during Digging Operation. Journal of Aerospace Engineering 9(1), 10–18 (1996)

Korzen, Z.: Mathematical modeling of the cutting process of strongly heterogeneous bulk materials with curvilinear edge tools. Studia Geotechnica et Mechanica 7(1), 27–54 (1985)

Lever, P.J., Wang, F.: Intelligent excavator control system for lunar mining system. J. Aerosp. Engrg. ASCE 8(1), 16–24 (1995)

Luengo, O., Singh, S., Cannon, H.: Modeling and identification of soil-tool interaction in automated excavation. In: Proc. IEEE/ RSJ Int. Conf. on Intelligent Robots and Sys., pp. 1900–1906. Institute of Electrical and Electronics Engineers, New York (1998)

Makarov, I.V.: Approximate calculation of the cutting forces in brittle materials. J. Min. Sci. 5, 54–60 (1992)

McCarthy, D.F.: Essential of soil mechanics and foundations: Basic geotechnics, 4th edn. Regents/Prentice-Hall, Englewood Cliffs (1993)

McKyes, E.: Soil cutting and tillage. Elsevier, New York (1985)

Osman, M.S.: The mechanics of soil cutting blades. J. Agric. Engrg. Res. 9(4), 313–328 (1964)

Reece, A.R.: The fundamental equation of earth-moving mechanics. Symposium on Earthmoving machinery. In: Proc., Instn. of Mech. Engrs., vol. 179 (3F), pp. 16–22 (1965)

Salcudean, S.E., Tafazoli, S., Lawrence, P.D., Chau, I.: Impedance control of a teleoperated mini excavator. In: Proc. of the 8th IEEE International Conference on Advanced Robotics (ICAR), Monterey, CA, USA, pp. 19–25 (1997)

Seward, D., Bradley, D., Mann, J., Goodwin, M.: Controlling an intelli-gent excavator for autonomous digging in difficult ground. In: Proc. 9th Int. Symp. on Automation and Robotics in Construction (ISARC), Tokyo, pp. 743–750 (1992)

Singh, S.: Learning to predict resistive forces during robotic excavation. In: IEEE International Conference on Proceedings Robotics and Automation 1995, pp. 2102–2107 (1995)

Singh, S.: Learning to predict resistive forces during robotic excavation. In: IEEE International Conference on Proceedings Robotics and Automation, pp. 2102–2107 (1995)

Singh, S.: State of the art in automation of earthmoving. J.Aerosp. Engrg. ASCE 10(4), 179–188 (1997)

Swick, W.C., Perumpral, J.V.: A model for predicting soiltool interaction. J. Terramechanics 25(1), 43–56 (1988)

Tan, C.P., Zweiri, Y., Althoefer, K., Seneviratne, L.: Online soil parameter estimation scheme based on Newton-Raphson method for autonomous excavation. IEEE/ASME Transactions on Mechatronics 10(2), 221–229 (2005)

Tan, C., Zweiri, Y., Althoefer, K., Seneviratne, L.: Online Soil-bucket Interaction Identification for Autonomous Excavation. In: Proceedings of the 2005 IEEE International Conference on Robotics and Automation, ICRA 2005, pp. 3576–3581 (2005)

Thakur, T.C., Godwin, R.J.: The mechanics of soil cutting by a rotating wire. J. Terramechanics 27(4), 291–305 (1990)

Timusk, M., Lipsett, M., Mechefske, C.: Fault detection using transient machine signals. Mechanical Systems and Signal Processing 22(7), 1724–1749 (2008)

Vähä, P.K., Skibniewski, M.J.: Dynamic Model of Excavator. Journal of Aerospace Engineering 6(2), 148–158 (1993)

Vahed, S., Althoefer, K., Seneviratne, L.D., Song, X., Dai, J.S., Lam, H.K.: Soil Estimation Based on Dissipation Energy During Autonomous Excavation. In: Proceedings of the 17th IFAC World Congress, pp. 13821–13826 (2008)

Vahed, S., Delaimi, H., Althoefer, K., Seneviratne, L.: On-line energy-based method for soil estimation and classification in autonomous excavation. In: IEEE/RSJ International Conference on Intelligent Robots and Systems, IROS 2007, pp. 554–559 (2007)

Vesic, A.S.: Synthesis of highway practice #42: Design of pile foundations. Washington, D.C. Nat. Coop. Hwy. Res. Prog. Transportation Research Board, National Research Council (1977)

Willman, B.M., Boles, W.W.: Soil-tool interaction theories as they apply to lunar soil simulant. J. Aerosp. Engrg. ASCE 8(2), 88–99 (1996)

Yao, Y., Zeng, D.C.: Investigation of the relationship between soil-metal friction and sliding speed. J. Terramechanics 27(4), 283–290 (1990)

Yin, Y., Grondin, G., Obaia, K., Elwi, A.: Fatigue life prediction of heavy mining equipment. Part 1: Fatigue load assessment and crack growth rate tests. Journal of Constructional Steel Research 63(11), 1494–1505 (2007)

Yong, R.N., Warkentin, B.P.: Introduction to soil behavior. McMillan, New York (1966)

Zelenin, A.N., Balovnev, V.I., Kerov, I.P.: Machines for moving the earth. Amerind Publishing, New Delhi (1985)

Author Index

Springer Series in Geomechanics and Geoengineering

Edited by W. Wu, R.I. Borja

Tejchman, J.
Shear Localization in
Granular Bodies with Micro-Polar
Hypoplasticity
316 p. 2008 [978-3-540-70554-3]

Iskander, M.
Modelling with Transparent Soils
331 p. 2010 [978-3-642-02500-6]

Tejchman, J. and Kozicki, J.
Experimental and Theoretical
Investigations of
Steel-Fibrous Concrete
289 p. 2010 [978-3-642-14602-2]

Iskander, M.
Behavior of Pipe Piles in Sand
267 p. 2010 [978-3-642-13107-3]

Wan, R., Alsaleh, M., and Labuz, J. (Eds.)
Bifurcations, Instabilities and Degradations
in Geomaterials
367 p. 2011 [978-3-642-18283-9]